Titles in This Series

Volume

5   D. A. Dawson, Editor
    Measure-valued processes, stochastic partial differential equations, and interacting systems
    1994

4   Hershy Kisilevsky and M. Ram Murty, Editors
    Elliptic curves and related topics
    1994

3   Rémi Vaillancourt and Andrei L. Smirnov, Editors
    Asymptotic methods in mechanics
    1993

2   Philip D. Loewen
    Optimal control via nonsmooth analysis
    1993

1   M. Ram Murty, Editor
    Theta functions
    1993

Volume 5

CRM PROCEEDINGS & LECTURE NOTES

Centre de Recherches Mathématiques
Université de Montréal

Measure-Valued Processes, Stochastic Partial Differential Equations, and Interacting Systems

D. A. Dawson
Editor

The Centre de Recherches Mathématiques (CRM) of the Université de Montréal was created in 1968 to promote research in pure and applied mathematics and related disciplines. Among its activities are special theme years, summer schools, workshops, postdoctoral programs, and publishing. The CRM is supported by the Université de Montréal, the Province of Québec (FCAR), and the Natural Sciences and Engineering Research Council of Canada. It is affiliated with the Institut des Sciences Mathématiques (ISM) of Montréal, whose constituent members are Concordia University, McGill University, the Université de Montréal, the Université du Québec à Montréal, and the Ecole Polytechnique.

American Mathematical Society
Providence, Rhode Island USA

The production of this volume was supported in part by the Fonds pour la Formation de Chercheurs et l'Aide à la Recherche (Fonds FCAR) and the Natural Sciences and Engineering Research Council of Canada (NSERC).

1991 *Mathematics Subject Classification.* Primary 60J80, 60J60; Secondary 60F15, 60G17.

Library of Congress Cataloging-in-Publication Data
Measure-valued processes, stochastic partial differential equations, and interacting systems/D. A. Dawson, editor.
    p. cm. — (CRM proceedings & lecture notes, ISSN 1065-8580; v. 5)
    ISBN 0-8218-6992-2
    1. Stochastic processes. 2. Stochastic partial differential equations. 3. Limit theorems (Probability theory) I. Dawson, Donald Andrew, 1937– . II. Université de Montréal, Centre de recherches mathématiques. III. Series.
QA274.M39 1994   93-48193
519.2—dc20   CIP

Copying and reprinting. Individual readers of this publication, and nonprofit libraries acting for them, are permitted to make fair use of the material, such as to copy an article for use in teaching or research. Permission is granted to quote brief passages from this publication in reviews, provided the customary acknowledgment of the source is given.

Republication, systematic copying, or multiple reproduction of any material in this publication (including abstracts) is permitted only under license from the American Mathematical Society. Requests for such permission should be addressed to the Manager of Editorial Services, American Mathematical Society, P.O. Box 6248, Providence, Rhode Island 02940-6248. Requests can also be made by e-mail to reprint-permission@math.ams.org.

The appearance of the code on the first page of an article in this publications (including abstracts) indicates the copyright owner's consent for copying beyond that permitted by Sections 107 or 108 of the U.S. Copyright Law, provided that the fee of $1.00 plus $.25 per page for each copy be paid directly to the Copyright Clearance Center, Inc., 222 Rosewood Drive, Danvers, Massachusetts 01923. This consent does not extend to other kinds of copying, such as copying for general distribution, for advertising or promotional purposes, for creating new collective works, or for resale.

        © Copyright 1994 by the American Mathematical Society. All rights reserved.
                    The American Mathematical Society retains all rights
                    except those granted to the United States Government.
                        Printed in the United States of America.

      ∞ The paper used in this book is acid-free and falls within the guidelines
                    established to ensure permanence and durability.
                          ♻ Printed on recycled paper.

                    This volume was typeset using $\mathcal{A}_{\mathcal{M}}\mathcal{S}$-TEX,
              the American Mathematical Society's TEX macro system,
      and submitted to the American Mathematical Society in camera-ready
                      form by the Centre de Recherches Mathématiques.

                        10 9 8 7 6 5 4 3 2 1   99 98 97 96 95 94

Contents

Preface
  D. A. Dawson vii

Superprocesses: The particle picture
  R. J. Adler 1

Une approche probabiliste de résolution d'équations non linéaires
  P. H. Bezandry, R. Ferland, G. Giroux et J.-C. Roberge 17

Variation of iterated Brownian motion
  K. Burdzy 35

The finite systems scheme: An abstract theorem and a new example
  J. T. Cox and A. Greven 55

On path properties of super-2 processes. I
  D. A. Dawson, K. J. Hochberg and V. Vinogradov 69

A type of interaction between superprocesses and branching particle systems
  E. B. Dynkin 83

Neutral allelic genealogy
  S. N. Ethier and T. Shiga 87

Superprocesses in catalytic media
  K. Fleischmann 99

A measure valued process arising from a branching particle system with changes of mass
  L. G. Gorostiza 111

Long time behavior of critical branching particle systems and applications
  L. G. Gorostiza and A. Wakolbinger 119

Newtonian particle mechanics and stochastic partial differential equations
  P. Kotelenez and K. Wang 139

Occupation time limit theorems for independent random walks
  T.-Y. Lee and B. Remillard 151

Large deviations and Boltzmann equation
  C. Léonard 165

A stochastic PDE arising as the limit of a long-range contact process, and
its phase transition
C. Mueller and R. Tribe   175

Some aspects of the Martin boundary of measure-valued diffusions
L. Overbeck   179

A Dirichlet form primer
B. Schmuland   187

Stationary distribution problem for interacting diffusion systems
T. Shiga   199

On clan-recurrence and -transience in time stationary branching Brownian
particle systems
A. Stoeckl and A. Wakolbinger   213

Tagged particle problem for an infinite hard core particle system in $\mathbb{R}^d$
H. Tanemura   221

A three level particle system and existence of general multilevel measure-
valued processes
Y. Wu   233

Preface

The objective of the CRM workshop and conference which were held during October, 1992 was to bring together researchers working in the areas of measure-valued processes, stochastic partial differential equations, and stochastic particle systems in order to develop an overview of recent developments and to identify and explore open problems which lie at the interfaces of these fields. Each of these areas had undergone profound development in recent years and it seemed to be an appropriate time for a workshop of this type which would emphasize their interrelations. The workshop included four series of lectures providing surveys of recent developments in these fields presented by R. T. Durrett, E. B. Dynkin, T. G. Kurtz, and E. A. Perkins. Much of the material of these lectures appears in their recent survey papers and monographs listed in the references below.

The papers in the present volume include contributions to some of the different major directions of research in these fields, explore the interface between them and describe some newly developing research problems and methodologies. One major direction of research which has flourished in recent years is the systematic development of the theory of measure-valued stochastic processes and their application to the study of limits of particle systems. Important classes of measure-valued processes include super-Brownian motion which arose from the theory of branching Brownian motions and the Fleming-Viot process which was motivated by problems of population genetics. Some new classes of measure-valued branching systems are described in the papers of Fleischmann (branching in a catalytic medium), Gorostiza (branching with change of mass), and Wu (multilevel branching). Path properties of multilevel super-Brownian motion are described in the paper of Dawson, Hochberg and Vinogradov, and path properties of iterated Brownian motion are studied in the paper of Burdzy. The paper of Dynkin is a complement to his Aisenstadt lectures in which he studies a very general class of measure-valued branching systems and the relation between such processes and a potential theory associated to a class of nonlinear partial differential equations. The paper of Adler demonstrates both the potential usefulness of computer visualization in the study of particle systems and the clustering phenomenon inherent in distributed branching systems. The paper of Ethier and Shiga is a contribution to the newly developing study of genealogical structures in the context of Fleming-Viot systems. Another important class of measure-valued processes are those which are absolutely continuous and have densities which satisfy stochastic partial differential equations. The papers of Kotelenez and Wang and of Mueller and Tribe explore two classes of stochastic partial differential equations describing measure-valued processes which arise as limits of interacting particle systems. The behavior of some classes of particle systems arising in statistical physics are studied in the papers of Bezandry, Ferland, Giroux and Roberge, and of Tanemura. The paper of Schmuland describes potential applications of the powerful method of Dirichlet forms to the study of stochastic partial differential equations and Fleming-Viot processes. Some new ideas and contributions to the study of the long time behavior and ergodic theory of measure-valued branching processes and systems of stochastic differential equations are given in the papers of Cox and Greven, of Gorostiza and Wakolbinger, Overbeck, Shiga, and of Stoeckl and Wakolbinger. Finally, various applications of large deviation methods in the study of particle systems are investigated in the papers of Lee and Remillard, and of Léonard.

We would like to thank all the participants of the conference and workshop and to authors of the present volume for their contribution to the successful realization of the objectives. We would also like to acknowledge NSERC for providing a conference grant. Finally, we would like to express our gratitude to the CRM for providing generous financial support as well as an ideal setting and highly efficient organization for the conference.

D. A. Dawson

References

R. Durrett, *Ten lectures on particle systems*, École d'Été de Probabilités de Saint-Flour XXIII, 1993.

E. B. Dynkin, *Aisenstadt lectures*, 1992.

S. N. Ethier and T.G. Kurtz, *Fleming-Viot processes in population genetics*, SIAM J. Control Opt. **31** (1992), 345-386.

E. A. Perkins, *On the martingale problem for interactive measure-valued branching diffusions*, Technical Report Series of the Laboratory for Research in Statistics and Probability, vol. 221, Carleton University, 1993.

Superprocesses: The Particle Picture

Robert J. Adler

0. Foreword

The material in this "paper" covers the last ten minutes of my talk at the Montréal conference. The talk had two parts: The first fifty minutes covered mathematics of the kind that both the writer and reader are most familiar with. This material is written up in [Adler 1992], and so does not appear again here. The last ten minutes comprised a brief slide show of some simulations of various "superprocesses", in dimensions one, two, and three. Some of these pictures, primarily for dimensions one and two, appear on the following pages.

I deeply believe that visualisation, along with the simulation that generates the objects we want to look at, is an important part of modern research in mathematics. Pictures will never replace mathematical proofs, but, as Don Geman once told me a decade ago, "It is much easier to prove a result once you are convinced that it is correct." The following collection of pictures is therefore offered in the hope that it will help others "see" what it is that superprocessors are currently studying, and perhaps lead to some new insights. As will be explained below, the programs that generated these pictures are freely avaliable to all who are interested. These programs, which allow one to see superprocesses developing in real time, generate far more insight than the fixed time snapshots shown below allow.

One more comment, before we start in ernest. I shall make no attempt below to carefully define the various processes nor precisely formulate results that I shall often describe. Neither shall I attempt to assign correct historical credit nor even give references. I shall assume that the reader is either familiar with the main definitions and results that I refer to, or has access to a copy of the encyclopædic review of [Dawson 1991], along with its all-encompassing bibliography.

1. "Superprocess" simulations

The quotation marks in the title of this subsection come from the fact that on a finite, and discrete, computer system it is obviously impossible to generate true superprocesses. *A fortiori*, there is no way to fully capture the fractal properties that make their sample path structure so fascinating. What one simulates is actually a discrete time, branching, random walk on a finite lattice. Nevertheless, I shall use the term "superprocess" in a loose sense in this paper, so that it will also cover these finite systems.

1991 *Mathematics Subject Classification.* Primary, 60J55, 60H15; Secondary, 60F17, 60F25.
Key words and phrases. Superprocesses, particle systems.
Research supported in part by US-Israel Binational Science Foundation (89-298), Israel Academy of Sciences (702-90).
This is the final form of the paper.

© 1994 American Mathematical Society
1065-8580/94 $1.00 + $.25 per page

Since the more interesting pictures appear in the two-dimensional setting, let us describe the simulation model there.

The action takes place on a $N_1 \times N_1$ lattice with periodic boundary conditions: i.e. a particle that wanders off the left hand side of the lattice immediately reappears at the right, ditto for top and bottom, etc. (In the examples, $N_1 = 51$.)

At time zero N_2 ($= 1,000$) particles are placed at the centre of the lattice: i.e. at the point $([N_1/2], [N_1/2])$. For the next N_3 iterations, each particle performs binary branching. (i.e. it either disappears or remains after having added another particle to the same site. The probability in each case is $1/2$, and each particle behaves independently of the others. After the branching, each particle takes N_4 steps of a symmetric random walk. This entire procedure (of N_3 branches and N_4 steps) is repeated N_5 times, and three pictures are generated.

The first two of these indicate the positions of the particles. The first gives a contour plot of the number of particles at each site. The second is a surface whose height is equal to this number. (At the left-hand side of this graph is a spike of height 10 to enable easy comparison between different graphs.)

As the above process develops, a count of the number of particles that ever visit a given site is kept. Set

$$L_n(i,j) = (\text{Number of particles that visit site } (i,j) \text{ up to the } n\text{-th stage})/N_2.$$

This is the local time process for the system. A plot of the local time appears after the previous two pictures.

This entire procedure is then repeated 5 times, so that there are five lines, each of three pictures, for each simulation. It is worth emphasising that the first line depicts the system after $N_3 \times N_5$ binary branchings, and $N_4 \times N_5$ steps of a random walk (for each of N_2 initial particles), and does not show the initial situation, in which all N_2 particles sit at the centre of the square.

There are also variants of this procedure in which attractive or dispersive interactions are allowed between the particles. These are all based on the local time, and a new parameter, β, that will indicate the strength of the interaction. The local time matrix L_n is used to change the drift of the random walk as follows:

Assume that a particle is about to take a step at the $(n+1)$-st stage. Before it takes this step, it calculates the horizontal and vertical "slopes" (i.e. first differences) of L_n at its position. If a slope is greater, in absolute value, than $1/\beta$, then it takes one step in the direction of $\text{sgn}(\beta)$ times the slope. Thus, when $\beta > 0$ particles will tend to cluster where they have been before, while when $\beta < 0$ they continually seek areas where they, or other particles, have not been in the past. After this interaction induced step has been taken, the particle takes the random step mentioned above. All particles move "together", so that the value of L_n is the same for all particles at the $(n+1)$-st stage.

Note that the processes thus obtained are no longer Markov, as the entire history of the process enters the drift via the local time. Furthermore, given the currently accepted usage of the term, they are also no longer superprocesses.

Following a few two-dimensional pictures, there are a few examples of the one-dimensional case. Here we present both the superprocess and the local time, and all parameters have the same meaning as before. Again, the initial state is generally N_2 particles at the center of the system, although also included are a couple of examples in which the initial state is a uniform distribution of particles throughout the system. As will be explained there, such an initial state offers very little in the way of additional information, and is visually confusing.

Finally, the paper closes with one three-dimensional system, just to give the reader a rough idea of what these look like. Unfortunately, without the benefit of colour, these are rather hard to interpret.

2. On choosing the parameters

The two parameters N_3 (# of branching generations), N_4 (# of random walk steps) are really the most important in terms of controlling what you are generating and seeing.

If $N_3 = 0$, there is no branching, and you are watching N_2 random walks on the torous, which will start diffusing like the heat equation, and soon go to white noise. This is an important first case to look at, since it gives one a basis for comparison.

In general, however, N_3 should be "large" in comparison to N_4. The reason for this is that, in the limit case, the random walk and branching scaling are different ($n^{-1/2}$ and n^{-1} respectively). A 5 : 1 ratio seems to give a nice balance on the 51×51 lattice. When the ratio is too small, the picture starts to look like a system of random walks, and when it is too high the branching takes over, and the system dies before any real movement occurs. (Remember, as $N_3 \to \infty$, the particles will disappear, since this is a *critical* branching process.)

A word on the choice of lattice size. Although there is a natural desire to take N_1 as large as possible, to see what happens on a bigger system, one should remember that (a) this is a self-similar process, so that you will only see more of the same thing, and (b) as N_1 increases the resolution of the graphs decreases rapidly.

Finally, an admission about how "typical" the simulations chosen for this collection are. They *are* typical, in all but one respect. All were chosen on the basis that throughout the simulation the number of particles grew neither too large, nor too small. In this sense, the examples chosen are somehow more akin to Fleming-Viot processes rather than superprocesses. Nevertheless, both of these classes processes are known to share many sample properties. The reason for this choice is a purely visual/didactic one: It is far easier to compare realisations of different branching and spreading mechanisms when the population sizes are inherently comparable.

3. On the programming

All programs were written in Fortran, and run on an IBM RS6000-320 workstation. The one and two-dimensional graphics were handled by the PGPLOT public domain graphics library of T. J. Pearson, and the three dimensional graphics by the commercial package DISSPLA. The former is quick but limited, the latter powerful but much slower.

Overall, the programs run fast enough that the two-dimensional systems shown below can be watched in real time. Two hundred generations with $N_1 = 51$, $N_2 = 1,000$, $N_3 = 5$, $N_4 = 1$ and $N_5 = 10$ (which means there are 20 snaphots) takes around 40 seconds to run. The three dimensional systems are much slower due to the slower speed of DISSPLA.

The one and two-dimensional programs are available for public use, and may be obtained by contacting me at ierhe01@technion.technion.ac.il. The three dimensional programs will work only on a computer that has a current version of DISSPLA installed.

ACKNOWLEDGEMENTS. Dr. Sergey Lyalko shared with me the teething pains of learning my way around AIX, PGPLOT, and the RS6000 which generated the pictures to follow. He also contributed heavily to the programming. His visualistions of three dimensional superprocesses, as already noted, are far more attractive than the one and two dimensional cases presented here.

References

R. J. Adler, *Superprocess local and intersection local times and their corresponding particle pictures*, Seminar on Stochastic Processes (E. Çinlar, K.L. Chung, M.J. Sharpe, eds.), Birkhäuser, 1993, pp. 1–42.

D. A. Dawson, *Measure-valued processes*, École d'éte de Probabilités de Saint-Flour, Lecture Notes in Math., vol. 1541, Springer, 1991, pp. 1–260.

ABSTRACT. We present a sequence of snapshots in time of discrete time, branching, random walks. In the high density, small lattice, limit, these processes converge to measure valued superprocesses. The snapshots, therefore, give us some idea of what "real" superprocesses look like.

FACULTY OF INDUSTRIAL ENGINEERING & MANAGEMENT, TECHNION — ISRAEL INSTITUTE OF TECHNOLOGY

E-mail address: ierhe01@technion.technion.ac.il

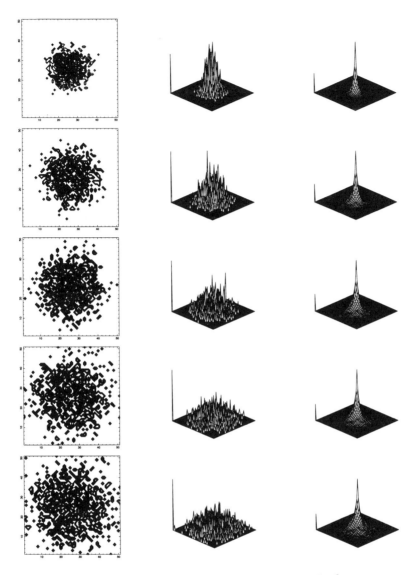

FIGURE 1. 2-d random walks. If we set $N_3 = 0$, there is no branching, and all we see are $N_2 = 1,000$ random walks on a torous. With $N_4 = 1$ and $N_5 = 50$, each picture shows the system after 50 steps by each particle. The mass of the superprocess should therefore spread out according to the heat kernel. The extent to which the mass spreads out in a more or less circular fashion indicates what sample fluctuations from 1,000 particles look like. Note that the local time is a much smoother process than the underlying superprocess. In fact, *its* contour lines (not shown here) are very close to perfect circles. Note also that after 250 steps, the superprocess is starting to look rather like white noise, as, in fact, it should.

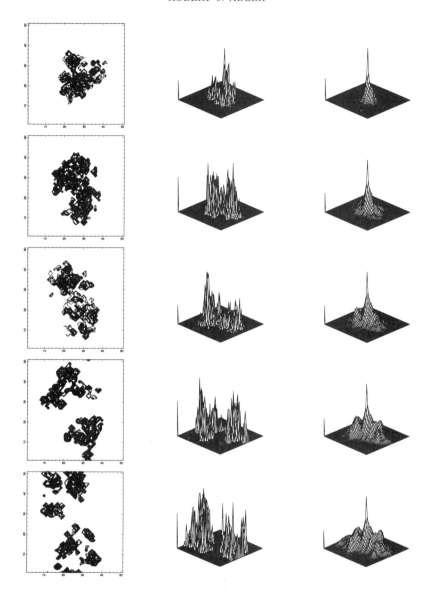

FIGURE 2. A simple 2-d superprocess: All parameters are as in Figure 1, except N_3 is now equal to 5, so that branching has been added to the previous random walk picture. Note how spatially disjoint "communities" of particles form very rapidly, as the mass of the superprocess spreads out. This is a result of weak "links" dying out. Of course, once such communities form, each one follows its own self-similar version of the original superprocess, albeit now with a different initial condition.

Note also that the local time process, while still much smoother than the superprocess itself, no longer has the spherical symmetry of the random walk situation.

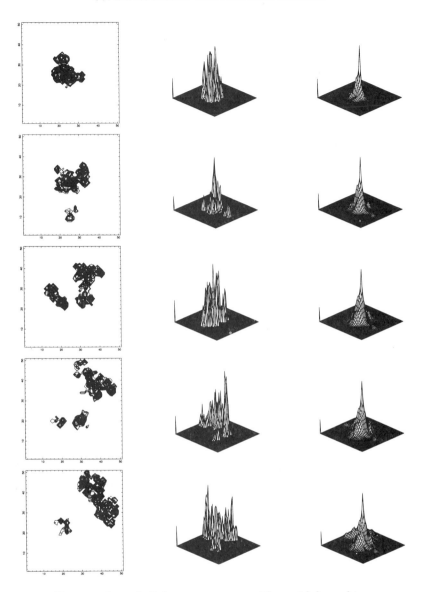

FIGURE 3. A 2-d superprocess with rapid branching: The same story as in Figure 2, but now $N_3 = 10$, so that the branching rate has been doubled. The result is a smaller number of communities, each one with a small spatial spread but a high population density. (Compare the sizes of the fixed spikes at the corner of the square.)

Since this and the previous two Figures give branching/walk ratios of 0,5 and 10, extrapolation should give you a good idea of what other parameter choices will do. Of course, when $N_3 = \infty$, the particles never move, and all one has is a critical Galton-Watson process at the centre of the square.

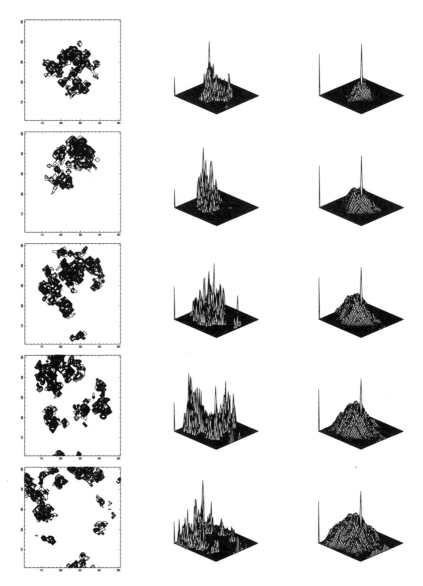

FIGURE 4. A superprocess with local repulsion: This is what I like to call the "goat process". The parameters are as for the "typical" case of Figure 2, but now the interaction parameter β is -20. Thus the particles, like nomadic goats, tend to avoid areas where they have been before; i.e. where the local time is high.

Note that the superprocess spreads out far more rapidly than before, the support seems far more disjointed, and the local time is much flatter. Careful observation shows that the local time spreads out from the centre like a two-dimensional wave, driving the particles before it.

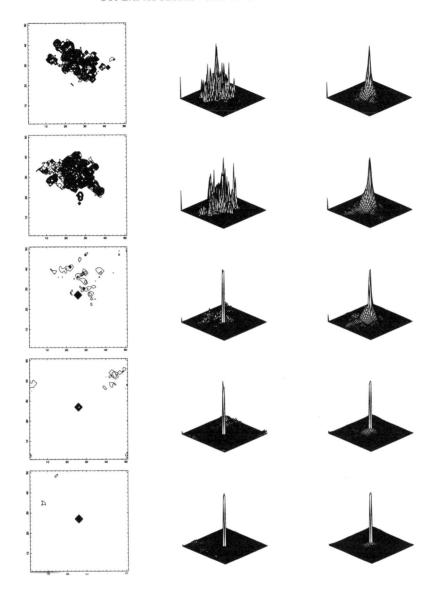

FIGURE 5. A superprocess with local attraction: This is the reverse of the previous case, with the interaction parameter β now set at 1.07, so that particles have a tendency to return to areas where the local time is high.

The first two pictures show the clustering that one would expect: This process spreads out slowly. The last two show that eventually everything gets stuck at the point of maximum local time, the origin. When $\beta < 1.06$ this phenomenon does not occur. When $\beta > 1.08$ it occurs almost immediately. Does this indicate the existence of a phase transition in the infinite density limit, or is it a finite sample phenomenon?

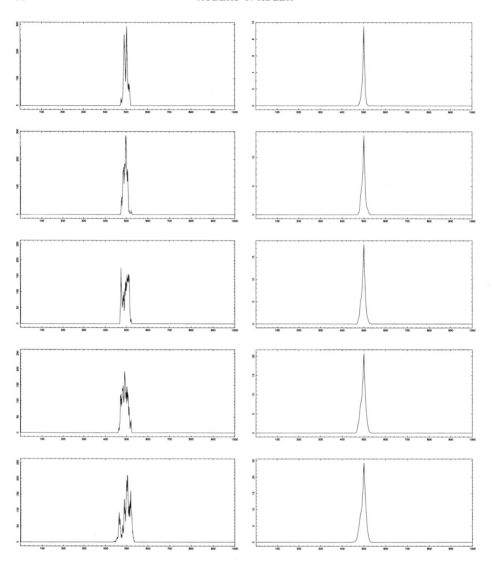

FIGURE 6. A 1-d superprocess: This is our first, rather unremarkable, example of a one-dimensional superprocess. There are $N_2 = 5,000$ initial particles, all initially placed at one site, and moving on a $N_1 = 1,000$ site circle. As before, there are $N_3 = 5$ branching generations in between each $N_4 = 1$ steps of the random walk. Here, as in all the one-dimensional cases bar Figure 8, each picture is after $N_5 = 100$ iterations.

Despite the rather jagged nature of the graphs, the superprocess in this case is known to have a smooth density, so that the pictures we see here should have, unlike the previous two-dimensional examples, a nice infinite density limit.

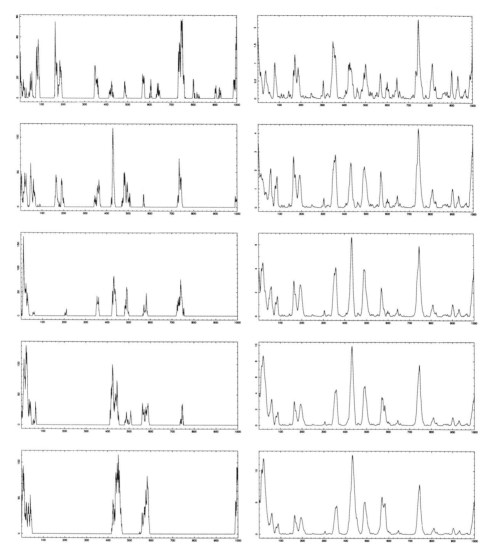

FIGURE 7. A 1-d superprocess with a "Lebesgue" initial condition: Everything is the same as in Figure 6, other than the fact that now the initial 5,000 particles are placed at 5 per site. What one sees is that, very rapidly, two types of regions develop. The lucky ones, where life was supported, and the others, where everything died. Within each one of the living regions, independent superprocesses develop.

The motivation of this example is to convince you that nothing was lost, in all of the previous examples, by looking at a delta function initial configuration.

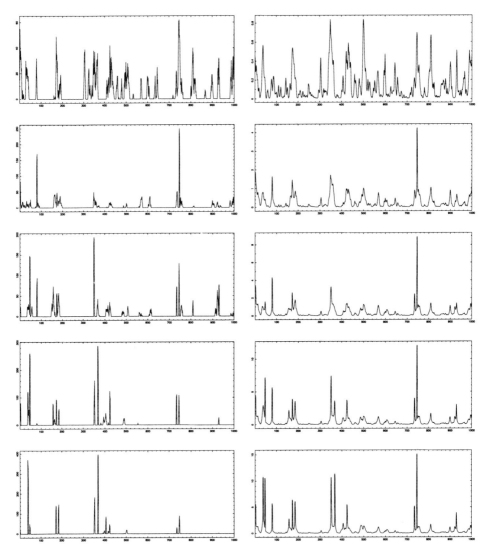

FIGURE 8. A 1-d superprocess with local attraction: This is the one-dimensional version of Figure 5, with $\beta = 5.0$. Each picture is now after $N_5 = 50$ iterations. The other difference is that the initial configuration was, as in the previous example, "Lebesgue". As a consequence, when the (extremely strong in this case) attractive interaction takes over, the fixation occurs at random points, rather than at the origin where, in Figure 5, the local time started off very high.

Once again, playing with the interaction parameter β leads one to believe that there may be a phase transition in β in the infinite density limit.

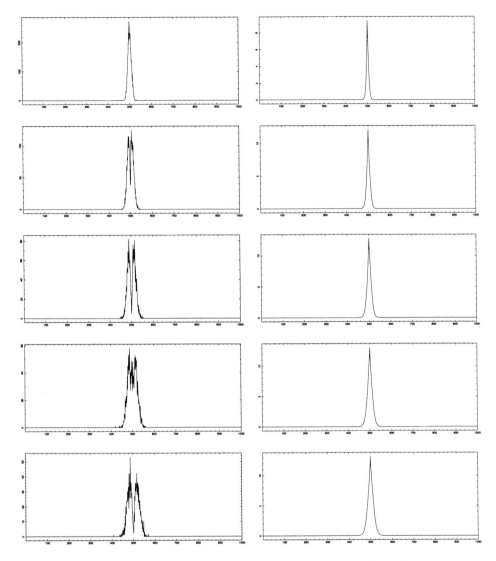

FIGURE 9. The catalyst process: In this process, branching occurs *only at the origin*, and at a very rapid rate. Elsewhere, there is the usual random walk. $N_3 = 250$ generations of branching occur at the origin, for every $N_4 = 1$ steps made by each particle. The rapid branching leads to almost zero mass at the origin, although the random walk continues to feed candidates for further branching into this single "catalyst" point.

In the limit, the density of the superprocess at the origin will, for any fixed time, be zero. But there are times at which mass falls into the origin, keeping the limit process random. (Away from the origin, there is only a deterministic heat kernel at work.)

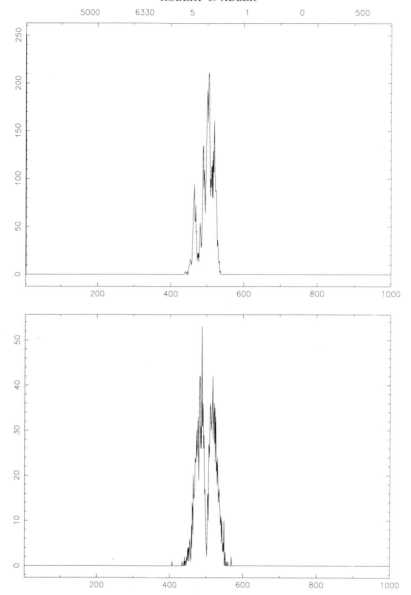

FIGURE 10. The catalyst process and the usual superprocess: The left hand picture is the last picture of Figure 6, enlarged. The right hand one is the last picture from the catalyst process.

In the former, the randomness comes from a random walk already in the regime of the strong law, and a branching mechanism spread throughout space. In the latter, there is the strong law random walk, and a far more rapid branching at the origin spreading heat kernel smoothed waves throughout space. Which is the smoother?

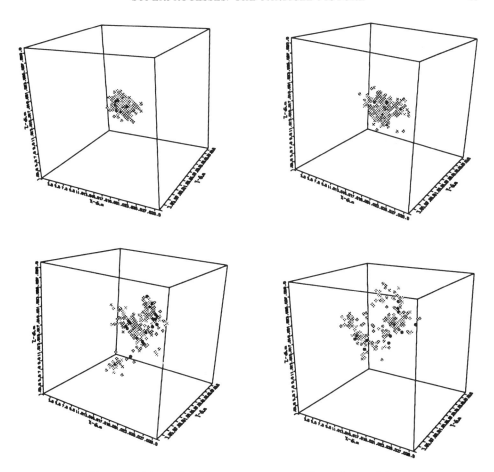

FIGURE 11. A 3-d superprocess: These four pictures (to be followed left to right, top to bottom) are of a 3-d superprocess ($N_3 = 5$, $N_4 = 1$, $N_5 = 50$) on a $31 \times 31 \times 31$ cubic lattice. Although the spatial development of this process is not really clear from this example, it has been included as a single example of what the 3-d situation is like. Watching the process develop temporally, with sites of high mass coloured, and the entire object rotating slowly, one gets a good feeling for the theorem that the support of the superprocess is two dimensional.

Une approche probabiliste de résolution d'équations non linéaires

Paul Hubert Bezandry, René Ferland,
Gaston Giroux et Jean-Claude Roberge

1. Introduction

L'étude des dynamiques de grands systèmes de particules en interaction mène à des équations d'évolution non linéaires. Maxwell et Boltzmann envisageaient un milieu dilué comme étant formé d'un grand nombre n de petites sphères de rayon $1/\sqrt{n}$. Ils obtenaient ainsi un processus de Markov Z^n à valeurs dans $(\mathbb{R}^6)^n$ où uniquement la préparation initiale des positions et des vitesses est aléatoire. Malheureusement, le générateur infinitésimal s'explicite peu. Pour contourner cette difficulté, une des simplifications qu'on rencontre est de substituer aux petites sphères des points infinitésimaux ; mais il faut alors inventer un mécanisme de rencontre de ces points. Cela est donné par une fonction d'intensité des sauts et un noyau d'interaction.

Dans son livre [Skorohod 1988], Skorohod étudie le cas de Maxwell ainsi modifié. Comme lui, nous étudions des systèmes où l'intensité des sauts est bornée mais nous travaillons avec des noyaux d'interaction plus généraux. Ce faisant, nous avons simplifié certaines démonstrations tout en modifiant certaines autres pour tenir compte de notre cadre plus général. Le principal propos de ce travail est de prouver l'existence d'une solution à l'équation non linéaire suivante :

$$(1.1) \quad \begin{cases} \dfrac{d}{dt}\langle \lambda_t, \varphi \rangle = \langle \lambda_t, \Gamma\varphi \rangle + \langle \lambda_t \otimes \lambda_t, \Lambda\varphi \rangle, \\ \lambda_0 = \mu. \end{cases}$$

Dans cette équation, λ_t et μ sont des mesures de probabilité sur $\mathbb{R}^3 \times \mathbb{R}^3$ et les crochets $\langle \cdot, \cdot \rangle$ sont utilisés pour désigner les intégrales. Les opérateurs Γ et Λ sont définis par les formules :

$$\Gamma\varphi(x,v) = v \cdot \nabla_x \varphi(x,v)$$

$$\Lambda\varphi(x,v;x',v') = \frac{1}{2}\int_{\mathbb{R}^6} [\varphi(x,\tilde{v}) - \varphi(x,v) + \varphi(x',\tilde{v}') - \varphi(x',v')] \, Q(v,v';d\tilde{v},d\tilde{v}').$$

La fonction φ est prise suffisamment régulière pour que $\Gamma\varphi$ soit bien défini et que les intégrales aient un sens. Dans la définition de Λ, Q est une probabilité de transition sur $\mathbb{R}^3 \times \mathbb{R}^3 \times \mathcal{B}(\mathbb{R}^3 \times \mathbb{R}^3)$ qui décrit le mécanisme d'évolution des vitesses des points

---

1991 *Mathematics Subject Classification.* Primary: 60K35, 60F99, 76P05 ; Secondary: 60G57.
La recherche du deuxième auteur est subventionnée par le CRSNG et le FCAR.
La recherche du troisième auteur est subventionnée par le CRSNG.
This is the final form of the paper.

infinitésimaux. *Nous supposerons tout le long de ce travail que Q est symétrique et que $\Lambda\varphi$ est continue bornée aussitôt que φ l'est.* On a alors le théorème suivant.

THÉORÈME 1.1. *Soit μ une probabilité sur $\mathbb{R}^3 \times \mathbb{R}^3$ et supposons que :*
(a) $\mu(\mathbb{R}^3 \times S_a) = 1$;
(b) $\int_{\mathbb{R}^3 \times \mathbb{R}^3} \|x\| \mu(dx,dv) < \infty$;
(c) $\forall v, v' \in S_a, Q(v,v'; S_a \times S_a) = 1$;
(d) *il existe une constante $L > 0$ telle que*

$$\int_{\mathbb{R}^3 \times \mathbb{R}^3} \|\tilde{v} - v\| Q(v,v'; d\tilde{v}, d\tilde{v}') \leq L(1 + \|v\| + \|v'\|)$$

et cela quels que soient v et v' dans $\mathbb{R}^3$.
Dans ces conditions, l'équation (1.1) possède une solution sur tout intervalle borné.
Dans l'énoncé du théorème, S_a est la boule fermée de rayon a dans $\mathbb{R}^3$.

2. Une loi des grands nombres

Dans cette section, on s'intéresse à l'équation ci-dessous qui est une version approchée de l'équation (1.1) :

(2.1) $\quad \begin{cases} \dfrac{d}{dt}\langle \lambda_t^N, \varphi \rangle = \langle \lambda_t^N, \Gamma^N \varphi \rangle + \langle \lambda_t^N \otimes \lambda_t^N, \Lambda \varphi \rangle, \\ \lambda_0^N = \mu. \end{cases}$

où $\Gamma^N \varphi(x,v) = N[\varphi(x + \frac{1}{N}v, v) - \varphi(x,v)]$ et $N > 0$ est un paramètre *fixé*. Moyennant nos hypothèses sur Q et μ, nous allons montrer, à l'aide d'une loi des grands nombres, que (2.1) possède une et une seule solution et nous utiliserons ensuite cette solution pour résoudre (1.1).

Cette idée est due à [Skorohod 1988] et le théorème 2.1 ci-dessous est analogue au résultat de ce dernier [Skorohod 1988, Theorem 1, p. 79]. Les hypothèses initiales du théorème 2.1 sont légèrement plus fortes que celles de Skorohod mais la conclusion également. En effet, nous obtenons la convergence en loi des processus empiriques alors que Skorohod démontre une convergence des mesures empiriques pour chaque instant t fixé. D'autre part, [Sznitman 1984] a déjà montré l'existence et l'unicité des solutions de (2.1). Mais ici, l'intensité des interactions est bornée (car N est fixé) et nos hypothèses initiales pour le théorème 2.1 sont donc moins fortes que celles de Sznitman car ces dernières sont conçues pour couvrir le cas où l'intensité est non bornée.

Commençons donc par associer à (2.1) une suite $\{W^{n,N}(t), t \in [0,T]\}$ de processus markoviens. Le n-ième processus prend ses valeurs dans $(\mathbb{R}^3 \times \mathbb{R}^3)^n$. On a donc

$$W^{n,N}(t) = (W_1^{n,N}(t), \ldots, W_n^{n,N}(t))$$

avec $W_j^{n,N}(t) = (X_j^{n,N}(t), V_j^n(t))$.

Le vecteur $X_j^{n,N}(t)$ est la composante position de $W_j^{n,N}(t)$ et $V_j^n(t)$ est sa composante vitesse. En fait, comme le montre l'écriture du générateur ci-dessous, l'évolution des vitesses ne dépend pas du paramètre N et la position initiale $X^{n,N}(0)$ non plus.

Le n-ième processus est gouverné par le générateur suivant :

$$G_n^N f(w_1, \ldots, w_n) = \mathcal{L}_n^N f(w_1, \ldots, w_n) + \mathcal{G}_n f(w_1, \ldots, w_n)$$

avec

$$\mathcal{L}_n^N f(w_1,\ldots,w_n) = N \sum_{j=1}^n \Gamma^{Nj} f(w_1,\ldots,w_n)$$

$$\mathcal{G}_n f(w_1,\ldots,w_n) = \frac{1}{n} \sum_{i \neq j} \Lambda^{ij} f(w_1,\ldots,w_n).$$

Les opérateurs Γ^{Nj} et Λ^{ij} sont analogues à Γ^N et Λ :

$$\Gamma^{Nj} f(w_1,\ldots,w_n) = f(w_1,\ldots,w_j+(v_j/N,0),\ldots,w_n) - f(w_1,\ldots,w_n)$$

$$\Lambda^{ij} f(w_1,\ldots,w_n) = \frac{1}{2} \int_{\mathbb{R}^6} \left[f^{ij}(w_1,\ldots,w_n) - f(w_1,\ldots,w_n) \right] Q(v_i,v_j;d\tilde{v},d\tilde{v}').$$

La fonction f^{ij} est obtenue de f en remplaçant les variables $w_i = (x_i, v_i)$ et $w_j = (x_j, v_j)$ par $\tilde{w}_i = (x_i, \tilde{v})$ et $\tilde{w}_j = (x_j, \tilde{v}')$ respectivement.

En dépit de cette lourdeur d'écriture pour $\mathcal{G}_n^N$, le comportement trajectoriel de $W^{n,N}$ est relativement simple. Initialement, le processus est dans l'état $W^{n,N}(0)$. Cet état demeure inchangé pour un temps exponentiel τ_1 (de paramètre $Nn + (n-1)$) à la fin duquel un "saut" est susceptible de survenir. Avec probabilité $Nn/(Nn + (n-1))$ le saut survient dans les composantes positions. Dans ce cas, le changement est le suivant : une composante j est choisie uniformément parmi les n possibles et le vecteur $X_j^{n,N}(0)$ est remplacé par $X_j^{n,N}(0) + V_j^n(0)/N$. Si le saut survient plutôt dans les vitesses, un couple (i,j) est choisi uniformément et les vitesses $(V_i^n(0), V_j^n(0))$ sont remplacées par $(\tilde{v}, \tilde{v}')$, un vecteur aléatoire dont la loi est donnée par $Q(V_i^n(0), V_j^n(0); d\tilde{v}, d\tilde{v}')$. Le nouvel état ainsi obtenu demeure inchangé pour un temps exponentiel τ_2 à la fin duquel un nouveau saut survient suivant le même schéma. Cette évolution se répète indéfiniment.

On peut supposer que tous les processus $\{W^{n,N}(t), t \in [0,T]\}$ sont définis sur un même espace probabilisé et que les trajectoires sont des fonctions continues à droite et pourvues de limites à gauche. La mesure aléatoire

$$\mu_t^{n,N} = \frac{1}{n} \sum_{j=1}^n \delta_{W_j^{n,N}(t)}$$

est la mesure empirique de $W^{n,N}(t)$ et $\{\mu_t^{n,N}, t \in [0,T]\}$ est le processus empirique associé à $\{W^{n,N}(t), t \in [0,T]\}$. Le reste de cette section est consacré à la démonstration du théorème suivant.

THÉORÈME 2.1. *Soit μ une probabilité sur $\mathbb{R}^3 \times \mathbb{R}^3$ et supposons que :*
(a) *il existe une constante $L > 0$ telle que*

$$\int_{\mathbb{R}^3 \times \mathbb{R}^3} \|\tilde{v} - v\| Q(v,v'; d\tilde{v}, d\tilde{v}') \leq L(1 + \|v\| + \|v'\|)$$

et cela quels que soient v et v' dans $\mathbb{R}^3$;
(b) $\sup_n \mathrm{E}\left[\frac{1}{n} \sum_{j=1}^n \|W_j^{n,N}(0)\|\right] < \infty$;
(c) $\mu_0^{n,N}$ *converge en loi vers μ.*

Dans ces conditions, les processus empiriques $\{\mu_t^{n,N}, t \in [0,T]\}$ convergent en loi vers un processus déterministe $\{\lambda_t^N, t \in [0,T]\}$ qui est l'unique solution de (2.1).

Avant de démontrer ce théorème, nous avons besoin de deux lemmes.

LEMME 2.2. *Sous les hypothèses du théorème 2.1, il existe une fonction réelle non décroissante $c_1(t)$ telle que*

$$\sup_n \mathrm{E}\left[\frac{1}{n}\sum_{j=1}^n \|W_j^{n,N}(t)\|\right] \leq c_1(t) < \infty.$$

DÉMONSTRATION. Regardons l'action du générateur G_n^N sur la fonction

$$f(w_1,\ldots,w_n) = \frac{1}{n}\sum_{j=1}^n \|w_j\|.$$

Tout d'abord, on a

$$\Gamma^{Nj}f(w_1,\ldots,w_n) = \frac{1}{n}\left\{\|w_j + (v_j/N, 0)\| - \|w_j\|\right\},$$

$$\Lambda^{ij}f(w_1,\ldots,w_n) = \frac{1}{2}\int_{\mathbb{R}^3\times\mathbb{R}^3}\frac{1}{n}\left\{\|\tilde{w}_i\| - \|w_i\| + \|\tilde{w}_j\| - \|w_j\|\right\}Q(v_i,v_j;d\tilde{v},d\tilde{v}')$$

où $\tilde{w}_i = (x_i,\tilde{v})$ et $\tilde{w}_j = (x_j,\tilde{v}')$. En prenant la valeur absolue et en utilisant l'hypothèse (a) du théorème 2.1 on trouve

$$|\Gamma^{Nj}f(w_1,\ldots,w_n)| \leq \frac{1}{n}\left\{\|v_j/N\|\right\},$$

$$|\Lambda^{ij}f(w_1,\ldots,w_n)| \leq \frac{1}{2}\int_{\mathbb{R}^3\times\mathbb{R}^3}\frac{1}{n}\left\{\|\tilde{v} - v_i\| + \|\tilde{v}' - v_j\|\right\}Q(v_i,v_j;d\tilde{v},d\tilde{v}')$$

$$\leq \frac{L}{n}\left(1 + \|v_i\| + \|v_j\|\right).$$

On en déduit que

$$|\mathcal{L}_n^N f(w_1,\ldots,w_n)| \leq \frac{1}{n}\sum_{j=1}^n \|v_j\| \leq f(w_1,\ldots,w_n),$$

$$|\mathcal{G}_n f(w_1,\ldots,w_n)| \leq \frac{L}{n^2}\sum_{i\neq j}(1 + \|v_i\| + \|v_j\|) \leq L + 2Lf(w_1,\ldots,w_n).$$

Si on pose maintenant $y^n(t) = \mathrm{E}[f(W^{n,N}(t))]$, nous savons que

$$y^n(t) = y^n(0) + \int_0^t \mathrm{E}[G_n^N f(W^{n,N}(s))]\,ds.$$

Les majorations ci-haut donnent que $\mathrm{E}[|G_n^N f(W^{n,N}(s))|] \leq (2L+1)y^n(s) + L$ et par conséquent

$$y^n(t) \leq y^n(0) + Lt + (2L+1)\int_0^t y^n(s)\,ds.$$

En appliquant le lemme de Gronwall et l'hypothèse (b) du théorème 2.1, on se rend compte que

$$c_1(t) = \left(\sup_n \mathrm{E}\left[\frac{1}{n}\sum_{j=1}^n \|W_j^{n,N}(0)\|\right] + Lt\right)e^{(2L+1)t}$$

est la fonction cherchée. □

LEMME 2.3. *Sous les hypothèses du théorème 2.1, il existe une constante M_T telle que*

$$\forall n, \forall m, \quad \Pr\left\{\sup_{0\leq t\leq T} \frac{1}{n}\sum_{j=1}^n \|W_j^{n,N}(t)\| > M_T 2^m\right\} \leq \frac{1}{2^m}.$$

DÉMONSTRATION. Considérons à la fonction $f(w_1,\ldots,w_n) = \frac{1}{n}\sum_{j=1}^n \|w_j\|$. Nous savons que $M_t^{n,N} = f(W^{n,N}(t)) - \int_0^t G_n^N f(W^{n,N}(s))\,ds$ est une martingale. L'inégalité

$$\sup_{0\leq t\leq T} \frac{1}{n}\sum_{j=1}^n \|W_j^{n,N}(t)\| \leq \sup_{0\leq t\leq T} |M_t^{n,N}| + \int_0^T |G_n^N f(W^{n,N}(s))|\,ds$$

entraîne que $\Pr\{\sup_{0\leq t\leq T} \frac{1}{n}\sum_{j=1}^n \|W_j^{n,N}(t)\| > c\}$ est majoré par la somme

$$\Pr\left\{\sup_{0\leq t\leq T} |M_t^{n,N}| > \frac{c}{2}\right\} + \Pr\left\{\int_0^T |G_n^N f(W^{n,N}(s))|\,ds > \frac{c}{2}\right\}.$$

Pour le premier terme, l'inégalité de Doob et la définition de $M_T^{n,N}$ donnent que

$$\Pr\left\{\sup_{0\leq t\leq T} |M_t^{n,N}| > \frac{c}{2}\right\} \leq \frac{2}{c} \mathrm{E}\big[|M_T^{n,N}|\big]$$
$$\leq \frac{2}{c}\left\{\mathrm{E}\left[\frac{1}{n}\sum_{j=1}^n \|W_j^{n,N}(T)\|\right] + \int_0^T \mathrm{E}\big[|G_n^N f(W^{n,N}(s))|\big]\,ds\right\}.$$

Pour le deuxième terme, l'inégalité de Tchebycheff donne

$$\Pr\left\{\int_0^T |G_n^N f(W^{n,N}(s))|\,ds > \frac{c}{2}\right\} \leq \frac{2}{c} \int_0^T \mathrm{E}\big[|G_n^N f(W^{n,N}(s))|\big]\,ds.$$

Or en démontrant le lemme 2.2, on a obtenu la majoration suivante :

$$\mathrm{E}\big[|G_n^N f(W^{n,N}(s))|\big] \leq \{L + (2L+1)c_1(T)\}.$$

Si on utilise cette majoration dans les inégalités précédentes, en prenant $M_T = 2[c_1(T) + 2T(L + (2L+1)c_1(T))]$, on obtient le résultat. □

DÉMONSTRATION (DU THÉORÈME 2.1). Nous procédons en trois étapes.

ÉTAPE 1. Nous démontrons que (2.1) possède au plus une solution. Soit φ une fonction mesurable bornée sur $\mathbb{R}^3 \times \mathbb{R}^3$ et λ^N, ν^N deux solutions de (2.1). Quel que soit $t \in [0, T]$, on a

$$\begin{aligned}
\left|\langle \lambda_t^N, \varphi \rangle - \langle \nu_t^N, \varphi \rangle\right| &= \left| \int_0^t \langle \lambda_s^N, \Gamma^N \varphi \rangle \, ds - \int_0^t \langle \nu_s^N, \Gamma^N \varphi \rangle \, ds \right. \\
&\quad \left. + \int_0^t \langle \lambda_s^N \otimes \lambda_s^N, \Lambda \varphi \rangle \, ds - \int_0^t \langle \nu_s^N \otimes \nu_s^N, \Lambda \varphi \rangle \, ds \right| \\
&\leq \int_0^t \left|\langle \lambda_s^N - \nu_s^N, \Gamma^N \varphi \rangle\right| ds + \int_0^t \left|\langle \lambda_s^N \otimes (\lambda_s^N - \nu_s^N), \Lambda \varphi \rangle\right| ds \\
&\quad + \int_0^t \left|\langle (\lambda_s^N - \nu_s^N) \otimes \nu_s^N, \Lambda \varphi \rangle\right| ds \\
&= \int_0^t \left|\langle \lambda_s^N - \nu_s^N, \Gamma^N \varphi \rangle\right| ds + \int_0^t \left|\langle \lambda_s^N - \nu_s^N, \psi_s \rangle\right| ds \\
&\quad + \int_0^t \left|\langle \lambda_s^N - \nu_s^N, \bar{\psi}_s \rangle\right| ds
\end{aligned}$$

où les fonctions ψ_s et $\bar{\psi}_s$ sont définies par :

$$\psi_s(x', v') = \int_{\mathbb{R}^6} \Lambda \varphi(x, v; x', v') \lambda_s^N(dx, dv),$$

$$\bar{\psi}_s(x, v) = \int_{\mathbb{R}^6} \Lambda \varphi(x, v; x'v') \nu_s^N(dx', dv').$$

Désignons par B_1 la boule unité fermée de l'espace des fonctions mesurables bornées sur $\mathbb{R}^3 \times \mathbb{R}^3$ pour la norme du supremum $\|\cdot\|_\infty$. On voit facilement que si $\varphi \in B_1$ alors $\frac{1}{2}\psi_s$, $\frac{1}{2}\bar{\psi}_s$ et $\frac{1}{2N}\Gamma^N \varphi$ appartiennent toutes trois à B_1. Puisque

$$\left|\langle \lambda_t^N, \varphi \rangle - \langle \nu_t^N, \varphi \rangle\right| \leq 2N \int_0^t \left|\left\langle \lambda_s^N - \nu_s^N, \frac{1}{2N}\Gamma^N \varphi \right\rangle\right| ds$$
$$+ 2 \int_0^t \left|\left\langle \lambda_s^N - \nu_s^N, \frac{1}{2}\psi_s \right\rangle\right| ds + 2 \int_0^t \left|\left\langle \lambda_s^N - \nu_s^N, \frac{1}{2}\bar{\psi}_s \right\rangle\right| ds,$$

il vient que

$$\left|\langle \lambda_t^N, \varphi \rangle - \langle \nu_t^N, \varphi \rangle\right| \leq (4 + 2N) \int_0^t \sup_{\varphi \in B_1} \left|\langle \lambda_s^N - \nu_s^N, \varphi \rangle\right| ds.$$

On en déduit que

$$\sup_{\varphi \in B_1} \left|\langle \lambda_t^N - \nu_t^N, \varphi \rangle\right| \leq (4 + 2N) \int_0^t \sup_{\varphi \in B_1} \left|\langle \lambda_s^N - \nu_s^N, \varphi \rangle\right| ds.$$

Si on pose $y(t) = \sup_{\varphi \in B_1} \left|\langle \lambda_t^N - \nu_t^N, \varphi \rangle\right|$, on a donc

$$0 \leq y(t) \leq (4 + 2N) \int_0^t y(s) \, ds; \quad y(0) = 0.$$

Par le lemme de Gronwall, $y(t) = 0$ pour tout t et donc $\lambda^N = \nu^N$.

ÉTAPE 2. Nous démontrons que la suite des processus $\{\mu_t^{n,N}, t \in [0,T]\}$ est relativement compacte pour la convergence en loi. D'après [Fernique 1991, théorème 4.4], il suffit pour cela de prouver que :
(1) il existe une suite $(K_m)_{m \geq 1}$ de sous-ensembles compacts de $\mathcal{M}_1^+(\mathbb{R}^3 \times \mathbb{R}^3)$ (les mesures de probabilité sur $\mathbb{R}^3 \times \mathbb{R}^3$ avec la topologie étroite) telle que

$$\forall n, \forall m, \quad \Pr\{\exists t \in [0,T] \mid \mu_t^{n,N} \notin K_m\} \leq \frac{1}{2^m};$$

(2) les processus réels $\{\langle \mu_t^{n,N}, \varphi \rangle, t \in [0,T]\}$ sont relativement compacts en loi et cela pour chaque fonction φ continue et bornée sur $\mathbb{R}^3 \times \mathbb{R}^3$.

La condition (1) résulte immédiatement du lemme 2.3 en prenant

$$K_m = \{\mu \in \mathcal{M}_1^+(\mathbb{R}^3 \times \mathbb{R}^3) \mid \langle \mu, \|w\| \rangle \leq M_T 2^m\}$$

où M_T est la constante qui apparaît dans ce lemme.

Pour avoir (2), nous appliquons un critère de Billingsley [Billingsley 1968, Theorem 15.5]. Les processus $\{\langle \mu_t^{n,N}, \varphi \rangle, t \in [0,T]\}$ sont relativement compacts lorsque les deux conditions suivantes sont satisfaites :
(i) $\lim_{R \uparrow \infty}(\sup_n \Pr\{\sup_{0 \leq t \leq T} |\langle \mu_t^{n,N}, \varphi \rangle| > R\}) = 0$;
(ii) $\forall \varepsilon > 0, \exists \delta > 0$ et $n_0 \geq 1$ tels que

$$\sup_{n \geq n_0} \Pr\left\{ \sup_{\substack{s,t \in [0,T] \\ |t-s| < \delta}} |\langle \mu_t^{n,N}, \varphi \rangle - \langle \mu_s^{n,N}, \varphi \rangle| \geq \varepsilon \right\} \leq \varepsilon.$$

La condition (i) est très facile à vérifier. Puique φ est bornée, on voit immédiatement que $\Pr\{\sup_{0 \leq t \leq T} |\langle \mu_t^{n,N}, \varphi \rangle| > R\} = 0$ aussitôt que $R > \|\varphi\|_\infty$. Pour démontrer (ii), on définit sur $D([0,T], \mathbb{R})$ le module V'' :

$$V''(f, \delta) = \sup\{|f(t) - f(r)| \wedge |f(r) - f(s)| ; 0 \leq s \leq r \leq t \leq T, |t-s| < \delta\}.$$

Il est bien connu que

$$(2.2) \qquad \sup_{\substack{s,t \in [0,T] \\ |t-s| < \delta}} |f(t) - f(s)| \leq 2V''(f, \delta) + \sup_{0 \leq t \leq T} |f(t) - f(t-)|.$$

Comme $\{W^{n,N}(t), t \in [0,T]\}$ est un processus de sauts dont au plus deux composantes changent au moment d'un saut, on voit que $|\langle \mu_t^{n,N}, \varphi \rangle - \langle \mu_{t-}^{n,N}, \varphi \rangle|$ est borné par $4\|\varphi\|_\infty/n$. Fixons $\varepsilon > 0$ et choisissons $n_0 \geq 1$ suffisament grand pour que $\|\varphi\|_\infty < n\varepsilon/8$ pour tout $n \geq n_0$. Pour ces entiers $n \geq n_0$, on a

$$\Pr\left\{ \sup_{0 \leq t \leq T} |\langle \mu_t^{n,N} \varphi \rangle - \langle \mu_{t-}^{n,N}, \varphi \rangle| > \frac{\varepsilon}{2} \right\} = 0$$

et (2.2) donne

$$\Pr\left\{ \sup_{\substack{s,t \in [0,T] \\ |t-s| < \delta}} |\langle \mu_t^{n,N} \varphi \rangle - \langle \mu_s^n, \varphi \rangle| > \varepsilon \right\} \leq \Pr\left\{ V''(\langle \mu^{n,N}, \varphi \rangle, \delta) > \frac{\varepsilon}{4} \right\}.$$

Pour démontrer (ii), il reste à prouver que

$$\lim_{\delta \downarrow 0} \sup_{n \geq n_0} \Pr\left\{ V''(\langle \mu^{n,N}, \varphi \rangle, \delta) > \frac{\varepsilon}{4} \right\} = 0.$$

Or, cette dernière relation est une conséquence de l'inégalité

(2.3) $\quad \mathrm{E}\big[(\langle\mu_t^{n,N},\varphi\rangle - \langle\mu_r^{n,N},\varphi\rangle)^2(\langle\mu_r^{n,N},\varphi\rangle - \langle\mu_s^{n,N},\varphi\rangle)^2\big] \leq C(t-s)^2$

où $0 \leq s \leq r \leq t \leq T$ (voir [Billingsley 1968, Theorem 15.6]). Pour démontrer (2.3), nous introduisons des martingales associées à $\{\mu_t^{n,N}, t \in [0,T]\}$. Notons $(\mathcal{F}_t^{n,N})$ la filtration induite par $\{W^{n,N}(t), t \in [0,T]\}$. En utilisant le générateur G_n^N, on peut montrer que les processus suivants :

$$M_t^{n,N} = \langle\mu_t^{n,N},\varphi\rangle - \int_0^t \langle\mu_s^{n,N}\dot\otimes\mu_s^{n,N}, \Lambda_1\varphi\rangle\, ds - \int_0^t \langle\mu_s^{n,N}, \Gamma_1^N\varphi\rangle\, ds$$

$$S_t^{n,N} = (M_t^{n,N})^2 - \frac{1}{n}\int_0^t \langle\mu_s^{n,N}\dot\otimes\mu_s^{n,N}, \Lambda_2\varphi\rangle\, ds - \frac{1}{n}\int_0^t \langle\mu_s^{n,N}, \Gamma_2^N\varphi\rangle\, ds$$

sont des $(\mathrm{Pr}, \mathcal{F}_t^{n,N})$-martingales. Dans l'écriture ci-haut, on a

$$\mu_s^{n,N}\dot\otimes\mu_s^{n,N} = \frac{1}{n^2}\sum_{i\neq j}\delta_{W_i^{n,N}(t)}\otimes\delta_{W_j^{n,N}(t)}$$

et

$$\Gamma_k^N\varphi(x,v) = N\left\{\varphi\left(x + \frac{1}{N}v, v\right) - \varphi(x,v)\right\}^k$$

$$\Lambda_k\varphi(x,v;x',v') = \frac{1}{2}\int_{\mathbb{R}^6}\{\varphi(x,\tilde{v}) - \varphi(x,v) + \varphi(x',\tilde{v}') - \varphi(x',v')\}^k Q(v,v';d\tilde{v},d\tilde{v}').$$

Utilisons ces martingales pour borner $\mathrm{E}\big[(\langle\mu_t^{n,N},\varphi\rangle - \langle\mu_r^{n,N},\varphi\rangle)^2 \mid \mathcal{F}_r^{n,N}\big]$. Étant donné la définition de $M_t^{n,N}$, cette espérance conditionnelle est presque sûrement majorée par

$$3\,\mathrm{E}\big[(M_t^{n,N})^2 - (M_r^{n,N})^2 \mid \mathcal{F}_r^{n,N}\big]$$
$$+ 3(t-r)\,\mathrm{E}\left[\int_r^t \{|\langle\mu_s^{n,N}\dot\otimes\mu_s^{n,N}, \Lambda_1\varphi\rangle|^2 + |\langle\mu_s^{n,N}, \Gamma_1\varphi\rangle|^2\}\, ds \,\Big|\, \mathcal{F}_r^{n,N}\right].$$

Mais $|\langle\mu_s^{n,N}\dot\otimes\mu_s^{n,N}, \Lambda_1\varphi\rangle|$ et $|\langle\mu_s^{n,N}, \Gamma_1\varphi\rangle|$ sont bornés respectivement par $2\|\varphi\|_\infty$ et $2N\|\varphi\|_\infty$. Le deuxième terme de la somme ci-haut est donc borné par $(4N^2 + 4)\|\varphi\|_\infty^2(t-r)$. D'autre part, le terme $\mathrm{E}\big[(M_t^{n,N})^2 - (M_r^{n,N})^2 \mid \mathcal{F}_r^{n,N}\big]$ se majore de façon analogue. Si on utilise la martingale $S_t^{n,N}$ cette fois-ci, on trouve

$$\mathrm{E}\big[(M_t^{n,N})^2 - (M_r^{n,N})^2 \mid \mathcal{F}_r^{n,N}\big]$$
$$= \frac{1}{n}\mathrm{E}\left[\int_r^t \{\langle\mu_s^{n,N}\dot\otimes\mu_s^{n,N}, \Lambda_2\varphi\rangle + \langle\mu_s^{n,N}, \Gamma_2^N\varphi\rangle\}\, ds \,\Big|\, \mathcal{F}_r^{n,N}\right].$$

À nouveau, $\langle\mu_s^{n,N}\dot\otimes\mu_s^{n,N}, \Lambda_2\varphi\rangle$ et $\langle\mu_s^{n,N}, \Gamma_2^N\varphi\rangle$ sont bornés par $8\|\varphi\|_\infty^2$ et $4N\|\varphi\|_\infty^2$. On obtient donc

(2.4) $\quad \mathrm{E}\big[(M_t^{n,N})^2 - (M_r^{n,N})^2 \mid \mathcal{F}_r^{n,N}\big] \leq \dfrac{8 + 4N}{n}\|\varphi\|_\infty^2(t-r).$

En combinant les deux majorations, on trouve que

$$\mathrm{E}\big[(\langle\mu_t^{n,N},\varphi\rangle - \langle\mu_r^{n,N},\varphi\rangle)^2 \mid \mathcal{F}_r^{n,N}\big] \leq C_1(t-r).$$

En prenant l'espérance, on obtient aussi

$$\mathrm{E}\big[(\langle \mu_t^{n,N}, \varphi\rangle - \langle \mu_r^{n,N}, \varphi\rangle)^2\big] \leq C_1(t-r)$$

et par conséquent pour $0 \leq s \leq r \leq t \leq T$:

$$\begin{aligned}
\mathrm{E}\big[&(\langle \mu_t^{n,N}, \varphi\rangle - \langle \mu_r^{n,N}, \varphi\rangle)^2 (\langle \mu_r^{n,N}, \varphi\rangle - \langle \mu_s^{n,N}, \varphi\rangle)^2\big] \\
&= \mathrm{E}\Big[(\langle \mu_r^{n,N}, \varphi\rangle - \langle \mu_s^{n,N}, \varphi\rangle)^2 \, \mathrm{E}\big[(\langle \mu_t^{n,N}, \varphi\rangle - \langle \mu_r^{n,N}, \varphi\rangle)^2 \mid \mathcal{F}_r^{n,N}\big]\Big] \\
&\leq C_1(t-r) \, \mathrm{E}\big[(\langle \mu_r^{n,N}, \varphi\rangle - \langle \mu_s^{n,N}, \varphi\rangle)^2\big] \\
&\leq C_1^2(t-r)(r-s) \\
&\leq C(t-s)^2.
\end{aligned}$$

ÉTAPE 3. Nous montrons la convergence en loi de $\{\mu_t^{n,N}, t \in [0,T]\}$ en identifiant le processus limite comme l'unique solution de (2.1). Les trajectoires du processus $\{\mu_t^{n,N}, t \in [0,T]\}$ sont des fonctions *cadlag* de $[0,T]$ dans $\mathcal{M}_1^+(\mathbb{R}^3 \times \mathbb{R}^3)$. Désignons par P^n la mesure de probabilité induite par $\{\mu_t^{n,N}, t \in [0,T]\}$ sur $D([0,T], \mathcal{M}_1^+(\mathbb{R}^3 \times \mathbb{R}^3))$ et par $\{U(t), t \in [0,T]\}$ le processus canonique des projections sur cet espace. Nous prouvons maintenant que tout point limite P^∞ de $\{P^n\}$ est concentré sur une trajectoire de $D([0,T], \mathcal{M}_1^+(\mathbb{R}^3 \times \mathbb{R}^3))$ à savoir la solution de (2.1). Cela terminera la démonstration du théorème.

Pour toute fonction φ continue et bornée sur $\mathbb{R}^3 \times \mathbb{R}^3$, considérons la variable suivante

$$M_t = \langle U(t), \varphi\rangle - \int_0^t \langle U(s), \Gamma^N \varphi\rangle \, ds - \int_0^t \langle U(s) \otimes U(s), \Lambda\varphi\rangle \, ds.$$

Prenons $\{P^{n_k}\}$ une sous-suite de $\{P^n\}$ qui converge étroitement vers P^∞ et montrons d'abord que

(2.5) $$\lim_k E^{n_k}\big[(M_t - M_0)^2\big] = E^\infty\big[(M_t - M_0)^2\big]$$

où E^n est l'espérance pour P^n. Soit la fonction $g(w) = (M_t(w) - M_0(w))^2$. Cette fonction est mesurable et bornée. De plus, elle est continue en toute trajectoire w de $C([0,T], \mathcal{M}_1^+(\mathbb{R}^3 \times \mathbb{R}^3))$. En effet, si $\{w^n\}$ converge vers w alors $\{w^n(s)\}$ converge étroitement vers $w(s)$ pour tout s (car w est continue en s). Il s'ensuit que $\langle w^n(s) \otimes w^n(s), \Lambda\varphi\rangle \to \langle w(s) \otimes w(s), \Lambda\varphi\rangle$ car $\Lambda\varphi$ est continue et bornée *par l'hypothèse faite sur Q dans l'introduction*. Une conclusion similaire s'applique à $\langle w^n(s), \Gamma^N \varphi\rangle$ puisque $\Gamma^N \varphi$ est continue et bornée. Le théorème de la convergence bornée donne alors que $g(w^n)$ converge vers $g(w)$. Par ailleurs, l'utilisation du critère de Billingsley à l'étape 2 montre non seulement la compacité relative mais aussi que $P^\infty(C([0,T], \mathcal{M}_1^+(\mathbb{R}^3 \times \mathbb{R}^3))) = 1$. L'ensemble des points de discontinuité de g est donc de mesure nulle pour P^∞. Cela entraîne (2.5) par le théorème des applications continues.

Montrons maintenant que

(2.6) $$\lim_k E^{n_k}\big[(M_t - M_0)^2\big] = 0.$$

À cet effet, soit $M_t^{(n)} = M_t - \frac{1}{n}\int_0^t \langle U(s), \bar{\Lambda}\varphi \rangle\, ds$ où $\bar{\Lambda}\varphi(x,v) = \Lambda\varphi(x,v;x,v)$. On a immédiatement que

$$E^{n_k}\left[(M_t - M_0)^2\right] \leq 2 E^{n_k}\left[(M_t^{(n_k)} - M_0^{(n_k)})^2\right] + \frac{2}{n_k{}^2} E^{n_k}\left[\left(\int_0^t \langle U(s), \bar{\Lambda}\varphi\rangle\, ds\right)^2\right].$$

Par (2.4), le premier terme du membre de droite est borné par $(8+4N)\|\varphi\|_\infty^2 t/n_k$ tandis que l'espérance dans le second terme est clairement bornée par une constante qui ne dépend que de φ et t. En prenant la limite sur k, on obtient (2.6).

Les relations (2.5) et (2.6) signifient que M_t est égal à M_0 presque sûrement pour P^∞. Puisque le processus $\{M_t,\, t \in [0,T]\}$ a des trajectoires continues à droite, il existe donc un ensemble $\Omega_0 \subset D([0,T], \mathcal{M}_1^+(\mathbb{R}^3 \times \mathbb{R}^3))$ avec $P^\infty(\Omega_0) = 1$ et $M_t(w) = M_0(w)$ quels que soient $t \in [0,T]$ et $w \in \Omega_0$. Compte-tenu de la définition de M_t, cela veut dire que les trajectoires de Ω_0 solutionnent (2.1). Finalement, l'hypothèse que $\mu_0^{n,N}$ converge en loi vers μ donne que $P^\infty(U(0) = \mu) = 1$. Par l'étape 1, $\Omega_0 \cap \{U(0) = \mu\}$ se réduit à une seule trajectoire w_0 et P^∞ est concentrée sur celle-ci. Cela montre à la fois que (2.1) possède une et une seule solution et que P^n converge étroitement vers P^∞. $\square$

REMARQUE. La démonstration précédente est en partie inspirée de [Dawson et Zheng 1991].

3. Existence des solutions

Revenons maintenant au théorème 1.1. Pour résoudre (1.1), nous démontrons comme à la section précédente, une loi des grands nombres. Nous considérons pour chaque n, un processus $\{Z^n(t),\, t \in [0,T]\}$ à valeurs dans $(\mathbb{R}^3 \times \mathbb{R}^3)^n$. La composante vitesse $V_j^n(t)$ de $Z_j^n(t)$ *est la même* que celle de $W_j^{n,N}(t)$, mais la composante position $X_j^n(t)$ est donnée par

$$X_j^n(t) = X_j^n(0) + \int_0^t V_j^n(s)\,ds.$$

On suppose que $Z^n(0) = W^{n,N}(0)$. Le processus $\{Z^n(t),\, t \in [0,T]\}$ est donc markovien et son générateur G_n est

$$G_n f(z_1, \ldots, z_n) = \mathcal{L}_n f(z_1, \ldots, z_n) + \mathcal{G}_n f(z_1, \ldots, z_n)$$

où

$$\mathcal{L}_n f(z_1, \ldots, z_n) = \sum_{j=1}^n v_j \cdot \nabla_{x_j} f(z_1, \ldots, z_n).$$

La mesure empirique de $Z^n(t)$ est notée μ_t^n. Le théorème 1.1 est alors un corollaire du résultat ci-dessous.

THÉORÈME 3.1. *Soit μ une probabilité sur $\mathbb{R}^3 \times \mathbb{R}^3$. On suppose qu'il existe une constante $a > 0$ telle que :*
  (a) $\mu(\mathbb{R}^3 \times S_a) = 1$;
  (b) $\forall v,\, v' \in S_a,\, Q(v,v'; S_a \times S_a) = 1$;
  (c) $\Pr\{V_j^n(0) \in S_a,\, j = 1, \ldots, n\} = 1$.
On suppose également que :
  (d) $\int_{\mathbb{R}^3 \times \mathbb{R}^3} \|x\|\mu(dx, dv) < \infty$;

(e) *il existe une constante $L > 0$ telle que*

$$\int_{\mathbb{R}^6} \|\tilde{v} - v\| Q(v, v'; d\tilde{v}, d\tilde{v}') \leq L(1 + \|v\| + \|v'\|)$$

et cela quels que soient v et v' dans $\mathbb{R}^3$;
(f) $\sup_n \mathrm{E}\big[\frac{1}{n} \sum_{j=1}^n \|X_j^n(0)\|\big] < \infty$;
(g) μ_0^n *converge en loi vers μ.*

Dans ces conditions, quel que soit $t > 0$, les mesures empiriques μ_t^n convergent en loi vers une probabilité λ_t et $\{\lambda_t, t \in [0, T]\}$ est une solution de (1.1).

Comme pour le théorème 2.1, nous avons besoin de quelques lemmes.

LEMME 3.2. *Sous les hypothèses du théorème 3.1, il existe une fonction réelle non décroissante $c_2(t)$ telle que*

$$\sup_n \mathrm{E}\bigg[\frac{1}{n} \sum_{j=1}^n \|Z_j^n(t)\|\bigg] \leq c_2(t) < \infty.$$

DÉMONSTRATION. On procède comme pour le lemme 2.2. On considère la fonction $f(z_1, \ldots, z_n) = \frac{1}{n} \sum_{j=1}^n \|z_j\|$. On sait déjà que

$$|\mathcal{G}_n f(z_1, \ldots, z_n)| \leq L + 2L f(z_1, \ldots, z_n).$$

D'autre part, on a

$$|\mathcal{L}_n f(z_1, \ldots, z_n)| = \bigg|\sum_{j=1}^n \frac{1}{n} \frac{(x_j \cdot v_j)}{\|z_j\|}\bigg| \leq \frac{1}{n} \sum_{j=1}^n \|z_j\| = f(z_1, \ldots, z_n).$$

Par conséquent, si on pose $y^n(t) = \mathrm{E}[f(Z^n(t))]$, nous obtenons la même relation qu'au lemme 2.2 soit

$$y^n(t) \leq y^n(0) + Lt + (2L+1) \int_0^t y^n(s) ds$$

et pour les mêmes raisons qu'avant, la fonction

$$c_2(t) = \bigg(\sup_n \mathrm{E}\bigg[\frac{1}{n} \sum_{j=1}^n \|Z_j^n(0)\|\bigg] + Lt\bigg) e^{(2L+1)t}$$

est la fonction cherchée. □

LEMME 3.3. *Sous les hypothèses du théorème 3.1, il existe une fonction réelle non décroissante $c_3(t)$ telle que*

$$\sup_n \mathrm{E}\bigg[\frac{1}{n} \sum_{j=1}^n \gamma(\|Z_j^n(t) - W_j^{n,N}(t)\|)\bigg] \leq \frac{c_3(t)}{N} < \infty,$$

où $\gamma(s) = s^2/(1+s^2)$.

DÉMONSTRATION. Posons $Y^{n,N}(t) = (Y_1^{n,N}(t), \ldots, Y_n^{n,N}(t))$ avec

$$Y_j^{n,N}(t) = (X_j^{n,N}(t), X_j^n(t), V_j^n(t)).$$

Le processus $\{Y^{n,N}(t), t \in [0, T]\}$ est markovien avec générateur :

$$H_n^N f(y_1, \ldots, y_n) = \mathcal{L}_n^N f(y_1, \ldots, y_n) + \mathcal{L}_n f(y_1, \ldots, y_n) + \mathcal{G}_n f(y_1, \ldots, y_n).$$

Bien entendu, les opérateurs $\mathcal{L}_n^N$, $\mathcal{L}_n$ et $\mathcal{G}_n$ agissent sur les composantes appropriées des vecteurs $y_j = (x_j^N, x_j, v_j)$. Considérons la fonction $f(y_1, \ldots, y_n) = \sum_{j=1}^n \gamma(\|x_j - x_j^N\|)/n$ et regardons l'action du générateur H_n^N sur celle-ci. Puisque f ne dépend pas des composantes v_j, on a $\mathcal{G}_n f = 0$. Pour $\mathcal{L}_n^N f$ et $\mathcal{L}_n f$, on a plutôt

$$\mathcal{L}_n f(y_1, \ldots, y_n) = \frac{1}{n} \sum_{j=1}^n \frac{2(x_j - x_j^N) \cdot v_j}{(1 + \|x_j - x_j^N\|^2)^2}$$

$$\mathcal{L}_n^N f(y_1, \ldots, y_n) = \frac{1}{n} \sum_{j=1}^n \frac{\|v_j\|^2/N - 2(x_j - x_j^N) \cdot v_j}{(1 + \|x_j - x_j^N - v_j/N\|^2)(1 + \|x_j - x_j^N\|^2)}.$$

On en déduit que

$$|\mathcal{L}_n f(y_1, \ldots, y_n) + \mathcal{L}_n^N f(y_1, \ldots, y_n)|$$
$$\leq \frac{1}{N}\left(\frac{1}{n}\sum_{j=1}^n \|v_j\|^2\right) + \frac{2}{n}\sum_{j=1}^n \frac{|(x_j - x_j^N) \cdot v_j|}{(1 + \|x_j - x_j^N\|^2)} g(y_1, \ldots, y_n)$$

avec

$$g(y_1, \ldots, y_n) = \left| \frac{1}{1 + \|x_j - x_j^N\|^2} - \frac{1}{1 + \|x_j - x_j^N - v_j/N\|^2} \right|.$$

Un calcul simple fournit la majoration

$$g(y_1, \ldots, y_n) \leq \frac{\|v_j/N\|(2\|x_j - x_j^N\| + \|v_j/N\|)}{(1 + \|x_j - x_j^N\|^2)(1 + \|x_j - x_j^N - v_j/N\|^2)}$$

et par conséquent on a

$$\frac{2}{n}\sum_{j=1}^n \frac{|(x_j - x_j^N) \cdot v_j|}{(1 + \|x_j - x_j^N\|^2)} g(y_1, \ldots, y_n)$$
$$\leq \frac{4}{N}\left(\frac{1}{n}\sum_{j=1}^n \|v_j\|^2 \gamma(\|x_j - x_j^N\|)\right) + \frac{2}{N^2}\left(\frac{1}{n}\sum_{j=1}^n \|v_j\|^3\right).$$

Puisque $X^n(0) = X^{n,N}(0)$, nous pouvons utiliser les majorations précédentes pour avoir que :

$$\mathrm{E}\left[\frac{1}{n}\sum_{j=1}^n \gamma(\|Z_j^n(t) - W_j^{n,N}(t)\|)\right] = \int_0^t \mathrm{E}[H_n^N f(Y^{n,N}(t))]\,ds$$
$$\leq \frac{4}{N}\int_0^t \mathrm{E}\left[\frac{1}{n}\sum_{j=1}^n \|V_j^n(s)\|^2 \gamma(\|X_j^n(s) - X_j^{n,N}(s)\|)\right]ds$$
$$+ \frac{1}{N}\int_0^t \mathrm{E}\left[\frac{1}{n}\sum_{j=1}^n \|V_j^n(s)\|^2\right]ds + \frac{2}{N^2}\int_0^t \mathrm{E}\left[\frac{1}{n}\sum_{j=1}^n \|V_j^n(s)\|^3\right]ds.$$

À cause des hypothèses (b) et (c), $\Pr\{V_j^n(s) \in S_a, j = 1, \ldots, n\} = 1$ et donc

$$\mathrm{E}\left[\frac{1}{n}\sum_{j=1}^{n}\gamma(\|Z_j^n(t) - W_j^{n,N}(t)\|)\right]$$
$$\leq \frac{a^2 t}{N} + \frac{2a^3 t}{N^2} + \frac{4a^2}{N}\int_0^t \mathrm{E}\left[\frac{1}{n}\sum_{j=1}^{n}\gamma(\|Z_j^n(s) - W_j^{n,N}(s)\|)\right] ds.$$

On applique encore le lemme de Gronwall et on trouve que $c_3(t) = (a^2 t + 2a^3 t)e^{4a^2 t}$ est la fonction cherchée. □

DÉMONSTRATION (DU THÉORÈME 3.1). Prenons une fonction φ continue et bornée sur $\mathbb{R}^3 \times \mathbb{R}^3$ et ayant un gradient $\nabla_x \varphi$ continu et borné. Pour tout $\varepsilon > 0$, on peut trouver une constante C_ε telle que

$$|\varphi(z) - \varphi(z')| \leq \varepsilon(1 + \|z\| + \|z'\|) + C_\varepsilon \gamma(\|z - z'\|),$$

où γ est la fonction du lemme 3.3 (voir [Skorohod 1988, p. 110]). On peut donc écrire les inégalités suivantes :

$$\mathrm{E}\big[|\langle \mu_t^n, \varphi\rangle - \langle \mu_t^{n,N}, \varphi\rangle|\big] \leq \mathrm{E}\left[\frac{1}{n}\sum_{j=1}^{n}|\varphi(Z_j^n(t)) - \varphi(W_j^{n,N}(t))|\right]$$
$$\leq \varepsilon\left(1 + \mathrm{E}\left[\frac{1}{n}\sum_{j=1}^{n}\|Z_j^n(t)\|\right] + \mathrm{E}\left[\frac{1}{n}\sum_{j=1}^{n}\|W_j^{n,N}(t)\|\right]\right)$$
$$+ C_\varepsilon \mathrm{E}\left[\frac{1}{n}\sum_{j=1}^{n}\gamma(\|Z_j^n(t) - W_j^{n,N}(t)\|)\right].$$

Les lemmes 2.2, 3.2 et 3.3 donnent alors

(3.1) $\quad \limsup_n \mathrm{E}\big[|\langle \mu_t^n, \varphi\rangle - \langle \mu_t^{n,N}, \varphi\rangle|\big] \leq \varepsilon(1 + c_1(t) + c_2(t)) + C_\varepsilon \dfrac{1}{N} c_3(t).$

Puisque $\langle \lambda_t^N, \varphi\rangle$ est une constante et que les variables $\langle \mu_t^{n,N}, \varphi\rangle$ sont bornées, on obtient comme conséquence du théorème 2.1 que

(3.2) $\quad \lim_n \mathrm{E}\big[|\langle \mu_t^{n,N}, \varphi\rangle - \langle \lambda_t^N, \varphi\rangle|\big] = 0.$

On en déduit l'inégalité

$$\limsup_n \mathrm{E}\big[|\langle \mu_t^n, \varphi\rangle - \langle \lambda_t^N, \varphi\rangle|\big] \leq \varepsilon\big(1 + c_1(t) + c_2(t)\big) + C_\varepsilon \frac{1}{N} c_3(t)$$

qui à son tour donne que

$$|\langle \lambda_t^N, \varphi\rangle - \langle \lambda_t^M, \varphi\rangle| \leq 2\varepsilon\big(1 + c_1(t) + c_2(t)\big) + C_\varepsilon \left(\frac{1}{N} + \frac{1}{M}\right) c_3(t).$$

La suite $\{\langle \lambda_t^N, \varphi\rangle, N \in \mathbb{N}\}$ est donc de Cauchy dans $\mathbb{R}$. D'autre part, c'est un corollaire du lemme 2.2 que $\sup_N \langle \lambda_t^N, \|z\|\rangle \leq c_1(t)$ et cette dernière inégalité permet facilement de montrer que $\{\lambda_t^N, N \in \mathbb{N}\}$ est tendu dans $\mathcal{M}_1^+(\mathbb{R}^3 \times \mathbb{R}^3)$. Ces deux conclusions suffisent pour prouver l'existence d'une mesure de probabilité λ_t vers

laquelle la suite $\{\lambda_t^N, N \in \mathbb{N}\}$ converge étroitement. Mais alors, les relations (3.1) et (3.2) donnent que
$$\lim_n \mathrm{E}\bigl[|\langle \mu_t^n, \varphi \rangle - \langle \lambda_t, \varphi \rangle|\bigr] = 0$$
et cela entraîne, d'après [Fernique 1992, théorème 2.2], que μ_t^n converge en loi vers λ_t.

Il reste maintenant à montrer que $\{\lambda_t, t \in [0, T]\}$ est une solution de (1.1). Si nous intégrons par rapport à t les équations (1.1) et (2.1), nous obtenons

(3.3) $\qquad \langle \lambda_t, \varphi \rangle = \langle \mu, \varphi \rangle + \int_0^t \{\langle \lambda_s, \Gamma\varphi \rangle + \langle \lambda_s \otimes \lambda_s, \Lambda\varphi \rangle\}\, ds$

(3.4) $\qquad \langle \lambda_t^N, \varphi \rangle = \langle \mu, \varphi \rangle + \int_0^t \{\langle \lambda_s^N, \Gamma^N\varphi \rangle + \langle \lambda_s^N \otimes \lambda_s^N, \Lambda\varphi \rangle\}\, ds.$

Nous savons que λ_t^N satisfait (3.4) et qu'elle converge étroitement vers λ_t ; nous voulons en déduire que λ_t satisfait (3.3). Observons d'abord que la convergence étroite de λ_s^N vers λ_s implique celle de $\lambda_s^N \otimes \lambda_s^N$ vers $\lambda_s \otimes \lambda_s$. Puisque $\Lambda\varphi$ est continue et bornée, on a
$$\lim_N \langle \lambda_s^N \otimes \lambda_s^N, \Lambda\varphi \rangle = \langle \lambda_s \otimes \lambda_s, \Lambda\varphi \rangle$$
quel que soit s. Il en résulte que
$$\lim_N \int_0^t \langle \lambda_s^N \otimes \lambda_s^N, \Lambda\varphi \rangle\, ds = \int_0^t \langle \lambda_s \otimes \lambda_s, \Lambda\varphi \rangle\, ds.$$

Il reste à montrer que

(3.5) $\qquad \displaystyle\lim_N \int_0^t \langle \lambda_s^N, \Gamma^N\varphi \rangle\, ds = \int_0^t \langle \lambda_s, \Gamma\varphi \rangle\, ds.$

Les processus Z^n et $W^{n,N}$ ont les mêmes composantes vitesses. Les hypothèses (b) et (c) donnent donc que $\Pr\{\mu_s^n(\mathbb{R}^3 \times S_a) = 1\} = \Pr\{\mu_s^{n,N}(\mathbb{R}^3 \times S_a) = 1\} = 1$. Comme μ_s^n et $\mu_s^{n,N}$ convergent en loi vers λ_s et λ_s^N, on en conclut que $\lambda_s^N(\mathbb{R}^3 \times S_a) = \lambda_s(\mathbb{R}^3 \times S_a) = 1$. Posons $C_\varphi = \sup_{x,v} \|\nabla_x \varphi(x, v)\|$ et observons que $|\Gamma^N\varphi(x, v)|$ et $|\Gamma\varphi(x, v)|$ sont tous deux majorés par $C_\varphi\|v\|$. Cela signifie que les intégrales $|\langle \lambda_s^N, \Gamma^N\varphi \rangle|$ et $|\langle \lambda_s, \Gamma\varphi \rangle|$ sont bornées par $C_\varphi a$. Si on montre que

(3.6) $\qquad \forall s \in [0, T], \quad \displaystyle\lim_N \langle \lambda_s^N, \Gamma^N\varphi \rangle = \langle \lambda_s, \Gamma\varphi \rangle$

on obtiendra (3.5) comme conséquence du théorème de la convergence bornée. Pour avoir (3.6), commençons par écrire l'inégalité suivante :

(3.7) $\qquad |\langle \lambda_s^N, \Gamma^N\varphi \rangle - \langle \lambda_s, \Gamma\varphi \rangle| \leq \langle \lambda_s^N, |\Gamma^N\varphi - \Gamma\varphi| \rangle + |\langle \lambda_s^N, \Gamma\varphi \rangle - \langle \lambda_s, \Gamma\varphi \rangle|.$

La fonction $\Gamma\varphi$ est continue et bornée sur $\mathbb{R}^3 \times S_a$. On peut donc construire une fonction continue et bornée ψ telle que $\langle \lambda_s^N, \Gamma\varphi \rangle = \langle \lambda_s^N, \psi \rangle$ et $\langle \lambda_s, \Gamma\varphi \rangle = \langle \lambda_s, \psi \rangle$. Le deuxième terme du membre de droite de (3.7) converge donc vers zéro car λ_s^N converge étroitement vers λ_s. D'autre part, la suite $\{\lambda_s^N\}$ est tendue. Par conséquent, pour tout $\varepsilon > 0$, il existe un compact K_ε tel que
$$\forall N, \quad \lambda_s^N(K_\varepsilon^c) < \frac{\varepsilon}{2(2C_\varphi a + 1)}.$$

Il s'ensuit que

$$\int_{K_\varepsilon^c} |\Gamma^N \varphi - \Gamma \varphi| \, d\lambda_s^N = \int_{K_\varepsilon^c \cap (\mathbb{R}^3 \times S_a)} |\Gamma^N \varphi - \Gamma \varphi| \, d\lambda_s^N \leq 2C_\varphi a \lambda_s^N(K_\varepsilon^c) < \frac{\varepsilon}{2}.$$

Finalement, la fonction $\Gamma^N \varphi$ converge uniformément vers $\Gamma \varphi$ sur le compact K_ε. Comme λ_s^N est une probabilité, on obtient pour N assez grand que

$$\int_{K_\varepsilon} |\Gamma^N \varphi - \Gamma \varphi| \, d\lambda_s^N < \frac{\varepsilon}{2}.$$

Le premier terme du membre de droite de (3.7) converge donc lui aussi vers zéro. Cela donne (3.6) et complète la preuve. $\square$

Nous terminons cette section en prolongeant le théorème 3.1 à une convergence des processus empiriques.

THÉORÈME 3.4. *Sous les hypothèses du théorème 3.1, les processus empiriques $\{\mu_t^n, t \in [0,T]\}$ convergent en loi vers le processus déterministe $\{\lambda_t, t \in [0,T]\}$.*

DÉMONSTRATION. Elle ressemble à celle du théorème 2.1. Il faut d'abord montrer que les processus $\{\mu_t^n, t \in [0,T]\}$ sont relativement compacts en loi. Nous utilisons le critère de Fernique *i.e.* nous vérifions les conditions (1) et (2) de l'étape 2 de la démonstration du théorème 2.1 (en prenant bien sûr dans (2) des fonctions φ continues bornées avec un gradient $\nabla_x \varphi$ continu et borné).

Pour la condition (1), nous commençons par établir l'inégalité suivante :

$$(3.8) \qquad \sup \mathrm{E}\left[\sup_{0 \leq t \leq T} \langle \mu_t^n, \|z\| \rangle \right] < \infty.$$

Les hypothèses (b) et (c) donnent que $\Pr\{\sup_{0 \leq s \leq t} \|V_j^n(s)\| \leq a\} = 1$. Puisque

$$\|X_j^n(t)\| \leq \|X_j^n(0)\| + t \sup_{0 \leq s \leq t} \|V_j^n(s)\|,$$

il vient presque sûrement que

$$\sup_{0 \leq t \leq T} \langle \mu_t^n, \|z\| \rangle \leq \frac{1}{n} \sum_{j=1}^n \|X_j^n(0)\| + (T+1)a$$

et (3.8) découle alors de l'hypothèse (f).

La propriété (1) résulte immédiatement de (3.8). En effet, il suffit de poser $M_T = \sup_n \mathrm{E}\left[\sup_{0 \leq t \leq T} \langle \mu_t^n, \|z\| \rangle\right]$, de prendre

$$K_m = \{\mu \in \mathcal{M}_1^+(\mathbb{R}^3 \times \mathbb{R}^3) \mid \langle \mu, \|z\| \rangle \leq M_T 2^m\}$$

et d'appliquer l'inégalité de Markov.

Pour établir la propriété (2), nous appliquons le critère de Billingsley. Un examen attentif de l'étape 2 de la démonstration du théorème 2.1 montrent que le raisonnement fait à ce moment-là s'applique encore ici. En fait, le seul résultat à démontrer est l'inégalité

$$\mathrm{E}\left[(\langle \mu_t^n, \varphi \rangle - \langle \mu_r^n, \varphi \rangle)^2 (\langle \mu_r^n, \varphi \rangle - \langle \mu_s^n, \varphi \rangle)^2\right] \leq C(t-s)^2.$$

Mais cela se fait comme avant en utilisant les martingales

$$M_t^n = \langle \mu_t^n, \varphi \rangle - \int_0^t \langle \mu_s^n \dot{\otimes} \mu_s^n, \Lambda_1 \varphi \rangle \, ds - \int_0^t \langle \mu_s^n, \Gamma\varphi \rangle \, ds$$

$$S_t^n = (M_t^n)^2 - \frac{1}{n} \int_0^t \langle \mu_s^n \dot{\otimes} \mu_s^n, \Lambda_2 \varphi \rangle \, ds$$

et le fait que $|\langle \mu_s^n \dot{\otimes} \mu_s^n, \Lambda_1 \varphi \rangle|$, $|\langle \mu_s^n, \Gamma\varphi \rangle|$ et $|\langle \mu_s^n \dot{\otimes} \mu_s^n, \Lambda_2 \varphi \rangle|$ sont presque sûrement bornés par $2\|\varphi\|_\infty$, $a \sup_{(x,v)} \|\nabla_x \varphi(x,v)\|$ et $8\|\varphi\|_\infty$ respectivement.

Pour terminer la démonstration, désignons par $\mathcal{C}$ l'ensemble des fonctions φ continues bornées avec un gradient $\nabla_x \varphi$ continu et borné. Par le théorème 3.1, nous savons que $\langle \mu_t^n, \varphi \rangle$ converge en loi vers $\langle \lambda_t, \varphi \rangle$ quels que soient $t \in [0,T]$ et $\varphi \in \mathcal{C}$. Mais comme la limite est non aléatoire, cela implique que le vecteur aléatoire $(\langle \mu_{t_1}^n, \varphi_1 \rangle, \ldots, \langle \mu_{t_k}^n, \varphi_k \rangle)$ converge en loi vers $(\langle \lambda_{t_1}, \varphi_1 \rangle, \ldots, \langle \lambda_{t_k}, \varphi_k \rangle)$ quels que soient $t_1, \ldots, t_k \in [0,T]$ et $\varphi_1, \ldots, \varphi_k \in \mathcal{C}$. Les processus $\{\langle \mu_t^n, \varphi \rangle, t \in [0,T]\}$ ont donc en loi au plus un point limite car tous les processus limites ont les mêmes marginales de dimension finie. La preuve est complète. □

Une fois le théorème 3.4 obtenu, il est naturel d'aborder l'étude des fluctuations. En négligeant le mouvement des particules (le terme $\Gamma\varphi$), un premier travail a d'abord été fait par McKean, puis par Tanaka et Uchiyama. Une analyse plus poussée fut par la suite entreprise dans [Bezandry, Fernique et Giroux 1993, Ferland, Fernique et Giroux 1992, Ferland et Roberge 1992] où on trouvera les références pertinentes. Il est permis de croire qu'une telle analyse puisse s'étendre au modèle étudié dans le présent exposé.

REMERCIEMENTS. Les auteurs tiennent à remercier Wolfgang Wagner pour ses remarques utiles qui ont permis de simplifier l'étape 3 de la preuve du théorème 2.1.

References

P. H. Bezandry, X. Fernique et G. Giroux, *A functional central limit theorem for a nonequilibrium model of interacting particles with unbounded intensity*, J. Statist. Phys. **72** (1993), 329-353.

P. Billingsley, *Convergence of probability measures*, John Wiley & Sons, New York, 1968.

D. A. Dawson et X. Zheng, *Law of large numbers and central limit theorem for unbounded jump mean-field models*, Adv. in Appl. Math. **12** (1991), 293-326.

R. Ferland, X. Fernique et G. Giroux, *Compactness of the fluctuations associated with some generalized nonlinear Boltzmann equations*, Canad. J. Math. **44** (1992), 1192-1205.

R. Ferland et J.-C. Roberge, *Binomial Boltzmann processes: Convergence of the fluctuations*, Transport Theory and Statist. Phys. (à paraître).

X. Fernique, *Convergence en loi de fonctions aléatoires continues ou cadlag, propriétés de compacités des lois*, Lecture Notes in Math., vol. 1485, Springer-Verlag, Berlin, Heidelberg et New York, 1991, pp. 178-195.

———, *Convergence en loi de variables aléatoires et de fonctions aléatoires, propriétés de compacités des lois*, Rapport de recherche no. 194, Département de mathématiques et d'informatique, Université du Québec à Montréal, Montréal, 1992.

A. V. Skorohod, *Stochastic equations for complex systems*, D. Reidel Publishing Company, Dordrecht, 1988.

A.-S. Sznitman, *Équations de type de Boltzmann, spatialement homogènes*, Z. Wahrsch. Verw. Gebiete **66** (1984), 559-592.

ABSTRACT. Par une loi des grands nombres, on démontre l'existence d'une solution pour une équation d'évolution non linéaire du type de Boltzmann.

Département de mathématiques et d'informatique, Université de Sherbrooke, Sherbrooke (Québec) Canada, J1K 2R1

E-mail address: bezandry@dmi.usherb.ca,

Département de mathématiques et d'informatique, Université du Québec à Montréal, Case postale 8888, Succursale "A", Montréal (Québec) Canada, H3C 3P8

E-mail address: ferland@math.uqam.ca

Département de mathématiques et d'informatique, Université de Sherbrooke, Sherbrooke (Québec) Canada, J1K 2R1

E-mail address: giroux@dmi.usherb.ca,

Département de mathématiques et d'informatique, Université de Sherbrooke, Sherbrooke (Québec) Canada, J1K 2R1

E-mail address: roberge@dmi.usherb.ca

Variation of Iterated Brownian Motion

Krzysztof Burdzy

1. Introduction and main results

Suppose that X^1, X^2 and Y are independent standard Brownian motions starting from 0 and let

$$(1.1) \qquad X(t) = \begin{cases} X^1(t) & \text{if } t \geq 0, \\ X^2(-t) & \text{if } t < 0. \end{cases}$$

We will consider the process

$$(1.2) \qquad \{Z(t) \stackrel{\text{df}}{=} X(Y(t)), t \geq 0\}$$

which we will call "iterated Brownian motion" or simply IBM. It can be proved that Z uniquely determines X and Y (see [Burdzy 1992] for a precise statement). A Law of Iterated Logarithm for IBM is also proved in [Burdzy 1992].

We consider IBM to be a process of independent interest but there exists an intriguing relationship between this process (strictly speaking its modification) and "squared Laplacian" which was discovered by [Funaki 1979]. So far, the probabilistic approach to bi-harmonic functions is much less successful than the probabilistic treatment of harmonic functions. [Krylov 1960] and [Hochberg 1978] attacked the problem using a signed finitely additive measure with infinite variation. [Mądrecki 1992] and [Mądrecki and Rybaczuk 1992] have a genuine probabilistic approach but their processes take values in an exotic space. Both models are used to define stochastic integrals for processes with "4-th order" scaling properties. Higher order variations of the process play an important role in Mądrecki and Rybaczuk's construction of the stochastic integral. It is no surprise that the 4-th variation of their process is a deterministic linear function. The quadratic variation of their process is, in a suitable sense, a Brownian motion. See (3.16) in Hochberg's paper for a result with similar intuitive content.

In this paper, we study higher order variations of IBM with view towards possible applications to the construction of the stochastic integral with respect to IBM. We prove that the 4-th variation of IBM is a deterministic linear function. This clearly means that the quadratic variation is infinite (although we do not prove this). We show that, in a weak sense, the "signed quadratic variation" of IBM is distributed like Brownian motion.

Suppose that $\Lambda = \{s = t_0 \leq t_1 \leq \cdots \leq t_n = t\}$ is a partition of $[s,t]$. The mesh of the partition Λ is defined as $|\Lambda| \stackrel{\text{df}}{=} \max_{1 \leq k \leq n} |t_k - t_{k-1}|$.

1991 *Mathematics Subject Classification*. Primary: 60J65; Secondary: 60G17.
Supported in part by NSF grant DMS 91-00244 and AMS Centennial Research Fellowship.
This is the final form of the paper.

THEOREM 1.
(i) *Fix some* $0 \leq s < t$. *The following limit exists in* L^p *for every* $p < \infty$.

(1.3) $$\lim_{|\Lambda| \to 0} \sum_{k=1}^{n} (Z(t_k) - Z(t_{k-1}))^4 = 3(t-s).$$

(ii) *Suppose in addition that* $t_k - t_{k-1} = (t-s)/n$ *for every* k. *Then*

(1.4) $$\lim_{|\Lambda| \to 0} \sum_{k=1}^{n} (Z(t_k) - Z(t_{k-1}))^3 = 0 \quad \text{in } L^p, \, p < \infty.$$

THEOREM 2. *Suppose that* $t_0 = 0$ *and* $t_k - t_{k-1} = 1/n$ *for* $k \geq 1$. *Let*

$$V_n(t_m) = \sum_{k=1}^{m} (Z(t_k) - Z(t_{k-1}))^2 \operatorname{sgn}(Z(t_k) - Z(t_{k-1})).$$

Extend V_n *continuously to* $[0, \infty)$ *by linear interpolation on each interval* $[t_{k-1}, t_k]$. *The processes* $\{V_n(s), s \geq 0\}$ *converge in distribution as* $n \to \infty$ *to a Brownian motion* $\{B(s), s \geq 0\}$ *with variance* $\operatorname{Var} B(s) = 3s$.

REMARKS. (i) The assumption that $t_k - t_{k-1} = t_j - t_{j-1}$ for all j and k is imposed for convenience in Theorem 1 (ii) and Theorem 2. The assumption seems to be unnecessary but it makes the calculations somewhat more manageable.

(ii) A heuristic argument suggests that the sequence $\{V_n(s)\}_{n \geq 1}$ for a fixed $s > 0$, has no subsequences converging in probability.

(iii) The models considered by [Funaki 1979] and [Mądrecki and Rybaczuk 1992] involve complex numbers. It might be worth having a look at the complex version of IBM. Suppose that Y is a standard one-dimensional Brownian motion, X is a two-sided complex (i.e., two-dimensional but written in complex notation) Brownian motion and $Z(t) = X(Y(t))$. Let $V_n(t_m) = \sum_{k=1}^{m} (Z(t_k) - Z(t_{k-1}))^2$, in the notation of Theorem 2. Then Theorem 2 holds for this complex analogue of quadratic variation. The limiting process B for V_n's is a complex (i.e., two-dimensional) Brownian motion with the quadratic variation 3 times as large as the standard one. This result may be proved just like Theorem 2 by using the method of moments.

The proof of Theorem 2 is based on estimates of moments of V_n's. The estimates are quite delicate and it would take enormous amount of space to write them down in all detail. We will carefully examine one crucial estimate and indicate how this can be generalized to other moments.

We would like to thank Ron Pyke for simple proofs of Lemmas 1 and 2 below.

2. Proofs

Throughout the paper, c will stand for a strictly positive and finite constant which may change the value from line to line.

We will need the following standard estimate. Let $a > 0$.

$$(2.1) \quad \int_a^\infty x^2 \frac{1}{\sqrt{2\pi t}} \exp(-x^2/2t)dx$$

$$= -x\sqrt{t/2\pi}\exp(-x^2/2t)\Big|_{x=a}^{x=\infty} + \int_a^\infty \sqrt{t/2\pi}\exp(-x^2/2t)dx$$

$$\leq a\sqrt{t/2\pi}\exp(-a^2/2t) + \int_a^\infty (x/a)\sqrt{t/2\pi}\exp(-x^2/2t)dx$$

$$= a\sqrt{t/2\pi}\exp(-a^2/2t) + (t/a)\sqrt{t/2\pi}\exp(-a^2/2t)$$

$$= (a + t/a)\sqrt{t/2\pi}\exp(-a^2/2t).$$

The next estimate may be derived in an analogous way using integration by parts.

$$(2.2) \quad \int_a^\infty x^4 \frac{1}{\sqrt{2\pi t}} \exp(-x^2/2t)dx \leq (a^3 + ta + t^2/a)\sqrt{t/2\pi}\exp(-a^2/2t).$$

LEMMA 1. *Suppose that for every (integer) $k \geq 1$, $k \neq 2$, the k-th moment of a random variable R is the same as that of a normal random variable U with mean 0 and variance σ^2. Then R and U have the same distribution.*

The point of the lemma is that we do not assume that the variances of R and U are identical. The lemma would follow immediately from known results (see [Durrett 1991, Theorem 3.9]) if we added this assumption.

PROOF. The $2k$-th moment μ_{2k} of U is equal to $\sigma^{2k}(2k-1)!!$. Thus

$$\limsup_{k\to\infty} \mu_{2k}^{1/2k}/2k = \limsup_{k\to\infty} (\sigma^{2k}(2k-1)!!)^{1/2k}/2k = 0.$$

[Durrett 1991] shows in the proof of Theorem 3.9 that this implies that the characteristic function φ_U has the following series expansion valid on the whole real line.

$$\varphi_U(t) = 1 + \sum_{k=1}^\infty \frac{t^k}{k!}\varphi_U^{(k)}(0).$$

The characteristic function φ_R of R is represented by an analogous series. For every $k \neq 2$, the k-th moment of R is the same as that of U so $\varphi_R^{(k)}(0) = \varphi_U^{(k)}(0)$ and it follows that the series for φ_R and φ_U may differ by at most one term. Hence

$$\varphi_R(t) = \varphi_U(t) + at^2$$

and

$$\varphi_R(t)/t^2 = \varphi_U(t)/t^2 + a$$

for all $t \neq 0$. Since $|\varphi_U(t)| \leq 1$ and $|\varphi_R(t)| \leq 1$,

$$0 = \lim_{t\to\infty} \varphi_R(t)/t^2 = \lim_{t\to\infty} \varphi_U(t)/t^2 + a = a.$$

It follows that $a = 0$ and, therefore, U and R have identical characteristic functions. □

LEMMA 2. *Let $f_\sigma(x)$ denote the centered normal density with standard deviation σ and let $\psi(x,\sigma) = f_\sigma(0) - f_\sigma(x)$. For all $k, n \geq 0$ there exists $c = c(k,n) < \infty$ such that*

$$\int_{-\infty}^\infty |x|^n \psi(x,\rho_1)\ldots\psi(x,\rho_k)f_\sigma(x)dx \leq c\rho_1^{-3}\ldots\rho_k^{-3}\sigma^{n+2k}$$

for all $\rho_1, \ldots, \rho_k \geq 0$ (c does not depend on ρ_j's or σ).

PROOF. Since $1 - e^{-y} \leq y$ for all $y \geq 0$,

$$\psi(x, \rho) = \frac{1}{\sqrt{2\pi}\rho}(1 - \exp(-x^2/2\rho^2)) \leq \frac{1}{2\sqrt{2\pi}}\frac{x^2}{\rho^3}.$$

Let ξ denote a standard normal random variable. Then the integral in the statement of the lemma equals

$$E|\sigma\xi|^n \prod_{j=1}^{k} \psi(\sigma\xi, \rho_j) \leq E|\sigma\xi|^{n+2k}(2\sqrt{2\pi})^{-k} \prod_{j=1}^{k} \rho_j^{-3} = c(k,n)\sigma^{n+2k} \prod_{j=1}^{k} \rho_j^{-3}. \quad \square$$

PROOF OF THEOREM 1 (i). We will only prove the convergence in L^p for $p = 2$. The general case may be treated in an analogous way.

Recall that $\Lambda = \{s = t_0 \leq t_1 \leq \cdots \leq t_n = t\}$ is a partition of $[s,t]$. Let $\Delta_i t \stackrel{\text{df}}{=} t_i - t_{i-1}$ and $\Delta_i Z \stackrel{\text{df}}{=} Z(t_i) - Z(t_{i-1})$. We have

$$(2.3) \quad \left[\sum_{i=1}^{n}(\Delta_i Z)^4 - 3(t-s)\right]^2 = \left[\sum_{i=1}^{n}((\Delta_i Z)^4 - 3\Delta_i t)\right]^2$$

$$= \sum_{i,j=1}^{n}((\Delta_i Z)^4 - 3\Delta_i t)((\Delta_j Z)^4 - 3\Delta_j t).$$

It will suffice to prove that the expectation of the above random variable goes to 0 as $|\Lambda|$ goes to 0.

Fix some $\alpha \in (0,1)$ and suppose that $i \neq j$. Fix some numbers $u_{i-1} < u_i < u_{j-1} < u_j$. We will compute

$$E((\Delta_i Z)^4 - 3\Delta_i t)((\Delta_j Z)^4 - 3\Delta_j t)$$

given

$$A_1 = A_1(u_{i-1}, u_i, u_{j-1}, u_j)$$
$$\stackrel{\text{df}}{=} \{Y(t_{i-1}) = u_{i-1}, Y(t_i) = u_i, Y(t_{j-1}) = u_{j-1}, Y(t_j) = u_j\}.$$

Given this condition, the processes $\{X(u_i + t) - X(u_i), t \geq 0\}$ and $\{X(u_i - t) - X(u_i), t \geq 0\}$ are independent standard Brownian motions. Given A_1, the random variable $(\Delta_i Z)^4 - 3\Delta_i t$ is defined in terms of the first process and $(\Delta_j Z)^4 - 3\Delta_j t$ is defined in terms of the second one. Since $E(X(s_1) - X(s_2))^4 = 3(s_1 - s_2)^2$, it follows that

$$(2.4) \quad E\Big(((\Delta_i Z)^4 - 3\Delta_i t)((\Delta_j Z)^4 - 3\Delta_j t) \,\Big|\, A_1\Big)$$
$$= E((\Delta_i Z)^4 - 3\Delta_i t \mid A_1) E((\Delta_j Z)^4 - 3\Delta_j t \mid A_1)$$
$$= E\Big((X(u_i) - X(u_{i-1}))^4 - 3\Delta_i t \,\Big|\, A_1\Big) E\Big((X(u_j) - X(u_{j-1}))^4 - 3\Delta_j t \mid A_1\Big)$$
$$= (3(u_i - u_{i-1})^2 - 3\Delta_i t)(3(u_j - u_{j-1})^2 - 3\Delta_j t).$$

The same argument works for any u_{i-1}, u_i, u_{j-1}, and u_j such that the interval with endpoints u_{i-1} and u_i is disjoint from the interval with endpoints u_{j-1} and u_j.

Suppose that $r > 2|\Lambda|^{\alpha/2}$. Let

$$A_2 = A_2(r)$$
$$\stackrel{\mathrm{df}}{=} \{|Y(t_i) - Y(t_{j-1})| = r, |Y(t_{i-1}) - Y(t_i)| < \Delta_i t^{\alpha/2},$$
$$|Y(t_j) - Y(t_{j-1})| < \Delta_j t^{\alpha/2}\}.$$

The increments $Y(t_{i-1}) - Y(t_i)$ and $Y(t_j) - Y(t_{j-1})$ are independent given A_2. If the event A_2 occurs then the interval with endpoints $Y(t_{i-1})$ and $Y(t_i)$ is disjoint from the interval with endpoints $Y(t_{j-1})$ and $Y(t_j)$. Hence we may integrate over suitable u_{i-1}, u_i, u_{j-1} and u_j in (2.4) to obtain

$$(2.5) \quad E\Big(\big((\Delta_i Z)^4 - 3\Delta_i t\big)\big((\Delta_j Z)^4 - 3\Delta_j t\big) \,\Big|\, A_2\Big)$$
$$= \int_{-\Delta_i t^{\alpha/2}}^{\Delta_i t^{\alpha/2}} (3u^2 - 3\Delta_i t)\frac{1}{\sqrt{2\pi\Delta_i t}}\exp(-u^2/2\Delta_i t)du \times$$
$$\times \int_{-\Delta_j t^{\alpha/2}}^{\Delta_j t^{\alpha/2}} (3v^2 - 3\Delta_j t)\frac{1}{\sqrt{2\pi\Delta_j t}}\exp(-v^2/2\Delta_j t)dv.$$

Since

$$\int_{-\infty}^{\infty} (3u^2 - 3\Delta_i t)\frac{1}{\sqrt{2\pi\Delta_i t}}\exp(-u^2/2\Delta_i t)du = 0,$$

(2.1) implies that for some $\beta > 0$ and small $\Delta_i t$,

$$(2.6) \quad \left|\int_{-\Delta_i t^{\alpha/2}}^{\Delta_i t^{\alpha/2}} (3u^2 - 3\Delta_i t)\frac{1}{\sqrt{2\pi\Delta_i t}}\exp(-u^2/2\Delta_i t)du\right|$$
$$= \left|2\int_{\Delta_i t^{\alpha/2}}^{\infty} (3u^2 - 3\Delta_i t)\frac{1}{\sqrt{2\pi\Delta_i t}}\exp(-u^2/2\Delta_i t)du\right|$$
$$\leq \int_{\Delta_i t^{\alpha/2}}^{\infty} 6u^2 \frac{1}{\sqrt{2\pi\Delta_i t}}\exp(-u^2/2\Delta_i t)du$$
$$\leq 6(\Delta_i t^{\alpha/2} + \Delta_i t/\Delta_i t^{\alpha/2})\sqrt{\Delta_i t/2\pi}\exp(-\Delta_i t^{\alpha}/2\Delta_i t)$$
$$\leq \exp(-\Delta_i t^{-\beta}).$$

This and (2.5) show that for small $|\Lambda|$

$$(2.7) \quad |E(((\Delta_i Z)^4 - 3\Delta_i t)((\Delta_j Z)^4 - 3\Delta_j t) \mid A_2)| \leq \exp(-\Delta_i t^{-\beta} - \Delta_j t^{-\beta}).$$

Let

$$A_3 = A_3(|\Lambda|)$$
$$\stackrel{\mathrm{df}}{=} \{|Y(t_i) - Y(t_{j-1})| > 2|\Lambda|^{\alpha/2}, |Y(t_{i-1}) - Y(t_i)| < \Delta_i t^{\alpha/2},$$
$$|Y(t_j) - Y(t_{j-1})| < \Delta_j t^{\alpha/2}\}.$$

It follows from (2.7) that

$$(2.8) \quad |E(((\Delta_i Z)^4 - 3\Delta_i t)((\Delta_j Z)^4 - 3\Delta_j t)\mathbf{1}_{A_3})|$$
$$\leq |E(((\Delta_i Z)^4 - 3\Delta_i t)((\Delta_j Z)^4 - 3\Delta_j t) \mid A_3)|$$
$$\leq \exp(-\Delta_i t^{-\beta} - \Delta_j t^{-\beta}).$$

Since $EX_t^8 = ct^4$,

(2.9) $\quad E((\Delta_i Z)^8 + 9(\Delta_i t)^2 \mid Y(t_{i-1}) - Y(t_i) = r) = cr^4 + 9(\Delta_i t)^2.$

An argument similar to that in (2.6) (except that we would use (2.2) rather than (2.1)) gives for small $\Delta_i t$ and some $\eta > 0$

(2.10) $\quad |E((\Delta_i Z)^8 + 9(\Delta_i t)^2)\mathbf{1}_{\{Y(t_{i-1})-Y(t_i)>\Delta_i^{\alpha/2}\}}|$
$$= \left|2\int_{\Delta_i t^{\alpha/2}}^{\infty}(cu^4 + 9(\Delta_i t)^2)\frac{1}{\sqrt{2\pi\Delta_i t}}\exp(-u^2/2\Delta_i t)du\right| \leq \exp(-\Delta_i t^{-\eta}).$$

Let
$$A_4 = A_4(|\Lambda|) \stackrel{df}{=} \{|Y(t_i) - Y(t_{j-1})| > 2|\Lambda|^{\alpha/2}\}$$
$$\cap \left[\{|Y(t_{i-1}) - Y(t_i)| > \Delta_i t^{\alpha/2}\} \cup \{|Y(t_j) - Y(t_{j-1})| > \Delta_j t^{\alpha/2}\}\right].$$

Then (2.10) yields

(2.11) $\quad |E(((\Delta_i Z)^4 - 3\Delta_i t)((\Delta_j Z)^4 - 3\Delta_j t)\mathbf{1}_{A_4})|$
$$\leq (E((\Delta_i Z)^4 - 3\Delta_i t)^2 \mathbf{1}_{A_4})^{1/2}(E((\Delta_j Z)^4 - 3\Delta_j t)^2 \mathbf{1}_{A_4})^{1/2}$$
$$\leq (E2((\Delta_i Z)^8 + 9(\Delta_i t)^2)\mathbf{1}_{A_4})^{1/2}(E2((\Delta_j Z)^8 + 9(\Delta_j t)^2)\mathbf{1}_{A_4})^{1/2}$$
$$\leq 2\exp(-\Delta_i t^{-\eta}/2 - \Delta_j t^{-\eta}/2).$$

We define $A_5 = A_5(|\Lambda|)$ to be $\{|Y(t_i) - Y(t_{j-1})| > 2|\Lambda|^{\alpha/2}\}$. Combining (2.8) and (2.11) yields for small $\Delta_i t$ and $\Delta_j t$

(2.12) $\quad |E(((\Delta_i Z)^4 - 3\Delta_i t)((\Delta_j Z)^4 - 3\Delta_j t)\mathbf{1}_{A_5})|$
$$\leq \exp(-\Delta_i t^{-\beta} - \Delta_j t^{-\beta}) + 2\exp(-\Delta_i t^{-\eta}/2 - \Delta_j t^{-\eta}/2)$$
$$\leq 3\exp(-2\Delta_i t^{-\beta} - \Delta_j t^{-\eta}).$$

Choose $\alpha, \gamma \in (0,1)$ such that $\delta \stackrel{df}{=} \alpha/2 - \gamma/2 > 0$. Let
$$A_6 = \{|Y(t_i) - Y(t_{j-1})| \leq 2|\Lambda|^{\alpha/2}\}$$
and suppose that $|t_i - t_{j-1}| > |\Lambda|^\gamma$. Taking the expectation on both sides of (2.9) gives
$$E((\Delta_i Z)^8 + 9(\Delta_i t)^2) = c\Delta_i t^2.$$

It is easy to see that
$$P(A_6) \leq c|\Lambda|^{\alpha/2}/|\Lambda|^{\gamma/2} = c|\Lambda|^\delta.$$

By the independence of increments of Y,
$$E(((\Delta_i Z)^8 + 9(\Delta_i t)^2)\mathbf{1}_{A_6}) = c\Delta_i t^2 P(A_6) = c\Delta_i t^2 |\Lambda|^\delta,$$
$$E(((\Delta_j Z)^8 + 9(\Delta_j t)^2)\mathbf{1}_{A_6}) = c\Delta_j t^2 |\Lambda|^\delta.$$

Hence

(2.13) $\quad |E(((\Delta_i Z)^4 - 3\Delta_i t)((\Delta_j Z)^4 - 3\Delta_j t)\mathbf{1}_{A_6})|$
$$\leq (E((\Delta_i Z)^4 - 3\Delta_i t)^2 \mathbf{1}_{A_6})^{1/2}(E((\Delta_j Z)^4 - 3\Delta_j t)^2 \mathbf{1}_{A_6})^{1/2}$$
$$\leq (E2((\Delta_i Z)^8 + 9(\Delta_i t)^2)\mathbf{1}_{A_6})^{1/2}(E2((\Delta_j Z)^8 + 9(\Delta_j t)^2)\mathbf{1}_{A_6})^{1/2}$$
$$\leq 2(c\Delta_i t^2 |\Lambda|^\delta)^{1/2}(c\Delta_j t^2 |\Lambda|^\delta)^{1/2}$$
$$= 2c\Delta_i t \Delta_j t |\Lambda|^\delta.$$

A similar application of the Schwarz inequality gives

(2.14) $$|E(((\Delta_i Z)^4 - 3\Delta_i t)((\Delta_j Z)^4 - 3\Delta_j t)| \leq c\Delta_i t \Delta_j t$$

for any i and j.

We conclude from (2.12) and (2.13) that for sufficiently small $|\Lambda|$

(2.15) $$\sum_{\substack{i,j=1 \\ |t_i - t_{j-1}| > |\Lambda|^\gamma}}^{n} |E((\Delta_i Z)^4 - 3\Delta_i t)((\Delta_j Z)^4 - 3\Delta_j t)|$$

$$\leq \sum_{\substack{i,j=1 \\ |t_i - t_{j-1}| > |\Lambda|^\gamma}}^{n} (3\exp(-2\Delta_i t^{-\beta} - \Delta_j t^{-\eta}) + 2c\Delta_i t \Delta_j t |\Lambda|^\delta) \leq c|\Lambda|^\delta.$$

As for the remaining terms, we use the estimate (2.14).

$$\sum_{\substack{i,j=1 \\ |t_i - t_{j-1}| \leq |\Lambda|^\gamma}}^{n} |E((\Delta_i Z)^4 - 3\Delta_i t)((\Delta_j Z)^4 - 3\Delta_j t)| \leq \sum_{\substack{i,j=1 \\ |t_i - t_{j-1}| \leq |\Lambda|^\gamma}}^{n} c\Delta_i t \Delta_j t \leq c|\Lambda|^\gamma.$$

This and (2.15) show that

(2.16) $$\sum_{i,j=1}^{n} |E((\Delta_i Z)^4 - 3\Delta_i t)((\Delta_j Z)^4 - 3\Delta_j t)| \to 0$$

as $|\Lambda| \to 0$. This completes the proof of (1.3) in the case $p = 2$. We can prove in a similar way that

(2.17) $$\lim_{|\Lambda| \to 0} \sum_{j_1,\ldots,j_p=1}^{n} \left| E \prod_{k=1}^{p} ((\Delta_{j_k} Z)^4 - 3\Delta_{j_k} t) \right| = 0$$

for any $p < \infty$. This can be used to show that the limit in (1.3) exists in L^p for every $p < \infty$. □

PROOF OF THEOREM 2. The proof will be based on the method of moments, i.e., we will show that the moments of V converge to the moments of B.

Recall that $t_0 = 0$ and $t_k - t_{k-1} = 1/n$ for $k \geq 1$. Fix some $0 \leq s_1 < s_2$. Let $\Theta = \Theta(n)$ be the set of all k such that $s_1 \leq t_{k-1} < t_k \leq s_2$. The set $\Theta(n)$ is non-empty for sufficiently large n. Recall that $\Delta_k t = t_k - t_{k-1}$, $\Delta_k Z = Z(t_k) - Z(t_{k-1})$, $\Delta_k Y = Y(t_k) - Y(t_{k-1})$ and let

$$\Delta_k^\pm Z^2 = (Z(t_k) - Z(t_{k-1}))^2 \operatorname{sgn}(Z(t_k) - Z(t_{k-1})).$$

We start with some estimates needed for computing the moments of the increments of V_n.

For every s, the distribution of $X(s)$ is normal so $EX^{2j}(s) = (2j-1)!!|s|^j$ [Durrett 1991, Excercise 3.18]. By conditioning on the value of $\Delta_k Y$ we obtain for some $d_j > 0$,

(2.18) $$E(\Delta_k Z)^{2j} = \int_{-\infty}^{\infty} EX^{2j}(s) \frac{1}{\sqrt{2\pi \Delta_k t}} \exp(-s^2/2\Delta_k t) ds$$

$$= \int_{-\infty}^{\infty} (2j-1)!!|s|^j \frac{1}{\sqrt{2\pi \Delta_k t}} \exp(-s^2/2\Delta_k t) ds = d_j (\Delta_k t)^{j/2}.$$

Hence, $E(\Delta_k Z)^{2j} < \infty$ for all $j < \infty$.

The main contribution in our moment estimates will come from the expectations of the form

$$(2.19) \qquad E\left(\sum_{k_1,\ldots,k_m \in \Theta} (\Delta_{k_1}^\pm Z^2)^2 (\Delta_{k_2}^\pm Z^2)^2 \ldots (\Delta_{k_m}^\pm Z^2)^2\right).$$

Suppose that $m = 2$. We have

$$\sum_{j,k\in\Theta} (\Delta_j^\pm Z^2)^2 (\Delta_k^\pm Z^2)^2 = \sum_{j,k\in\Theta} \Delta_j Z^4 \Delta_k Z^4$$

$$= \left[\sum_{j\in\Theta}((\Delta_j Z)^4 - 3\Delta_j t)\right]^2 - \sum_{j,k\in\Theta} 9\Delta_j t \Delta_k t + 2 \sum_{j,k\in\Theta} 3\Delta_j Z^4 \Delta_k t.$$

It is easy to check that $d_2 = 3$ in (2.18). Thus

$$E 3\Delta_j Z^4 \Delta_k t = 9\Delta_j t \Delta_k t$$

and, therefore,

$$E \sum_{j,k\in\Theta} (\Delta_j^\pm Z^2)^2 (\Delta_k^\pm Z^2)^2 = E\left[\sum_{j\in\Theta}((\Delta_j Z)^4 - 3\Delta_j t)\right]^2 + \sum_{j,k\in\Theta} 9\Delta_j t \Delta_k t.$$

The expectation on the right hand side goes to 0 as n goes to ∞, by (2.16). Hence

$$(2.20) \qquad \lim_{n\to\infty} E \sum_{j,k\in\Theta} (\Delta_j^\pm Z^2)^2 (\Delta_k^\pm Z^2)^2 = \lim_{n\to\infty} \sum_{j,k\in\Theta} 9\Delta_j t \Delta_k t = 9(s_2 - s_1)^2.$$

In order to estimate the expectations in (2.19) for $m \geq 3$, we use induction. We will treat only the case $m = 3$.

$$\sum_{i,j,k\in\Theta} (\Delta_i^\pm Z^2)^2 (\Delta_j^\pm Z^2)^2 (\Delta_k^\pm Z^2)^2 = \sum_{i,j,k\in\Theta} \Delta_i Z^4 \Delta_j Z^4 \Delta_k Z^4$$

$$= \left[\sum_{j\in\Theta}((\Delta_j Z)^4 - 3\Delta_j t)\right]^3 + \sum_{i,j,k\in\Theta} 27\Delta_i t \Delta_j t \Delta_k t$$

$$- 3\sum_{i,j,k\in\Theta} 9\Delta_i Z^4 \Delta_j t \Delta_k t + 3\sum_{i,j,k\in\Theta} 3\Delta_i Z^4 \Delta_j Z^4 \Delta_k t.$$

By (2.17), (2.18) and (2.20),

$$\lim_{n\to\infty} E\left[\sum_{j\in\Theta}((\Delta_j Z)^4 - 3\Delta_j t)\right]^3 = 0,$$

$$\lim_{n\to\infty} E\left[3\sum_{i,j,k\in\Theta} 9\Delta_i Z^4 \Delta_j t \Delta_k t\right] = \lim_{n\to\infty} E\left[3\sum_{i,j,k\in\Theta} 27\Delta_i t \Delta_j t \Delta_k t\right] = 81(s_2 - s_1)^3,$$

$$\lim_{n\to\infty} E\left[3\sum_{i,j,k\in\Theta} 3\Delta_i Z^4 \Delta_j Z^4 \Delta_k t\right] = \lim_{n\to\infty} (s_2 - s_1) E\left[3\sum_{i,j\in\Theta} 3\Delta_i Z^4 \Delta_j Z^4\right]$$

$$= 81(s_2 - s_1)^3.$$

Hence,
$$\lim_{n\to\infty} E \sum_{i,j,k\in\Theta} (\Delta_i^\pm Z^2)^2 (\Delta_j^\pm Z^2)^2 (\Delta_k^\pm Z^2)^2$$
$$= \lim_{n\to\infty} E \sum_{i,j,k\in\Theta} 27 \Delta_i t \Delta_j t \Delta_k t = 27(s_2 - s_1)^3.$$

In the same way, using induction, we may prove that

(2.21) $$\lim_{n\to\infty} E\left(\sum_{k_1,\ldots,k_m \in \Theta} (\Delta_{k_1}^\pm Z^2)^2 (\Delta_{k_2}^\pm Z^2)^2 \ldots (\Delta_{k_m}^\pm Z^2)^2 \right) = 3^m (s_2 - s_1)^m.$$

Suppose that $q_1, \ldots, q_m \geq 1$ are integers and at least one of them is strictly greater than 1. By Hölder's inequality and (2.18),

$$E[(\Delta_{k_1}^\pm Z^2)^{2q_1} (\Delta_{k_2}^\pm Z^2)^{2q_2} \ldots (\Delta_{k_m}^\pm Z^2)^{2q_m}]$$
$$\leq (E(\Delta_{k_1}^\pm Z^2)^{2mq_1})^{1/m} \ldots (E(\Delta_{k_m}^\pm Z^2)^{2mq_m})^{1/m}$$
$$\leq (d_{2mq_1} (\Delta_{k_1} t)^{mq_1})^{1/m} \ldots (d_{2mq_m} (\Delta_{k_m} t)^{mq_m})^{1/m}$$
$$\leq c(1/n)^{q_1 + \cdots + q_m},$$

and, therefore,

(2.22) $$\lim_{n\to\infty} E\left(\sum_{k_1,\ldots,k_m \in \Theta} (\Delta_{k_1}^\pm Z^2)^{2q_1} (\Delta_{k_2}^\pm Z^2)^{2q_2} \ldots (\Delta_{k_m}^\pm Z^2)^{2q_m} \right)$$
$$\leq \lim_{n\to\infty} n^m c(1/n)^{q_1 + \cdots + q_m} = 0.$$

The absolute value of the difference between
$$\sum_{k_1,\ldots,k_m \in \Theta} (\Delta_{k_1}^\pm Z^2)^2 (\Delta_{k_2}^\pm Z^2)^2 \ldots (\Delta_{k_m}^\pm Z^2)^2$$
and
$$\sum_{\substack{k_1,\ldots,k_m \in \Theta \\ k_1,\ldots,k_m \text{ distinct}}} (\Delta_{k_1}^\pm Z^2)^2 (\Delta_{k_2}^\pm Z^2)^2 \ldots (\Delta_{k_m}^\pm Z^2)^2$$
is bounded by a finite sum of the expressions of the form
$$\sum_{k_1,\ldots,k_i \in \Theta} (\Delta_{k_1}^\pm Z^2)^{2q_1} (\Delta_{k_2}^\pm Z^2)^{2q_2} \ldots (\Delta_{k_i}^\pm Z^2)^{2q_i}$$
where at least one of the q_j's is greater than 1. It follows from (2.21) and (2.22) that

(2.23) $$\lim_{n\to\infty} E\left(\sum_{\substack{k_1,\ldots,k_m \in \Theta \\ k_1,\ldots,k_m \text{ distinct}}} (\Delta_{k_1}^\pm Z^2)^2 (\Delta_{k_2}^\pm Z^2)^2 \ldots (\Delta_{k_m}^\pm Z^2)^2 \right) = 3^m (s_2 - s_1)^m.$$

Next we tackle the expectations of the form
$$E\left(\sum_{k_1,\ldots,k_m \in \Theta} (\Delta_{k_1}^\pm Z^2)^{q_1} (\Delta_{k_2}^\pm Z^2)^{q_2} \ldots (\Delta_{k_m}^\pm Z^2)^{q_m} \right),$$
where $q_1, \ldots, q_m \geq 1$ are integers but they are not necessarily even. We will tacitly assume that the sum is taken over indices which are pairwise distinct (the other

terms appear in sums with different exponents q_k). We will illustrate the method by analyzing in detail only one sum, namely,

$$E\left(\sum_{\substack{k_1,\ldots,k_4\in\Theta \\ k_1<k_2<k_3<k_4}} (\Delta_{k_1}^{\pm}Z^2)^2\Delta_{k_2}^{\pm}Z^2(\Delta_{k_3}^{\pm}Z^2)^2\Delta_{k_4}^{\pm}Z^2\right).$$

Note that the indices in the last sum are ordered — the sum with unordered indices may be obtained by adding a finite number of sums with ordered indices. Let

$$A_1 = A_1(u_{k_1-1}, u_{k_1}, u_{k_2-1}, u_{k_2}, u_{k_3-1}, u_{k_3}, u_{k_4-1}, u_{k_4})$$
$$\stackrel{\mathrm{df}}{=} \{Y(t_{k_i-1}) = u_{k_i-1}, Y(t_{k_i}) = u_{k_i},\ i=1,2,3,4\}.$$

Let A_2 denote the event that there exists a number a such that for each i, the interval with endpoints u_{k_i-1} and u_{k_i} is either contained in (a, ∞) or in $(-\infty, a)$ and that each half-line contains at least one of these intervals. Suppose for a moment that u_{k_i-1} and u_{k_i} are such that A_2 holds, for example, the intervals corresponding to $i = 1, 2$ are in (a, ∞) and the other two are contained in the other half-line. The same argument that leads to (2.4) gives in the present case

(2.24) $\quad E((\Delta_{k_1}^{\pm}Z^2)^2\Delta_{k_2}^{\pm}Z^2(\Delta_{k_3}^{\pm}Z^2)^2\Delta_{k_4}^{\pm}Z^2 \mid A_1)$
$$= E((\Delta_{k_1}^{\pm}Z^2)^2\Delta_{k_2}^{\pm}Z^2 \mid A_1)E((\Delta_{k_3}^{\pm}Z^2)^2\Delta_{k_4}^{\pm}Z^2 \mid A_1).$$

Both conditional expectations on the right hand side are equal to zero since the random variables have symmetric (conditional) distributions. If the event A_2 is realized in some other way, the conditional expectation on the left hand side of (2.24) may be factored in some other way such that at least one conditional expectation on the right hand side is equal to 0 because of symmetry of the involved distribution. Hence,

(2.25) $\quad E((\Delta_{k_1}^{\pm}Z^2)^2\Delta_{k_2}^{\pm}Z^2(\Delta_{k_3}^{\pm}Z^2)^2\Delta_{k_4}^{\pm}Z^2\mathbf{1}_{A_2}) = 0.$

For arbitrary values of u_i's,

(2.26) $\quad E((\Delta_{k_1}^{\pm}Z^2)^2\Delta_{k_2}^{\pm}Z^2(\Delta_{k_3}^{\pm}Z^2)^2\Delta_{k_4}^{\pm}Z^2 \mid A_1)$
$$\leq (E((\Delta_{k_1}^{\pm}Z^2)^4 \mid A_1))^{1/2}(E((\Delta_{k_2}^{\pm}Z^2)^2 \mid A_1))^{1/2}$$
$$\times (E((\Delta_{k_3}^{\pm}Z^2)^4 \mid A_1))^{1/2}(E((\Delta_{k_4}^{\pm}Z^2)^2 \mid A_1))^{1/2}$$
$$= (EX^8(|u_{k_1-1} - u_{k_1}|))^{1/2}(EX^4(|u_{k_2-1} - u_{k_2}|))^{1/2}$$
$$\times (EX^8(|u_{k_3-1} - u_{k_3}|))^{1/2}(EX^4(|u_{k_4-1} - u_{k_4}|))^{1/2}$$
$$\leq c|u_{k_1-1} - u_{k_1}|^2|u_{k_2-1} - u_{k_2}||u_{k_3-1} - u_{k_3}|^2|u_{k_4-1} - u_{k_4}|.$$

Note that

(2.27) $\quad E((\Delta_{k_1}^{\pm}Z^2)^2\Delta_{k_2}^{\pm}Z^2(\Delta_{k_3}^{\pm}Z^2)^2\Delta_{k_4}^{\pm}Z^2$
$$\mid A_1(u_{k_1-1}, u_{k_1}, u_{k_2-1}, u_{k_2}, u_{k_3-1}, u_{k_3}, u_{k_4-1}, u_{k_4}))$$
$$= -E((\Delta_{k_1}^{\pm}Z^2)^2\Delta_{k_2}^{\pm}Z^2(\Delta_{k_3}^{\pm}Z^2)^2\Delta_{k_4}^{\pm}Z^2$$
$$\mid A_1(u_{k_1-1}, u_{k_1}, u_{k_2}, u_{k_2-1}, u_{k_3-1}, u_{k_3}, u_{k_4-1}, u_{k_4}))$$

because when we exchange the roles of u_{k_2-1} and u_{k_2}, we change, in a sense, the sign of $\Delta_{k_2}^{\pm} Z^2$. For the same reason we have

$$(2.28) \quad E((\Delta_{k_1}^{\pm} Z^2)^2 \Delta_{k_2}^{\pm} Z^2 (\Delta_{k_3}^{\pm} Z^2)^2 \Delta_{k_4}^{\pm} Z^2 \\ \mid A_1(u_{k_1-1}, u_{k_1}, u_{k_2-1}, u_{k_2}, u_{k_3-1}, u_{k_3}, u_{k_4-1}, u_{k_4})) \\ = -E((\Delta_{k_1}^{\pm} Z^2)^2 \Delta_{k_2}^{\pm} Z^2 (\Delta_{k_3}^{\pm} Z^2)^2 \Delta_{k_4}^{\pm} Z^2 \\ \mid A_1(u_{k_1-1}, u_{k_1}, u_{k_2-1}, u_{k_2}, u_{k_3-1}, u_{k_3}, u_{k_4}, u_{k_4-1})).$$

Let $\rho = \rho(r_1, r_2, r_3, r_4) = |r_1| + |r_2| + |r_3| + |r_4|$. We claim that

$$(2.29) \quad |E((\Delta_{k_1}^{\pm} Z^2)^2 \Delta_{k_2}^{\pm} Z^2 (\Delta_{k_3}^{\pm} Z^2)^2 \Delta_{k_4}^{\pm} Z^2)| \\ \leq \left| \int \left[|r_1|^2 |r_2| |r_3|^2 |r_4| P(Y(t_{k_1}) - Y(t_{k_1-1}) \in dr_1) P(Y(t_{k_2}) - Y(t_{k_2-1}) \in dr_2) \right. \right. \\ \left. \times P(Y(t_{k_3}) - Y(t_{k_3-1}) \in dr_3) P(Y(t_{k_4}) - Y(t_{k_4-1}) \in dr_4) \right] \\ \mathbf{1}_{|r_5| \leq \rho} \mathbf{1}_{|r_6| \leq \rho} \left[P(Y(t_{k_2-1}) - Y(t_{k_1}) \in dr_5) P(Y(t_{k_3-1}) - Y(t_{k_2}) \in dr_6) \right. \\ \left. - P(Y(t_{k_2-1}) - Y(t_{k_1}) \in d(r_5 + r_2)) P(Y(t_{k_3-1}) - Y(t_{k_2}) \in d(r_6 - r_2)) \right] \\ \left. \mathbf{1}_{|r_7| \leq \rho} \left[P(Y(t_{k_4-1}) - Y(t_{k_3}) \in dr_7) - P(Y(t_{k_4-1}) - Y(t_{k_3}) \in d(r_7 + r_4)) \right] \right|.$$

The terms in the first pair of large square brackets are justified by (2.26). The terms in the second and third pair of large square brackets come from (2.27)–(2.28). The presence of indicator functions follows from (2.25).

We will now estimate

$$(2.30) \quad \left| \int \mathbf{1}_{|r_5| \leq \rho} \mathbf{1}_{|r_6| \leq \rho} [P(Y(t_{k_2-1}) - Y(t_{k_1}) \in dr_5) P(Y(t_{k_3-1}) - Y(t_{k_2}) \in dr_6) \\ - P(Y(t_{k_2-1}) - Y(t_{k_1}) \in d(r_5 + r_2)) P(Y(t_{k_3-1}) - Y(t_{k_2}) \in d(r_6 - r_2))] \right|.$$

Recall $\psi(x, \sigma)$ from Lemma 2. Since $t_{k_2-1} - t_{k_1} = (k_2 - 1 - k_1)/n$, the standard deviation of $Y(t_{k_2-1}) - Y(t_{k_1})$ is equal to $((k_2 - 1 - k_1)/n)^{1/2}$. Hence

$$(2.31) \quad |P(Y(t_{k_2-1}) - Y(t_{k_1}) \in dr_5) - P(Y(t_{k_2-1}) - Y(t_{k_1}) \in d(r_5 + r_2))| \\ \leq (\psi(r_5, ((k_2 - 1 - k_1)/n)^{1/2}) + \psi(r_5 + r_2, ((k_2 - 1 - k_1)/n)^{1/2})) dr_5 \\ \leq 2\psi(2\rho, ((k_2 - 1 - k_1)/n)^{1/2}) dr_5$$

provided $|r_5| \leq \rho$. Similarly,

$$(2.32) \quad |P(Y(t_{k_3-1}) - Y(t_{k_2}) \in dr_6) - P(Y(t_{k_3-1}) - Y(t_{k_2}) \in d(r_6 - r_2))| \\ \leq 2\psi(2\rho, ((k_3 - 1 - k_2)/n)^{1/2}) dr_6$$

assuming $|r_6| \leq \rho$. We will assume temporarily that $k_j - k_{j-1} > 1$ for all j in order to avoid normal variables with zero variance and in order to be able to use Lemma 2. We will get rid of this assumption later.

For any reals a, Δa, b and Δb,

$$|ab - (a + \Delta a)(b + \Delta b)| \leq |a \Delta b| + |b \Delta a| + |\Delta a \Delta b|.$$

This, (2.31) and (2.32) imply

$$|P(Y(t_{k_2-1}) - Y(t_{k_1}) \in dr_5)P(Y(t_{k_3-1}) - Y(t_{k_2}) \in dr_6)$$
$$- P(Y(t_{k_2-1}) - Y(t_{k_1}) \in d(r_5 + r_2))P(Y(t_{k_3-1}) - Y(t_{k_2}) \in d(r_6 - r_2))|$$
$$\leq 2\psi(2\rho, ((k_2 - 1 - k_1)/n)^{1/2})dr_5 P(Y(t_{k_3-1}) - Y(t_{k_2}) \in dr_6)$$
$$+ 2\psi(2\rho, ((k_3 - 1 - k_2)/n)^{1/2})dr_6 P(Y(t_{k_2-1}) - Y(t_{k_1}) \in dr_5)$$
$$+ 2\psi(2\rho, ((k_2 - 1 - k_1)/n)^{1/2})dr_5 2\psi(2\rho, ((k_3 - 1 - k_2)/n)^{1/2})dr_6.$$

The integral in (2.30) is, therefore, bounded by

$$(2.33) \quad \left| \int \mathbf{1}_{|r_5| \leq \rho} \mathbf{1}_{|r_6| \leq \rho} \right.$$
$$[2\psi(2\rho, ((k_2 - 1 - k_1)/n)^{1/2})dr_5 P(Y(t_{k_3-1}) - Y(t_{k_2}) \in dr_6)$$
$$+ 2\psi(2\rho, ((k_3 - 1 - k_2)/n)^{1/2})dr_6 P(Y(t_{k_2-1}) - Y(t_{k_1}) \in dr_5)$$
$$\left. + 2\psi(2\rho, ((k_2 - 1 - k_1)/n)^{1/2})dr_5 2\psi(2\rho, ((k_3 - 1 - k_2)/n)^{1/2})dr_6] \right|$$
$$\leq c\psi(2\rho, ((k_2 - 1 - k_1)/n)^{1/2})\rho^2((k_3 - 1 - k_2)/n)^{-1/2}$$
$$+ c\psi(2\rho, ((k_3 - 1 - k_2)/n)^{1/2})\rho^2((k_2 - 1 - k_1)/n)^{-1/2}$$
$$+ c\psi(2\rho, ((k_2 - 1 - k_1)/n)^{1/2})\rho^2 \psi(2\rho, ((k_3 - 1 - k_2)/n)^{1/2}).$$

We can prove in the same way that

$$(2.34) \quad \left| \int \mathbf{1}_{|r_7| \leq \rho} [P(Y(t_{k_4-1}) - Y(t_{k_3}) \in dr_7) \right.$$
$$\left. - P(Y(t_{k_4-1}) - Y(t_{k_3}) \in d(r_7 + r_4))] \right| \leq c\psi(2\rho, ((k_4 - 1 - k_3)/n)^{1/2})\rho.$$

We now substitute this estimate and (2.33) into (2.29) to obtain

$$(2.35) \quad |E((\Delta_{k_1}^\pm Z^2)^2 \Delta_{k_2}^\pm Z^2 (\Delta_{k_3}^\pm Z^2)^2 \Delta_{k_4}^\pm Z^2)|$$
$$\leq \left| \int |r_1|^2 |r_2| |r_3|^2 |r_4| P(Y(t_{k_1}) - Y(t_{k_1-1}) \in dr_1) P(Y(t_{k_2}) - Y(t_{k_2-1}) \in dr_2) \right.$$
$$P(Y(t_{k_3}) - Y(t_{k_3-1}) \in dr_3) P(Y(t_{k_4}) - Y(t_{k_4-1}) \in dr_4)$$
$$[c\psi(2\rho, ((k_2 - 1 - k_1)/n)^{1/2})\rho^2((k_3 - 1 - k_2)/n)^{-1/2}$$
$$+ c\psi(2\rho, ((k_3 - 1 - k_2)/n)^{1/2})\rho^2((k_2 - 1 - k_1)/n)^{-1/2}$$
$$+ c\psi(2\rho, ((k_2 - 1 - k_1)/n)^{1/2})\rho^2 \psi(2\rho, ((k_3 - 1 - k_2)/n)^{1/2})]$$
$$\left. c\psi(2\rho, ((k_4 - 1 - k_3)/n)^{1/2})\rho \right|.$$

By multiplying out the expression in brackets on the right hand side we obtain three terms under the integral sign. The first one is equal to

$$(2.36) \quad \int |r_1|^2 |r_2||r_3|^2 |r_4| c\psi(2\rho, ((k_2 - 1 - k_1)/n)^{1/2}) \rho^2 ((k_3 - 1 - k_2)/n)^{-1/2}$$

$$c\psi(2\rho, ((k_4 - 1 - k_3)/n)^{1/2}) \rho P(Y(t_{k_1}) - Y(t_{k_1-1}) \in dr_1)$$
$$P(Y(t_{k_2}) - Y(t_{k_2-1}) \in dr_2) P(Y(t_{k_3}) - Y(t_{k_3-1}) \in dr_3)$$
$$P(Y(t_{k_4}) - Y(t_{k_4-1}) \in dr_4)$$
$$= \int |r_1|^2 |r_2||r_3|^2 |r_4| c\psi(2(|r_1| + |r_2| + |r_3| + |r_4|), ((k_2 - 1 - k_1)/n)^{1/2})$$
$$(|r_1| + |r_2| + |r_3| + |r_4|)^2 ((k_3 - 1 - k_2)/n)^{-1/2}$$
$$c\psi(2(|r_1| + |r_2| + |r_3| + |r_4|), ((k_4 - 1 - k_3)/n)^{1/2})(|r_1| + |r_2| + |r_3| + |r_4|)$$
$$P(Y(t_{k_1}) - Y(t_{k_1-1}) \in dr_1) P(Y(t_{k_2}) - Y(t_{k_2-1}) \in dr_2)$$
$$P(Y(t_{k_3}) - Y(t_{k_3-1}) \in dr_3) P(Y(t_{k_4}) - Y(t_{k_4-1}) \in dr_4).$$

Note that

$$\psi(2(|r_1| + |r_2| + |r_3| + |r_4|), ((k_2 - 1 - k_1)/n)^{1/2}) \leq \sum_{j=1}^{4} \psi(8|r_j|, ((k_2 - 1 - k_1)/n)^{1/2})$$

and the standard deviation of $Y(t_{k_j}) - Y(t_{k_j-1})$ is equal to $n^{-1/2}$. This and Lemma 2 can be used to show that (2.36) is bounded by

$$c((k_2 - 1 - k_1)/n)^{-3/2} ((k_3 - 1 - k_2)/n)^{-1/2} ((k_4 - 1 - k_3)/n)^{-3/2} n^{-13/2}.$$

The other terms in (2.35) may be treated in a similar way so

$$(2.37) \quad |E((\Delta_{k_1}^{\pm} Z^2)^2 \Delta_{k_2}^{\pm} Z^2 (\Delta_{k_3}^{\pm} Z^2)^2 \Delta_{k_4}^{\pm} Z^2)|$$
$$\leq c((k_2 - 1 - k_1)/n)^{-3/2} ((k_3 - 1 - k_2)/n)^{-1/2} ((k_4 - 1 - k_3)/n)^{-3/2} n^{-13/2}$$
$$+ c((k_3 - 1 - k_2)/n)^{-3/2} ((k_2 - 1 - k_1)/n)^{-1/2} ((k_4 - 1 - k_3)/n)^{-3/2} n^{-13/2}$$
$$+ c((k_2 - 1 - k_1)/n)^{-3/2} ((k_3 - 1 - k_2)/n)^{-3/2} ((k_4 - 1 - k_3)/n)^{-3/2} n^{-15/2}.$$

Now we discuss our temporary assumption that $k_j - k_{j-1} > 1$. If $k_4 - k_3 = 1$ then the last term in large square brackets in (2.29) should be replaced by 1. If $k_4 - k_3 = 2$ then the effect of the same term on our estimate is that of a multiplicative constant (see, e.g., (2.34)). It follows that the terms corresponding to $k_j - k_{j-1} = 1$ contribute to our sums as much as those corresponding to $k_j - k_{j-1} = 2$ (up to a multiplicative constant). Having this in mind, we may write

$$(2.38) \quad \left| E\left(\sum_{\substack{k_1, \ldots, k_4 \in \Theta \\ k_1 < k_2 < k_3 < k_4}} (\Delta_{k_1}^{\pm} Z^2)^2 \Delta_{k_2}^{\pm} Z^2 (\Delta_{k_3}^{\pm} Z^2)^2 \Delta_{k_4}^{\pm} Z^2 \right) \right|$$

$$\leq \sum_{\substack{k_1, \ldots, k_4 \in \Theta \\ k_1 < k_2 < k_3 < k_4 \\ k_j - k_{j-1} > 1}} \left[c(\tfrac{1}{n}(k_2 - 1 - k_1))^{-3/2} (\tfrac{1}{n}(k_3 - 1 - k_2))^{-1/2} (\tfrac{1}{n}(k_4 - 1 - k_3))^{-3/2} n^{-13/2} \right.$$

$$+ c(\tfrac{1}{n}(k_3 - 1 - k_2))^{-3/2} (\tfrac{1}{n}(k_2 - 1 - k_1))^{-1/2} (\tfrac{1}{n}(k_4 - 1 - k_3))^{-3/2} n^{-13/2}$$

$$\left. + c(\tfrac{1}{n}(k_2 - 1 - k_1))^{-3/2} (\tfrac{1}{n}(k_3 - 1 - k_2))^{-3/2} (\tfrac{1}{n}(k_4 - 1 - k_3))^{-3/2} n^{-15/2} \right].$$

We have

(2.39)
$$\sum_{\substack{k_1,\ldots,k_4 \in \Theta \\ k_1 < k_2 < k_3 < k_4 \\ k_j - k_{j-1} > 1}} c(\tfrac{1}{n}(k_2 - 1 - k_1))^{-3/2}(\tfrac{1}{n}(k_3 - 1 - k_2))^{-1/2}(\tfrac{1}{n}(k_4 - 1 - k_3))^{-3/2} n^{-13/2}$$

$$\leq cn^{-3} \sum_{k_1=1}^{n} \sum_{j_1=1}^{\infty} j_1^{-3/2} \sum_{j_2=1}^{n} j_2^{-1/2} \sum_{j_3=1}^{\infty} j_3^{-3/2}$$

$$\leq cn^{-3} n n^{1/2} = cn^{-3/2}.$$

Similar bounds hold for other terms in (2.38) so that

$$\left| E\left(\sum_{\substack{k_1,\ldots,k_4 \in \Theta \\ k_1 < k_2 < k_3 < k_4}} (\Delta_{k_1}^{\pm} Z^2)^2 \Delta_{k_2}^{\pm} Z^2 (\Delta_{k_3}^{\pm} Z^2)^2 \Delta_{k_4}^{\pm} Z^2 \right) \right| \leq cn^{-3/2} + cn^{-3/2} + cn^{-2}$$

and, therefore,

$$\lim_{n \to \infty} \left| E\left(\sum_{\substack{k_1,\ldots,k_4 \in \Theta \\ k_1 < k_2 < k_3 < k_4}} (\Delta_{k_1}^{\pm} Z^2)^2 \Delta_{k_2}^{\pm} Z^2 (\Delta_{k_3}^{\pm} Z^2)^2 \Delta_{k_4}^{\pm} Z^2 \right) \right| = 0.$$

Now we will explain how this result may be generalized. Suppose that q_j is either equal to 1 or 2 for $j = 1, \ldots, m$. The expectation

$$|E((\Delta_{k_1}^{\pm} Z^2)^{q_1} (\Delta_{k_2}^{\pm} Z^2)^{q_2} \ldots (\Delta_{k_m}^{\pm} Z^2)^{q_m})|$$

may be bounded as in (2.37) by a product of factors corresponding to $(\Delta_{k_j}^{\pm} Z^2)^{q_j}$. If $q_j = 1$ and $j > 1$ then the factor is of the form

$$c((k_j - 1 - k_{j-1})/n)^{-3/2} n^{-2}$$

and when $q_j = 2$, $j > 1$, then the factor is

$$c((k_j - 1 - k_{j-1})/n)^{-1/2} n^{-3/2}.$$

For $j = 1$, the factor is n^{-1} or $n^{-1/2}$, depending on whether $q_j = 2$ or 1. Summing as in (2.39) shows that each factor corresponding to $(\Delta_{k_j}^{\pm} Z^2)^{q_j}$ contributes $cn^{-1/2}$ to the sum of expectations provided $j > 1$. The contribution from the first factor is either c or $cn^{1/2}$. Hence

(2.40) $$\left| E\left(\sum_{k_1,\ldots,k_m \in \Theta} (\Delta_{k_1}^{\pm} Z^2)^{q_1} (\Delta_{k_2}^{\pm} Z^2)^{q_2} \ldots (\Delta_{k_m}^{\pm} Z^2)^{q_m} \right) \right| \leq cn^{(-m+2)/2}.$$

If $m > 3$ then

(2.41) $$\lim_{n \to \infty} \left| E\left(\sum_{k_1,\ldots,k_m \in \Theta} (\Delta_{k_1}^{\pm} Z^2)^{q_1} (\Delta_{k_2}^{\pm} Z^2)^{q_2} \ldots (\Delta_{k_m}^{\pm} Z^2)^{q_m} \right) \right| = 0.$$

Suppose that some q_j's are greater than 2 and at least one of them is odd (we need this assumption to prove (2.25)). The only part of the proof that will be affected by this change in the assumptions is that the powers of r_j's in (2.29) will increase. If every q_j is equal to $2p_j + 1$ or $2p_j + 2$ for some integer $p_j \geq 0$ then instead of (2.40) we will have

$$(2.42) \quad \left| E\left(\sum_{k_1,\ldots,k_m \in \Theta} (\Delta_{k_1}^{\pm} Z^2)^{q_1} (\Delta_{k_2}^{\pm} Z^2)^{q_2} \ldots (\Delta_{k_m}^{\pm} Z^2)^{q_m} \right) \right| \leq cn^{-p_1-\cdots-p_m+(-m+2)/2}.$$

Next we will analyse the fourth moment of V_n's. We have

$$(2.43) \quad \left(\sum_{k \in \Theta} \Delta_k^{\pm} Z^2 \right)^4 = 3 \sum_{\substack{j,k \in \Theta \\ j,k \text{ distinct}}} (\Delta_j^{\pm} Z^2)^2 (\Delta_k^{\pm} Z^2)^2 + \sum_{k \in \Theta} (\Delta_k^{\pm} Z^2)^4$$

$$+ c \sum_{\substack{j,k \in \Theta \\ j \neq k}} (\Delta_j^{\pm} Z^2)^3 \Delta_k^{\pm} Z^2 + c \sum_{\substack{j,k,m \in \Theta \\ m \neq j \neq k < m}} (\Delta_j^{\pm} Z^2)^2 \Delta_k^{\pm} Z^2 \Delta_m^{\pm} Z^2$$

$$+ c \sum_{\substack{i,j,k,m \in \Theta \\ i < j < k < m}} \Delta_i^{\pm} Z^2 \Delta_j^{\pm} Z^2 \Delta_k^{\pm} Z^2 \Delta_m^{\pm} Z^2.$$

It follows from (2.22), (2.23), (2.41) and (2.42) that

$$\lim_{n \to \infty} E\left(3 \sum_{\substack{j,k \in \Theta \\ j,k \text{ distinct}}} (\Delta_j^{\pm} Z^2)^2 (\Delta_k^{\pm} Z^2)^2 \right) = 27(s_2 - s_1)^2,$$

$$\lim_{n \to \infty} \left| E\left(\sum_{k \in \Theta} (\Delta_k^{\pm} Z^2)^4 \right) \right| = 0,$$

$$\lim_{n \to \infty} \left| E\left(c \sum_{\substack{j,k \in \Theta \\ j \neq k}} (\Delta_j^{\pm} Z^2)^3 \Delta_k^{\pm} Z^2 + c \sum_{\substack{j,k,m \in \Theta \\ m \neq j \neq k < m}} (\Delta_j^{\pm} Z^2)^2 \Delta_k^{\pm} Z^2 \Delta_m^{\pm} Z^2 \right. \right.$$

$$\left. \left. + c \sum_{\substack{i,j,k,m \in \Theta \\ i<j<k<m}} \Delta_i^{\pm} Z^2 \Delta_j^{\pm} Z^2 \Delta_k^{\pm} Z^2 \Delta_m^{\pm} Z^2 \right) \right| = 0.$$

Hence

$$(2.44) \quad \lim_{n \to \infty} E\left(\sum_{k \in \Theta} \Delta_k^{\pm} Z^2 \right)^4 = 27(s_2 - s_1)^2.$$

More generally, suppose that $m > 1$ is an integer. Then

$$\left(\sum_{k \in \Theta} \Delta_k^{\pm} Z^2 \right)^{2m} = (2m-1)!! \sum_{\substack{k_1,\ldots,k_m \in \Theta \\ k_1,\ldots,k_m \text{ distinct}}} (\Delta_{k_1}^{\pm} Z^2)^2 (\Delta_{k_2}^{\pm} Z^2)^2 \ldots (\Delta_{k_m}^{\pm} Z^2)^2$$

$$+ \sum_{\substack{2(q_1+\cdots+q_j)=2m \\ q_1,\ldots,q_j \geq 2}} c(q_1,\ldots,q_j) \sum_{k_1,\ldots,k_m \in \Theta} (\Delta_{k_1}^{\pm} Z^2)^{2q_1} (\Delta_{k_2}^{\pm} Z^2)^{2q_2} \ldots (\Delta_{k_j}^{\pm} Z^2)^{2q_j}$$

$$+ \sum_{q_1+\cdots+q_j=2m} c(q_1,\ldots,q_j) \sum_{k_1,\ldots,k_j \in \Theta} (\Delta_{k_1}^{\pm} Z^2)^{q_1} (\Delta_{k_2}^{\pm} Z^2)^{q_2} \ldots (\Delta_{k_j}^{\pm} Z^2)^{q_j},$$

where q_i's are positive integers and in the last sum, at least two q_i's are odd. By (2.23)

$$\lim_{n\to\infty} E\left((2m-1)!! \sum_{\substack{k_1,\ldots,k_m\in\Theta \\ k_1,\ldots,k_m \text{ distinct} \\ 2m=q}} (\Delta^{\pm}_{k_1} Z^2)^2 (\Delta^{\pm}_{k_2} Z^2)^2 \ldots (\Delta^{\pm}_{k_m} Z^2)^2\right)$$

$$= (2m-1)!! 3^m (s_2 - s_1)^m.$$

We have

$$\lim_{n\to\infty} E\Bigg(\sum_{\substack{2(q_1+\cdots+q_j)=2m \\ q_1,\cdots,q_j \geq 2}} c(q_1,\ldots,q_j) \sum_{k_1,\ldots,k_m\in\Theta} (\Delta^{\pm}_{k_1} Z^2)^{2q_1} (\Delta^{\pm}_{k_2} Z^2)^{2q_2} \ldots (\Delta^{\pm}_{k_j} Z^2)^{2q_j}$$
$$+ \sum_{q_1+\cdots+q_j=2m} c(q_1,\ldots,q_j) \sum_{k_1,\ldots,k_j\in\Theta} (\Delta^{\pm}_{k_1} Z^2)^{q_1} (\Delta^{\pm}_{k_2} Z^2)^{q_2} \ldots (\Delta^{\pm}_{k_j} Z^2)^{q_j}\Bigg) = 0$$

by (2.22), (2.41) and (2.42). We conclude that

$$(2.45) \qquad \lim_{n\to\infty} E\left(\sum_{k\in\Theta} \Delta^{\pm}_k Z^2\right)^{2m} = (2m-1)!! 3^m (s_2 - s_1)^m$$

for $m > 1$. Note that

$$(2.46) \qquad \lim_{n\to\infty} E\left(\sum_{k\in\Theta} \Delta^{\pm}_k Z^2\right)^{2m+1} = \lim_{n\to\infty} 0 = 0$$

since the random variables under the expectation have symmetric distributions.

We see from (2.45)–(2.46) that the moments of $\widetilde{V}_n(s_2) - \widetilde{V}_n(s_1) \stackrel{\mathrm{df}}{=} \sum_{k\in\Theta} \Delta^{\pm}_k Z^2$ (except possibly the second moment) converge as n goes to infinity to the moments of the normal distribution with mean 0 and variance $3(s_2 - s_1)$. The difference

$$|\widetilde{V}_n(s_2) - \widetilde{V}_n(s_1) - (V_n(s_2) - V_n(s_1))|$$

is bounded by $|\Delta^{\pm}_j Z^2 + \Delta^{\pm}_k Z^2|$ for appropriate j and k. Since

$$\lim_{n\to\infty} E|\Delta^{\pm}_j Z^2 + \Delta^{\pm}_k Z^2|^p = 0$$

for every $p < \infty$ and every choice of j and k, an application of Minkowski's inequality shows that the moments of $V_n(s_2) - V_n(s_1)$ have the same limits as those of $\widetilde{V}_n(s_2) - \widetilde{V}_n(s_1)$.

It is perhaps appropriate to explain why we have not proved the convergence of the second moments of $V_n(s_2) - V_n(s_1)$. In order to do it we would have to have very accurate estimates of

$$E\left(\sum_{j,k\in\Theta} \Delta^{\pm}_j Z^2 \Delta^{\pm}_k Z^2\right)$$

which cannot be found using our method.

A simple modification of the proof of Theorem 4.5.5 of [Chung 1974] shows that every subsequence of $\{V_n(s_2) - V_n(s_1)\}_{n\geq 1}$ has a further subsequence which converges in distribution to a random variable which has the same moments (with possible exception of variance) as the centered normal with variance $3(s_2 - s_1)$.

Lemma 1 implies that there is only one distribution which has the same moments of order greater than 2 as $N(0, 3(s_2 - s_1))$ and so $\{V_n(s_2) - V_n(s_1)\}_{n \geq 1}$ converges in distribution to $N(0, 3(s_2 - s_1))$.

We will indicate how one can prove that the finite-dimensional distributions of V_n also converge to those of Brownian motion $B(s)$ with variance $3s$. In order to prove that a pair of random variables has a two-dimensional normal distribution it suffices to show that all linear combinations of the random variables are normal. One can show this by finding the moments of all linear combinations. Let us fix some $0 \leq s_1 < s_2 \leq u_1 < u_2$ and let $\Theta(s)$ and $\Theta(u)$ be the obvious analogues of Θ. Let a and b be arbitrary real numbers. In order to find

$$\lim_{n \to \infty} E \left(a \sum_{k \in \Theta(s)} \Delta_k^{\pm} Z^2 + b \sum_{k \in \Theta(u)} \Delta_k^{\pm} Z^2 \right)^m$$

we might calculate

$$\lim_{n \to \infty} E \left[\left(a \sum_{k \in \Theta(s)} \Delta_k^{\pm} Z^2 \right)^{m_1} \left(b \sum_{k \in \Theta(u)} \Delta_k^{\pm} Z^2 \right)^{m_2} \right]$$

for various values of m_1 and m_2. Doing this would require estimates of

$$E \Bigg(\sum_{k_1, \ldots, k_m \in \Theta(s)} (\Delta_{k_1}^{\pm} Z^2)^{q_1} (\Delta_{k_2}^{\pm} Z^2)^{q_2} \ldots (\Delta_{k_m}^{\pm} Z^2)^{q_m}$$
$$\times \sum_{k_1, \ldots, k_i \in \Theta(u)} (\Delta_{k_1}^{\pm} Z^2)^{p_1} (\Delta_{k_2}^{\pm} Z^2)^{p_2} \ldots (\Delta_{k_i}^{\pm} Z^2)^{p_i} \Bigg).$$

It can be shown that these expectations converge to the desired limits. The method of proof is a routine adaptation of the one used in the case of the one-dimensional distributions of V_n. We omit the details as they are tedious. The case of m-dimensional distributions, $m > 2$, can be dealt with in the same way.

Since the the finite-dimensional distributions of V_n converge as $n \to \infty$ to those of B it remains to check that the distributions of V_n are tight. It will suffice to show that there exists $c < \infty$ such that

(2.47) $$E(V_n(s_2) - V_n(s_1))^4 \leq c(s_2 - s_1)^2$$

for all s_1, s_2 and all n (see [Billingsley 1968, (12.51)]). It follows easily from the scaling properties of Brownian motions X and Y that Z has the "4-th order" scaling properties while V_n's have the Brownian scaling properties. In other words, the distribution of $\{\sqrt{n/m} V_n(s \cdot m/n), s \geq 0\}$ is the same as that of $\{V_m(s), s \geq 0\}$. It follows that (2.47) holds for all n if it holds for a single n. We will show that (2.47) holds for $n = 1$. Suppose that $s_1 = 0$. Then we want to show that

$$E(V_1(s_2))^4 \leq c s_2^2.$$

This is true for $s_2 = 1$ and some c since V_1 has all moments. It is easy to extend it to all $s_2 \in [0, 1]$ since V_1 is (by definition) a linear function on $[0, 1]$. Suppose that s_2 is an integer greater than 1. Then

$$E(V_1(s_2))^4 = E(\sqrt{s_2} V_{s_2}(s_2/s_2))^4 = s_2^2 E(V_{s_2}(1))^4$$

by the scaling property of V_n's. We have proved in (2.44) that

$$\limsup_{s_2 \to \infty} E(V_{s_2}(1))^4 < c < \infty$$

so

(2.48) $$E(V_1(s_2))^4 \leq cs_2^2$$

for all integer s_2. It is elementary to extend the inequality to non-integer s_2 using the fact that V_1 is a linear function between integers. Now we use another invariance property of V_n's, namely, that of translation invariance of their increments. The distribution of $\{V_1(s) - V_1(0), s \geq 0\}$ is the same as that of $\{V_1(s+a) - V_1(a), s \geq 0\}$ provided $a > 0$ is an integer. This and (2.48) imply that

$$E(V_1(s_2) - V_1(s_1))^4 \leq c(s_2 - s_1)^2$$

for integer numbers s_1 and all $s_2 > s_1$. Extending the result to all s_1 is easy.

This concludes the proof of the weak convergence of V_n's to B. □

PROOF OF THEOREM 1 (ii). We will prove the theorem only for $s = 0$ and $t = 1$. The estimate (2.42) obtained in the proof of Theorem 2 will be used to prove (1.4). We will apply the estimate with $s_1 = 0$ and $s_2 = t$ (recall the notation form the previous proof). We want to show that

$$\lim_{n \to \infty} E \left(\sum_{j=1}^n (\Delta_j Z)^3 \right)^p = 0.$$

It will suffice to show that

(2.49) $$\lim_{n \to \infty} \left| E \left(\sum_{k_1, \ldots, k_m \in \Theta} (\Delta_{k_1} Z)^{3q_1} (\Delta_{k_2} Z)^{3q_2} \ldots (\Delta_{k_m} Z)^{3q_m} \right) \right| = 0$$

for integer $q_j \geq 1$ such that $q_1 + \cdots + q_m = p$. The sum extends over distinct k_j's. If all q_j's are even then (2.49) follows from (2.22).

Assume that at least one of q_j's is odd. Let us adopt the following convention for integer m

$$(\Delta_k^\pm Z^2)^{m+1/2} = (\Delta_k Z)^{2m+1}.$$

One can verify that (2.42) remains true if some of q_j's are not integers but have the form $k + 1/2$ for some integer k. Such q_j's should be treated as odd integers for the purpose of the decomposition $q_j = 2p_j + 1$ used in (2.42). We obtain from (2.42)

(2.50) $$\left| E \left(\sum_{k_1, \ldots, k_m \in \Theta} (\Delta_{k_1} Z)^{3q_1} (\Delta_{k_2} Z)^{3q_2} \ldots (\Delta_{k_m} Z)^{3q_m} \right) \right|$$
$$\leq cn^{-r_1 - \cdots - r_m + (-m+2)/2}$$

where $3q_j/2 = 2r_j + 2$ if $3q_j/2$ is an even integer and $3q_j/2 = 2r_j + 1$ otherwise. If $m > 2$ then (2.50) implies (2.49). If $m = 2$ and p is large then $r_1 + r_2 > 0$ and again (2.50) implies (2.49). □

References

P. Billingsley, *Convergence of probability measures*, Wiley, New York, 1968.

K. Burdzy, *Some path properties of iterated Brownian motion*, Seminar on Stochastic Processes 1992 (E. Çinlar, K. L. Chung and M. Sharpe, eds.), Birkhäuser, Boston, 1993, pp. 67–87.

R. Durrett, *Probability: Theory and examples*, Wadsworth, Belmont, CA, 1991.

T. Funaki, *Probabilistic construction of the solution of some higher order parabolic differential equations*, Proc. Japan Acad. **55** (1979), 176–179.

K. J. Hochberg, *A signed measure on path space related to Wiener measure*, Ann. Probab. **6** (1978), 433–458.

V. Yu. Krylov, *Some properties of the distribution corresponding to the equation $\partial u/\partial t = (-1)^{q+1} \partial^{2q} u/\partial x^{2q}$*, Soviet Math. Dokl. **1** (1960), 760–763.

A. Mądrecki, *The 4-a-stable motions, construction, properties and applications* (1992) (preprint).

A. Mądrecki and M. Rybaczuk, *New Feynman-Kac type formula*, Reports Math. Phys (1992) (to appear).

DEPARTMENT OF MATHEMATICS, UNIVERSITY OF WASHINGTON, GN-50, SEATTLE, WA 98195, U.S.A.

E-mail address: burdzy@math.washington.edu

The Finite Systems Scheme: An Abstract Theorem and a New Example

J. Theodore Cox and Andreas Greven

1. Introduction and main results

In this we paper compare the long term behavior of finite and infinite systems of interacting components. Our motivation is to better understand the following two notions:

(i) Large finite systems are well approximated by infinite systems, which are often easier to analyse mathematically.

(ii) Computer simulations of large finite systems can be used to understand the behavior of infinite systems.

In the sequence of papers [Cox and Greven 1990], [Cox and Greven 1991], [Dawson and Greven 1993a], and [Cox, Greven and Shiga 1993], a scheme to compare finite and infinite systems was developed and applied to the following examples: critical branching random walk, the voter model, the contact process, and certain systems of interacting diffusions, including the Wright-Fisher stepping stone model and continuous state branching models with migration. Our aim here is to present an *abstract theorem* which tries to present a unified approach to the basic phenomenon these examples exhibit. We will sketch an application of the theorem to a new example, a system of interacting measure-valued diffusions, the interacting Fleming-Viot process. We begin by recalling one of the examples already treated, then formulate the abstract theorem, and finally apply it to the new example.

Interacting diffusions. We define the Markov processes $x(t) = (x_i(t))_{i \in \mathbb{Z}^d} \in [0,1]^{\mathbb{Z}^d}$ and $x^N(t) = (x_i^N(t))_{i \in \Lambda_N} \in [0,1]^{\Lambda_N}$ (where $\Lambda_N = (-N, N]^d \cap \mathbb{Z}^d$) through the following systems of stochastic differential equations:

$$(1.1) \quad dx_i(t) = \left[\sum_{j \in \mathbb{Z}^d} a(i,j) x_j(t) - x_i(t) \right] dt + \sqrt{g(x_i(t))}\, dw_i(t),$$

$$i \in \mathbb{Z}^d, \quad x(0) \in [0,1]^{\mathbb{Z}^d},$$

and

$$(1.2) \quad dx_i^N(t) = \left[\sum_{j \in \mathbb{Z}^d} a^N(i,j) x_j^N(t) - x_i^N(t) \right] dt + \sqrt{g(x_i^N(t))}\, dw_i(t),$$

$$i \in \Lambda_N, \quad x^N(0) \in [0,1]^{\Lambda_N}.$$

1991 *Mathematics Subject Classification.* Primary: 60K35; Secondary: 60J60.
This is the final form of the paper.

The ingredients of the above equations are the following:
- A Lipschitz continuous function $g : [0,1] \to \mathbb{R}^+$ satisfying

(1.3) $$g > 0 \text{ on } (0,1) \text{ and } g(0) = g(1) = 1.$$

- An irreducible transition kernel $a(\cdot, \cdot)$ on $\mathbb{Z}^d \times \mathbb{Z}^d$ satisfying

(1.4) $$a(i,j) \geq 0, \quad a(i,j) = a(0, j-i), \quad \sum_{i \in \mathbb{Z}^d} a(0,i) = 1,$$

and the kernel $a^N(\cdot, \cdot)$ on $\Lambda_N \times \Lambda_N$ defined by

(1.5) $$a^N(i,j) = \sum_{k \in \mathbb{Z}^d} a(i, j + 2Nk).$$

- A family $\{w_i(t)\}_{i \in \mathbb{Z}^d}$ of independent standard Brownian motions on $\mathbb{R}$.

Results of [Shiga and Shimizu 1980] imply that these systems uniquely determine Feller diffusion processes $x(t)$, $x^N(t)$.

The long term behavior of $x^N(t)$ is easy to understand. Since $\sum_{i \in \Lambda_N} x_i^N(t)$ is a bounded martingale, it is not hard to see that if $\mathcal{L}(x^N(0))$ is homogeneous, and $\theta = Ex_0^N(0)$, then

(1.6) $$\mathcal{L}(x^N(t)) \Rightarrow (1-\theta)\delta_{\mathbf{0}^N} + \theta \delta_{\mathbf{1}^N} \text{ as } t \to \infty.$$

Here $\mathcal{L}$ denotes law, $\Rightarrow$ denotes weak convergence, $\delta_{\mathbf{0}^N}$ ($\delta_{\mathbf{1}^N}$) is the unit point mass at the element of $[0,1]^{\Lambda_N}$ with all coordinates equal to 0 (respectively, 1).

The behavior of the infinite system is different, and depends on the kernel $\hat{a}(\cdot, \cdot)$ defined by

(1.7) $$\hat{a}(i,j) = \frac{a(i,j) + a(j,i)}{2}.$$

It has been shown (see [Cox and Greven 1994] and [Shiga 1992]) that if $\mathcal{L}(x(0))$ is homogeneous and shift ergodic, and $\theta = Ex_0(0)$, then for $\hat{a}(\cdot, \cdot)$ transient,

(1.8) $$\mathcal{L}(x(t)) \Rightarrow \nu_\theta \text{ as } t \to \infty,$$

where ν_θ is a homogeneous, shift ergodic, invariant measure with

(1.9) $$E_\theta^\nu x_0 = \theta.$$

On the other hand, if $\hat{a}(\cdot, \cdot)$ is recurrent, then

(1.10) $$\mathcal{L}(x(t)) \Rightarrow (1-\theta)\delta_{\mathbf{0}} + \theta \delta_{\mathbf{1}},$$

where $\delta_{\mathbf{0}}$ ($\delta_{\mathbf{1}}$) is the unit point mass at the element of $[0,1]^{\mathbb{Z}^d}$ which is identically 0 (respectively, 1). For transient $\hat{a}(\cdot, \cdot)$, (1.6) and (1.8) show clearly that $x(t)$ and $x^N(t)$ behave differently in the limit $t \to \infty$. It would seem at first sight that the ergodic theorem for the infinite systems says little about the long term behavior of large finite systems, and that computer simulations of finite systems can say little about infinite systems. We shall see that this is not really the case, provided that we observe the finite systems over an appropriate time scale.

We will now compare the behavior of $x(t)$ and $x^N(t)$ as *both* N and t tend to infinity. To do this we introduce the following elements of the *finite systems scheme*:
- The time scale $\beta(N) = (2N)^d$.

- The empirical density

(1.11) $$\Theta^N(x) = |\Lambda_N|^{-1} \sum_{i \in \Lambda_N} x_i.$$

- The rescaled density process

(1.12) $$Z^N(t) = \Theta^N(x^N(t\beta(N))).$$

- The associated one-dimensional diffusion process $Z(t)$ defined by

(1.13) $$dZ(t) = \sqrt{g^*(Z(t))}\, dw(t),$$

where $w(t)$ is a standard Brownian motion, and $g^* : [0,1] \to \mathbb{R}^+$ is the function

(1.14) $$g^*(\theta) = E^{\nu_\theta} g(x_0).$$

(By Lemma 4 of [Cox, Greven and Shiga 1993], g^* is Lipschitz continuous and satisfies (1.3), hence $Z(t)$ is well defined.) The probability transition kernel of $Z(t)$ is denoted $Q_t(\cdot,\cdot)$.

The following theorem is a special case of Theorem 1 of [Cox, Greven and Shiga 1993].

THEOREM 1. *Let $\mu^N = \mathcal{L}(x^N(0))$, $N = 1, 2, \ldots$, be homogeneous, and assume that $\mu^N \Rightarrow \mu$ as $N \to \infty$, where μ is a homogenous, shift-ergodic measure on $[0,1]^{\mathbb{Z}^d}$. Let $\theta = E^\mu x_0$.*

(a) If $\hat{a}(\cdot,\cdot)$ is transient, then as $N \to \infty$,

(1.15) $$\mathcal{L}((Z^N(t))_{t \geq 0}) \Rightarrow \mathcal{L}((Z(t))_{t \geq 0}), \quad Z(0) = \theta.$$

Furthermore, for any sequence $T(N) \to \infty$ with $T(N)/\beta(N) \to s \in [0,\infty]$,

(1.16) $$\mathcal{L}(x^N(T(N))) \Rightarrow \int_{[0,1]} Q_s(\theta, d\theta')\nu_{\theta'}.$$

(b) If $\hat{a}(\cdot,\cdot)$ is recurrent, then for any $T(N) \to \infty$,

(1.17) $$\mathcal{L}(x^N(T(N))) \Rightarrow (1-\theta)\delta_0 + \theta\delta_1.$$

Let us first interpret the transient case. If we look at the system in a time scale $T(N) \ll N^d$, then the finite and infinite systems display the same behavior – both look like ν_θ. Thus computer simulations of systems on Λ_N, run up to times of smaller order than N^d, should provide reliable information on the evolution of the infinite system. If $T(N) \gg N^d$, then since $Z(\infty)$ equals 0 with probability $1-\theta$ and 1 with probability θ, we see the finite systems effect (1.6). In the intermediate case, where $T(N) \sim N^d$, we see the system "diffuse" through the equilibrium states of the infinite system, and this diffusion is governed by the evolution of the empirical density.

The function g^* can be explicitly calculated in the case where g has the special form $g(\theta) = c\theta(1-\theta)$. The map $g \to g^*$ is a complicated, nonlinear map, and has only been analysed completely in the mean-field setting of [Dawson and Greven 1993a] (see [Baillon, Clément, Greven and den Hollander 1993]). An interesting open problem is to relate the accessibility of the boundary points for the diffusion $Z(t)$ to properties of g.

In the case $\hat{a}(\cdot,\cdot)$ is recurrent, the system is less sensitive to its size if we study only local observables, and in all time scales clusters. However, it is possible to obtain more detailed information about global observables such as the empirical density or the cluster formation, and then certain critical time scales become important; such results depend in a sensitive way on finer properties of $\hat{a}(\cdot,\cdot)$. We omit any further discussion of these ideas, and instead refer the interested reader to [Cox and Greven 1991], [Dawson and Greven 1993a], [Cox, Greven and Shiga 1993], and [Klenke 1993].

Abstract theorem. In order to be able to formulate an abstract theorem we have to first set up the appropriate scenario. As we do so we will refer back to the interacting diffusion example just described in order to make things more concrete. We divide our assumptions into four groups. But first some general notation.

Let E be a countable set, let I be a polish space, and let $X(t)$ be a Markov processes with state space I^E (endowed with the product topology) and transition semigroup operators $S(t)$. Let $\{E_N, N = 1, 2, \ldots\}$ be a sequence of finite subsets of E which increase to E, and for $N = 1, 2, \ldots$ let $X^N(t)$ be Markov processes with state space I^{E_N} and transition semigroup operators $S^N(t)$. For $X \in I^E$ define the element $X|_{E_N} \in I^{E_N}$ by the restriction $(X|_{E_N})_i = X_i, i \in E_N$. We assume there are extension operators $\phi_N : I^{E_N} \to I^E$ satisfying $(\phi_N X^N)|_{E_N} = X^N$. We allow the ϕ_N to act on measures in the obvious way. If μ is a measure on I^E, and μ^N is a measure on I^{E_N}, $N = 1, 2, \ldots$, we will write $\mu^N \Rightarrow \mu$ as $N \to \infty$ for $\phi_N \mu^N \Rightarrow \mu$. A function $f : I^E \to \mathbb{R}$ is called tame if it is bounded, continuous, and depends on only finitely many coordinates. We will write $\mu^N - \nu^N \Rightarrow 0$ as $N \to \infty$ to mean that for all tame functions f, $\langle \mu^N, f \rangle - \langle \nu^N, f \rangle \to 0$.

REMARK. In some cases the "correct" state space for $X(t)$ is some proper subset of I^E. For simplicity we will ignore this possibility, since it presents no essential difficulty.

I: The $X^N(\cdot)$ are finite versions of $X(\cdot)$. This first set of assumptions makes precise the statement that the $X^N(\cdot)$ can be considered finite versions of $X(\cdot)$, and "identifies" certain classes of appropriate initial distributions $\mathcal{M}$ and $\mathcal{M}^N$ for $X(t)$ and $X^N(t)$.

(A1) For each $X(0) \in I^E$, letting $X^N(0) = X(0)|_{E_N}$,
$$\mathcal{L}(X^N(t)) \Rightarrow \mathcal{L}(X(t)) \quad \text{as} \quad N \to \infty$$
for all $t \geq 0$.

(A2) There are sets of measures $\mathcal{M}$ and $\mathcal{M}^N$ on I^E, respectively I^{E_N}, such that $\phi_N(\mathcal{M}_N) \subset \mathcal{M}$, and these sets are closed under the actions of the appropriate semigroups: for all $t \geq 0$, $\mu^N \in \mathcal{M}^N$ implies $\mu^N S^N(t) \in \mathcal{M}^N$, and $\mu \in \mathcal{M}$ implies $\mu S(t) \in \mathcal{M}$.

INTERACTING DIFFUSION EXAMPLE. Here $I = [0,1]$, $E = \mathbb{Z}^d$, $E_N = \Lambda_N$, $(\phi_N X^N)_j = X_i^N$ where $i \in \Lambda_N$ and $i = j \mod 2N$, and $X(t)$ is defined by (1.1), $X^N(t)$ is defined by (1.2). The matrix $\hat{a}(\cdot,\cdot)$ is assumed to be transient. $\mathcal{M}$ is the set of homogeneous measures on $I^{\mathbb{Z}^d}$, and $\mathcal{M}_N$ is the set of homogeneous measures on I^{Λ_N}. A coupling argument (see Prop. 2.3 of [Cox, Greven and Shiga 1993]) can be used to obtain (A1).

REMARK. There are other choices for E and E_N, as in [Dawson and Greven 1993a] and [Fleischmann and Greven 1992].

II: *Basic ergodic properties of the infinite system.* The next assumptions are concerned with the equilibrium states and the ergodic properties of the infinite systems. Let $\mathcal{I}$ denote the set of invariant measures for $X(t)$: i.e., all $\mu \in \mathcal{M}$ such that $\mu S(t) = \mu$ for all $t \geq 0$.

(A3) There is is a polish space $\mathcal{J}$ and a set of probability measures $\{\nu_\theta, \theta \in \mathcal{J}\}$ on I^E such that $\{\nu_\theta, \theta \in \mathcal{J}\} \subset \mathcal{I}$. For $\theta \in \mathcal{J}$ let $\mathcal{M}_\theta$ be the set all $\mu \in \mathcal{M}$ such that $\mu S(t) \Rightarrow \nu_\theta$ as $t \to \infty$. (Typically, $\{\nu_\theta, \theta \in \mathcal{J}\}$ is the set of extreme points of $I \cap \mathcal{M}$.)

(A4) There are functions $\Theta^N : I^{E_N} \to \mathcal{J}$ such that for all $\mu \in \mathcal{M}$, the limit

$$\Theta^N(X|_{E_N}) \to \Theta(X) \text{ as } N \to \infty \text{ a.s. } \mu$$

exists. Furthermore, we require that if $\Theta(X) = \theta$ a.s. μ, then $\mu \in \mathcal{M}_\theta$, and for all $\mu \in \mathcal{M}$,

$$\mu = \int_\mathcal{J} \mu_\theta \, \Gamma(d\theta),$$

where Γ is the law of $\Theta(X)$ with respect to μ, and $\mu_\theta \in \mathcal{M}_\theta$, $\theta \in \mathcal{J}$.

INTERACTING DIFFUSION EXAMPLE. $\mathcal{J} = [0,1]$, and

$$\Theta^N(X) = |\Lambda_N|^{-1} \sum_{i \in \Lambda_N} X_i.$$

The ν_θ are obtained by taking weak limits as $t \to \infty$ of $\delta_\theta S(t)$, $\theta \in [0,1]$. $\mathcal{M}_\theta$ is the set of homogeneous measures such that $\theta(X) = \lim_{N \to \infty} \Theta^N(X) = \theta$ a.s. μ. Every homogeneous measure μ can be decomposed as required in (A4) by choosing

$$\mu_\theta(\cdot) = \mu(\cdot \mid \Theta(X) = \theta).$$

III: *Time scale.* Here we impose certain regularity properties of the Θ^N, and identify a fundamental time scale $\beta(N)$ connected with the Θ^N. We will write $\bar\Theta^N(t)$ for $\Theta^N(X^N(t))$.

(A5) Let $T(N)$ be any sequence tending to infinity, suppose $\mathcal{L}(X^N(0)) \in \mathcal{M}^N$, and $\mathcal{L}(x(0)) \in \mathcal{M}_\theta$ for some $\theta \in \mathcal{J}$. Assume further that along some subsequence $N(k)$,

$$\mathcal{L}(X^{N(k)}(T(N(k)))) \Rightarrow \mathcal{L}(X) \text{ as } k \to \infty.$$

Then

$$\mathcal{L}(\Theta^{N(k)}(X^{N(k)})) \Rightarrow \mathcal{L}(\Theta(x)).$$

(A6) There is a sequence $\beta(N) \to \infty$ as $N \to \infty$ such that if $\mathcal{L}(X^N(0)) \in \mathcal{M}^N$, and $\mathcal{L}(X^N(0)) \Rightarrow \mathcal{L}(X(0)) \in \mathcal{M}$, then the family of measures $\mathcal{L}(\bar\Theta^N(t\beta(N))_{t\geq 0})$ on $D_\mathcal{J}[0,\infty)$ is tight. Furthermore, all weak limit points $(\theta_s)_{s\geq 0}$ should have the property: $Ef(\theta_s)$ is continuous in s for all bounded continuous functions f on $\mathcal{J}$.

INTERACTING DIFFUSION EXAMPLE. See Lemma 2.2 of [Cox, Greven, Shiga 1993] for (A5). With $\beta(N) = (2N)^d$, (A6) is easily obtained from (1.2) by martingale considerations.

IV: Compactness and strengthened convergence properties. Here we impose some needed compactness, and also require a strengthened form of (A1), as well as a strengthened Feller property for $X(t)$.

(A7) Fix $T < \infty$, and let $\{t_N\}$ satisfy $t_N \leq T\beta(N)$ for N. If $\mathcal{L}(X^N(0)) \in \mathcal{M}^N$, then the family $\{\mathcal{L}(X^N(t_N))\}$ is tight, and all weak limit points belong to $\mathcal{M}$.

(A8) If $\mu, \mu_N, N = 1, 2, \ldots \in \mathcal{M}$, and $\mu_N \Rightarrow \mu$ and $t_N \to \infty$ as $N \to \infty$, then
$$\mu_N S(t_N) - \mu S(t_N) \Rightarrow 0.$$

(A9) There is a sequence $L(N) \to \infty$ such that if $t_N \leq L(N)$ and $\mu^N \in \mathcal{M}_N$, then
$$\mu^N S^N(t_N) - (\phi_N \mu^N) S(t_N) \Rightarrow 0.$$

INTERACTING DIFFUSION EXAMPLE. (A7) holds trivially since I^E is compact. Both (A8) and (A9) can be proved using coupling techniques (see [Cox, Greven and Shiga 1993]).

REMARK. The most difficult properties to check for the interacting diffusions are (A4) and (A8). In the case of the contact process, property (A6) and the identification of $\beta(N)$ is nontrivial.

We are now ready to state our comparison theorem. We assume that $X(t)$, $X^N(t), N = 1, 2, \ldots$ satisfy (A1)–(A9).

THEOREM 2. *Assume $\mathcal{L}(X(0)) \in \mathcal{M}_\theta$ for some $\theta \in \mathcal{J}$. Let $\{N_k\}$ be any sequence such that as $k \to \infty$,*

$$(1.18) \quad \mathcal{L}((\bar{\Theta}^{N_k}(s\beta(N_k)))_{s\geq 0}) \Rightarrow \mathcal{L}((\theta_s)_{s\geq 0}) \quad \text{in} \quad D_\mathcal{J}[0, \infty).$$

If $T(N) \to \infty$, and $T(N)/\beta(N) \to s \in [0, \infty)$ as $N \to \infty$, then

$$(1.19) \quad \mathcal{L}(X^{N_k}(s\beta(N_k))) \Rightarrow \int_\mathcal{J} P(\theta_s \in d\theta)\nu_\theta.$$

The existence of a sequence $\{N_k\}$ and a process $(\theta_s)_{s\geq 0}$ satisfying (0.18) is guaranteed by (A6). The limit $(\theta_s)_{s\geq 0}$ depends in general on the particular sequence $\{N_k\}$. If it doesn't, then the theorem implies

$$(1.20) \quad \mathcal{L}(X^N(s\beta(N))) \Rightarrow \int_\mathcal{J} P(\theta_s \in d\theta)\nu_\theta \quad \text{as} \quad N \to \infty.$$

It is possible, in some cases, to verify this lack of dependence on $\{N_k\}$, thus obtaining (1.20). In others cases it seems necessary to prove convergence of $\mathcal{L}(\bar{\Theta}^N(s\beta(N)))$ and $\mathcal{L}(X^N(s\beta(N)))$ *simultaneously*.

INTERACTING FLEMING-VIOT PROCESSES. We first recall the definition of the basic *noninteracting* Fleming-Viot process $\theta(t)$. For $z \in \mathcal{P}([0, 1])$ let $Q_z(\cdot, \cdot)$ be the signed measure on $[0, 1] \times [0, 1]$

$$Q_z(du, dv) = z(du)\delta_u(dv) - z(du)z(dv).$$

For functions F acting on measures given by

$$(1.21) \quad F(z) = \left(\int_{[0,1]} f(v) z(dv)\right)^n, \quad f \in C[0, 1], \quad n \in \mathbb{N}^+,$$

define

$$\text{(1.22)} \qquad \frac{\partial F(z)}{\partial z}(u) = \lim_{\epsilon \to 0} \frac{F(z + \epsilon \delta_u) - F(z)}{\epsilon}, \quad u \in [0,1].$$

Finally, for $D > 0$ define the operator L_D acting on functions F of the form (1.21) by

$$\text{(1.23)} \qquad (L_D F)(z) = \frac{D}{2} \int_{[0,1]} \int_{[0,1]} \frac{\partial^2 F(z)}{\partial z \partial z}(u,v) Q_z(du, dv).$$

With these definitions, the basic Fleming-Viot process $(\theta(t))_{t \geq 0}$ with fluctuation constant D is the $D_{\mathcal{P}([0,1])}[0, \infty)$ process which satisfies:

$$\text{(1.24)} \qquad \left(F(\theta(t)) - F(\theta(0)) - \int_0^t L_D(F)(\theta(s)) ds \right)_{t \geq 0}$$

is a martingale for all functions F of the form (1.21). See the monograph [Dawson 1991] for a comprehensive account of the basic facts concerning the Fleming-Viot process, including proofs of existence and uniqueness of (more general versions of) $\theta(t)$. The Fleming-Viot process is a model for the evolution, through resampling, of genetic types in a closed population.

We wish to consider a countable number of such populations which interact through migration. Our interacting Fleming-Viot process $Y(t)_{t \geq 0}$ will be a Markov process with state space $\mathcal{E} = (\mathcal{P}([0,1]))^{\mathbb{Z}^d}$; we write $Y(t) = (Y_\xi(t))_{\xi \in \mathbb{Z}^d}$ with $Y_\xi(t) \in \mathcal{P}([0,1])$. The idea behind the model is this. Suppose we have genetic types which are labelled with elements $u \in [0,1]$. Then the population at colony ξ is described by the empirical distribution of types, which is an element in $\mathcal{P}([0,1])$. The dynamics of the model are obtained by taking a diffusion limit of a discrete model allowing for the *migration* of individuals between colonies and *resampling* within colonies of the types from one generation to the next. Mathematically, the process $Y(t)$ is defined as the solution to a martingale problem as follows.

Let $\mathcal{A}$ be the algebra generated by functions F of the form

$$\text{(1.25)} \qquad F(x) = \prod_{i=1}^n \int_0^1 u^{a_i} x_{\xi_i}(du), \quad u \in [0,1], \quad a_i \in \mathbb{N}^+, \quad x \in \mathcal{E}.$$

Partial derivatives of a function $F \in \mathcal{A}$ are defined by

$$\text{(1.26)} \qquad \frac{\partial F(x)}{\partial x_\xi}(u) = \lim_{\epsilon \to 0} \frac{F(x^\epsilon(u)) - F(x)}{\epsilon}$$

where $x^\epsilon(u)$ is defined by

$$\text{(1.27)} \qquad x^\epsilon_{\xi'}(u) = \begin{cases} x_{\xi'} & \text{if } \xi' \neq \xi, \\ x_\xi + \epsilon \delta_u & \text{if } \xi' = \xi. \end{cases}$$

There is an irreducible transition kernel $a(\cdot, \cdot)$ on $\mathbb{Z}^d \times \mathbb{Z}^d$ satisfying

$$\text{(1.28)} \qquad a(\xi, \xi') \geq 0, \quad a(\xi, \xi') = a(0, \xi' - \xi), \quad \sum_{\xi \in \mathbb{Z}^d} a(0, \xi) = 1.$$

The operator L acting on functions in $\mathcal{A}$ is defined by

$$
\begin{aligned}
(LF)(x) = & \sum_{\xi,\xi' \in \mathbb{Z}^d} a(\xi,\xi') \int_{[0,1]} \frac{\partial F(x)}{\partial x_\xi}(u)\,(x_{\xi'}(du) - x_\xi(du)) \\
& + \frac{D}{2} \sum_{\xi \in \mathbb{Z}^d} \int_{[0,1]} \int_{[0,1]} \frac{\partial^2 F(x)}{\partial x_\xi \partial x_\xi}(u,v)\,Q_{x_\xi}(du,dv).
\end{aligned}
\tag{1.29}
$$

We define the process $(Y(t))_{t \geq 0}$ to be the $D_{\mathcal{E}}[0,\infty)$ process satisfying:

$$
F(Y(t)) - F(Y(0)) - \int_0^t (LF)(Y(s))\,ds
\tag{1.30}
$$

is a martingale for all $F \in \mathcal{A}$. We call $Y(t)$ a system of interacting Fleming-Viot processes. Existence and uniqueness questions for this martingale problem are settled in [Dawson, Greven and Vaillancourt 1993].

If we replace $\mathbb{Z}^d$ above with Λ_N (as defined in the interacting diffusion example) we can define an algebra $\mathcal{A}_N$ of functions in the obvious way. Replacing $a(\cdot,\cdot)$ by $a^N(\cdot,\cdot)$ as in (1.4) we get an operator L_N on $\mathcal{A}_N$ which then allows us to define as in (1.29) the finite system $Y^N(t)$ of interacting Fleming-Viot processes uniquely as the solution to the appropriate martingale problem. $(Y^N(t))_{t \geq 0}$ is a $D_{\mathcal{E}_N}[0,\infty)$ process, where $\mathcal{E}_N = (\mathcal{P}([0,1]))^{\Lambda_N}$.

The ergodic theory of the finite systems $Y^N(t)$ is very simple. For $u \in [0,1]$ let $\bar{\delta}_u \in \mathcal{E}$ be defined by setting $(\bar{\delta}_u)_\xi = \delta_u$. Let $\bar{\delta}_u^N$ be the corresponding element of $\mathcal{E}_N$. Suppose that for some fixed $\theta \in \mathcal{P}([0,1])$, $Y_\xi^N(0)$ has mean measure θ for all $\xi \in \Lambda_N$. (By this we mean that $E\langle Y_\xi^N(0), f\rangle = \langle \theta, f \rangle$ for all $f \in C[0,1]$.) Then $Y^N(t)$ tends to the traps $\bar{\delta}_u^N$ as $t \to \infty$. That is,

$$
\mathcal{L}(Y^N(t)) \Rightarrow \int_{[0,1]} \delta_{\bar{\delta}_u^N}\,\theta(du) \text{ as } t \to \infty.
\tag{1.31}
$$

For the infinite system we must consider the symmetrized kernel $\hat{a}(\xi,\xi') = (a(\xi,\xi') + a(\xi',\xi))/2$. The following results were obtained in [Dawson, Greven and Vaillancourt 1993]. If $\hat{a}(\cdot,\cdot)$ is recurrent, and $\mathcal{L}(Y(0))$ is homogeneous, with each coordinate $Y_\xi(0)$ having mean measure θ, then

$$
\mathcal{L}(Y(t)) \Rightarrow \int_{[0,1]} \delta_{\bar{\delta}_u}\,\theta(du) \text{ as } t \to \infty.
\tag{1.32}
$$

If $\hat{a}(\cdot,\cdot)$ is transient, then there exists a set of extremal, homogeneous invariant measures $\{\nu_\theta, \theta \in \mathcal{P}([0,1])\}$ which are mixing and satisfy

$$
E^{\nu_\theta}\langle Y_\xi, f\rangle = \langle \theta, f\rangle \text{ for all } f \in C[0,1].
\tag{1.33}
$$

Furthermore, for every homogenous, shift-ergodic initial distribution $\mathcal{L}(Y(0))$ such that each coordinate $Y_\xi(0)$ has mean measure θ,

$$
\mathcal{L}(Y(t)) \Rightarrow \nu_\theta \text{ as } t \to \infty.
\tag{1.34}
$$

Note that as with the interacting diffusions, the long term behavior of the finite and infinite systems is different when the $\hat{a}(\cdot,\cdot)$ is transient.

To apply the finite systems scheme to the interacting Fleming-Viot process we introduce the following. Define the continuous time kernel

$$\hat{a}_t(\xi, \xi') = e^{-t} \sum_{n=0}^{\infty} t^n \hat{a}^{(n)}(\xi, \xi')/n!,$$

the time scale $\beta(N) = (2N)^d$, the $\mathcal{P}([0,1])$-valued variables

$$\Theta^N(Y) = |\Lambda_N|^{-1} \sum_{\xi \in \Lambda_N} Y_\xi,$$

and let $\theta(t)$ be the basic Fleming-Viot process with fluctuation constant

$$D^* = \frac{D}{1 + D \int_0^\infty \hat{a}_{2t}(0,0) dt}$$

and probability transition function $Q_t(\cdot, \cdot)$.

THEOREM 3. *Let $\mu^N = \mathcal{L}(Y^N(0))$, $N = 1, 2, \ldots$, be homogeneous, and assume that $\mu^N \Rightarrow \mu$ as $N \to \infty$, where μ is a homogeneous, shift-ergodic measure on $\mathcal{E}$, where, under μ, each coordinate Y_ξ has mean measure $\theta \in \mathcal{P}([0,1])$.*

(a) *If $\hat{a}(\cdot, \cdot)$ is transient, then as $N \to \infty$,*

(1.35) $$\mathcal{L}((\Theta^N(Y^N(t\beta(N))))_{t \geq 0}) \Rightarrow \mathcal{L}((\theta(t))_{t \geq 0}), \quad \theta(0) = \theta.$$

Furthermore, for any sequence $T(N) \to \infty$ with $T(N)/\beta(N) \to s \in [0, \infty]$,

(1.36) $$\mathcal{L}(Y^N(T(N))) \Rightarrow \int_{\mathcal{P}([0,1])} Q_s(\theta, d\theta') \nu_{\theta'}.$$

(b) *If $\hat{a}(\cdot, \cdot)$ is recurrent, then for any $T(N) \to \infty$,*

(1.37) $$\mathcal{L}(Y^N(t)) \Rightarrow \int_{[0,1]} \delta_{\bar{\delta}_u} \theta(du).$$

REMARK. A proof of Theorem 2 could be given by using duality and adapting the techniques used to handle the voter model in [Cox and Greven 1990]. Another possibility is to proceed as in [Cox, Greven and Shiga 1993]. Shiga (private communication) has also given a proof of Theorem 2. In Section 2 we will sketch a proof based on Theorem 1.

2. Proof of Theorem 1

The proof we present here is a rigorous version of the following simple idea. Choose $L(N)$ as in (A9), and let $T'(N) = T(N) - L(N)$. By the Markov property we may write

$$\mathcal{L}(X^N(T(N))) = \int P(\bar{\Theta}^N(T'(N)) \in d\theta') \mu_{\theta'}^N S^N(L(N))$$

where $\mu_{\theta'}^N = \mathcal{L}(X^N(T'(N)) \mid \bar{\Theta}^N(T'(N)) = \theta')$. Now we know that $\bar{\Theta}^N(t)$ will change little over the time interval $[T'(N), T(N)]$, so $P(\bar{\Theta}^N(T'(N)) \in d\theta')$ should converge to $P(\theta(s) \in d\theta')$. Furthermore, $L(N)$ has been chosen so that we should have

$$\mu_{\theta'}^N S^N(L(N)) \approx \mu_{\theta'}^N S(L(N)) \Rightarrow \nu_\theta.$$

Putting these pieces together we have

$$\mathcal{L}(X^N(T(N))) \Rightarrow \int P(\theta(s) \in d\theta') \nu_{\theta'}.$$

Here is the formal argument:

STEP 1. By (A6) we can choose a sequence $\{N_k\}$ such that

(2.1) $$\mathcal{L}((\bar{\Theta}^{N_k}(s\beta(N_k)))_{s\geq 0}) \Rightarrow \mathcal{L}((\theta_s)_{s\geq 0})$$

for some $D_{\mathcal{J}}[0,\infty)$ process $(\theta_s)_{s\geq 0}$. Now choose a squence $L(N)$ as in (A9) such that $L(N) = o(\beta(N))$ as $N \to \infty$. By the tightness in (A6), for the same limit process $(\theta_s)_{s\geq 0}$,

(2.2) $$\mathcal{L}((\bar{\Theta}^{N_k}(s\beta(N_k) - L(N_k)))_{s\geq 0}) \Rightarrow \mathcal{L}((\theta_s)_{s\geq 0}).$$

From now on we work with a *fixed* s.

By assumption (A7) we may choose a subsequence $\{\tilde{N}_k\}$ of $\{N_k\}$ so that

(2.3) $$\mathcal{L}(X^{\tilde{N}_k}(s\beta(\tilde{N}_k) - L(\tilde{N}_k))) \Rightarrow \mu(s) \in \mathcal{M}.$$

Using (A4) we decompose $\mu(s)$ in the form

(2.4) $$\mu(s) = \int_{\mathcal{J}} \mu_\rho \Gamma(d\rho),$$

where Γ is the law under $\mu(s)$ of $\Theta(X)$, and $\mu_\rho \in \mathcal{M}_\rho$ for all $\rho \in \mathcal{J}$. In view of (A2) and (A5), (2.1) implies that $\Gamma(\cdot) = P(\theta_s \in \cdot)$. Thus we have established

(2.5) $$\mu(s) = \int_{\mathcal{J}} \mu_\rho P(\theta_s \in d\rho).$$

STEP 2. Our goal in this step is to prove

(2.6) $$\mu(s)S(t) \Rightarrow \int_{\mathcal{J}} \nu_\rho P(\theta_s \in d\rho) \text{ as } t \to \infty.$$

To do this, let f be a tame function on I^E, and compute using (2.5)

(2.7) $$\begin{aligned}\langle \mu(s)S(t), f \rangle &= \left\langle \left(\int_{\mathcal{J}} \mu_\rho P(\theta_s \in d\rho)\right) S(t), f \right\rangle \\ &= \left\langle \int_{\mathcal{J}} \mu_\rho P(\theta_s \in d\rho), S(t)f \right\rangle \\ &= \int_{\mathcal{J}} \langle \mu_\rho, S(t)f \rangle P(\theta_s \in d\rho) \\ &= \int_{\mathcal{J}} \langle \mu_\rho S(t), f \rangle P(\theta_s \in d\rho).\end{aligned}$$

But since $\mu_\rho \in \mathcal{M}\rho$,

(2.8) $$\langle \mu_\rho S(t), f \rangle \to \langle \nu_\rho, f \rangle \text{ as } t \to \infty \text{ for all } \rho \in \mathcal{J},$$

and we get (2.6) by the bounded convergence theorem.

STEP 3. In this step we show that

(2.9) $$\mathcal{L}(X^{\tilde{N}_k}(s\beta(\tilde{N}_k))) \Rightarrow \int_{\mathcal{J}} \nu_\rho P(\theta_s \in d\rho).$$

Given this, since the limit is independent of the subsequence $\{\tilde{N}_k\}$, convergence must hold along $\{N_k\}$. Since s was arbitrary, this is enough to prove (2.9). Introduce the following abbreviations:

(2.10)
$$\bar{\mu}_k = \mathcal{L}(X^{\tilde{N}_k}(s\beta(\tilde{N}_k) - L(\tilde{N}_k)) \in \mathcal{M}^{\tilde{N}_k},$$
$$\hat{\mu}_k = \phi_{\tilde{N}_k}\bar{\mu}_k(s) \in \mathcal{M}.$$

Since by assumption $\hat{\mu}_k \Rightarrow \mu(s)$ as $k \to \infty$, we can conclude from (A8) that

(2.11)
$$\hat{\mu}_k S(L(\tilde{N}_k)) - \mu(s)S(L(\tilde{N}_k))) \Rightarrow 0.$$

By (A9) we have

(2.12)
$$\bar{\mu}_k S^{\tilde{N}_k}(L(\tilde{N}_k)) - \hat{\mu}_k S(L(\tilde{N}_k)) \Rightarrow 0.$$

Combining (2.6) with (2.11) and (2.12) gives

(2.13)
$$\bar{\mu}_k S^{\tilde{N}_k}(L(\tilde{N}_k)) \Rightarrow \int_J \nu_\rho P(\theta_s \in d\rho).$$

By the Markov property, the left-hand side of (2.13) is the same as the left-hand side of (2.9), so we are done.

3. Sketch of proof of Theorem 2

We will discuss on the transient case (a), since (b) is much simpler and somewhat different qualitatively. Our proof breaks naturally into two parts. First we need to check the conditions (A1)-(A9). Then we need to prove convergence, not just tightness, of the density process.

STEP 1. Let us start by identifying the objects in the abstract theorems. We have $I = J = \mathcal{P}([0,1])$, $E = \mathbb{Z}^d$, $E_N = \Lambda_N$, $\beta(N) = (2N)^d$, $\mathcal{M}$ ($\mathcal{M}^N$) are the homogeneous measures on $I^{\mathbb{Z}^d}$ (respectively, I^{Λ_N}). For $\theta \in \mathcal{P}([0,1])$ let $\mathcal{M}_\theta$ be the set of all $\mu \in \mathcal{M}$ which are (spatially) ergodic and such that $E^\mu \langle Y_0, f \rangle = \langle \theta, f \rangle$, $f \in C[0,1]$. Let

$$\Theta^N(Y) = |\Lambda_N|^{-1} \sum_{\xi \in \Lambda_N} Y_\xi, \quad \Theta(Y) = \lim_{N \to \infty} \Theta^N(Y),$$

and let $(\theta(s))_{s \geq 0}$ be the basic Fleming-Viot process with initial state θ. Now we consider the conditions (A1)–(A9).

(A1) This follows from the existence and uniqueness theorem for $Y(t)$, Theorem 0.0 in [Dawson, Greven and Vaillancourt 1993].

(A2) This is obvious from the form of L.

(A3) See Theorem 0.1 in [Dawson, Greven and Vaillancourt 1993].

(A4) The first part of (A4) is the ergodic theorem for stationary random fields, see Section 6.4 of [Kreugel 1985]. For the second part we note that since $\mathcal{E}$ is a Polish space, we can obtain the μ_θ by conditioning, and again refer to Theorem 1.1 of [Dawson, Greven and Vaillancourt 1993].

(A5) This can be reduced to the situation for interacting Fisher-Wright diffusions using the fact that for all $A \in \mathcal{B}([0,1])$, $\langle Y(t), 1_A \rangle_{t \geq 0}$ is a system of interacting Fisher-Wright diffusions.

(A6) The key fact here is that for fixed $f \in C[0,1]$, with

$$Z^N(t) = \langle \Theta^N(Y^N(t\beta(N))), f \rangle,$$

$(Z^N(t))_{t \geq 0}$ is a bounded martingale with increasing process

$$D \int_0^t |\Lambda_N|^{-1} \sum_{\xi \in \Lambda_N} EV_f(Y^N(s\beta(N))) \, ds,$$

where $V_f(z) = \langle z, f^2 \rangle - \langle z, f \rangle^2$. From here one proceeds as in the proof of (4.1) of [Cox, Greven and Shiga 1993].

(A7) Since $[0, 1]$ is compact, $\mathcal{P}([0,1])$ is compact in the weak topology, and hence $\mathcal{E}$ is compact in the product topology. Hence $\mathcal{P}(\mathcal{E})$ is compact in the weak topology of measures on $\mathcal{E}$.

(A8) It is this condition that requires the most work. It is possible to construct a *coupling* as in [Cox, Greven and Shiga 1993] (but see also Section 4c of [Dawson, Greven and Vaillancourt 1993]), and proceed as in the proof of Corollary 1 there. Another approach is to work directly with the dual process.

(A9) A somewhat stronger condition than (A9) can be obtained using either coupling or duality, as is done for interacting diffusions in Proposition 2.3 of [Cox, Greven and Shiga 1993].

STEP 2. The problem here is to show that all possible weak limits $(\theta_s)_{s \geq 0}$ from (A6) must be one and the same, namely, the basic Fleming-Viot process $(\theta(s))_{s \geq 0}$ with fluctuation constant D^*. There are several possible approaches to this problem. One is to use martingale methods, based on the approach used in [Cox, Greven and Shiga 1993], supplemented with duality. A less elegant (indeed, rather tedious) method is to show that if $F_1, \ldots, F_m$ are functions of form (1.25), and $0 \leq s_1 < \ldots < s_m$, then as $N \to \infty$,

$$(3.1) \qquad E \prod_{l=1}^m F_l(\Theta^N(s_l \beta(N))) \to E \prod_{l=1}^m F_l(\theta(s_l)).$$

To do this one uses duality to convert both sides of (3.1) into statements concerning the dual processes. This approach was used for the voter model (which has a simpler dual) in [Cox 1989]. To carry out the details some modifications of the coalescing random walk results derived in [Cox 1989], as well as some extensions of similar results in [Cox and Griffeath 1990], need to be made. Otherwise, the basic method is the same.

References

J.-B. Baillon, Ph. Clément, A. Greven and F. den Hollander, *On the attracting orbit of a nonlinear transformation arising from renormalization of hierarchically interacting diffusions*, Canad. J. Math. (to appear).

J.T. Cox, *Coalescing random walks and voter model consensus times on the torus in $\mathbb{Z}^d$*, Ann. Probab. **17** (1989), 1333–1366.

J.T. Cox and A. Greven, *On the long term behavior of some finite particle systems*, Probab. Theor. Relat. Fields **85** (1990), 195–237.

_____, *On the long term behavior of some finite particle systems: a critical dimension example*, Random Walks, Brownian Motion and Interacting Particle Systems (R. Durrett and H. Kesten, eds.), Birkhäuser, 1991, pp. 203–213.

_____, *Ergodic theorems for infinite systems of interacting diffusions*, Ann. Probab. (1994) (to appear).

J.T. Cox, A. Greven and T. Shiga, *Finite and infinite systems of interacting diffusions*, Preprint (1993).

J.T. Cox and D. Griffeath, *Mean field asymptotics for the planar stepping stone model*, Proc. London Math. Soc. **61** (1990), 189–208.

D.A. Dawson, *Measure-valued Markov processes*, École d'éte de probabilités de Saint Flour, 1991.

D.A. Dawson and A. Greven, *Multiple time scale analysis of interacting diffusions*, Prob. Theor. Relat. Fields **95** (1993), 467–508.

_____, *Models of interacting diffusions: multiple time scale phenomena, phase transition and patterns of cluster formation*, Prob. Theor. Relat. Fields **96** (1993), 435–474.

D.A. Dawson, A. Greven and J. Vaillancourt, *Equilibria and quasi equilibria for infinite collections of interacting Fleming-Viot processes*, Preprint (1993).

K. Fleischmann and A. Greven, *Diffusive clustering in an infinite system of hierarchically interacting diffusions*, Probab. Theor. Relat. Fields (to appear).

A. Klenke, *Wechselwirkende Diffusionen auf der hierarchischen Gruppe-Vergleich endlicher Fall versus unendlicher Fall* (1993), Diplomarbeit, Göttingen.

U. Krengel, *Ergodic theorems* (1985), De Gruyter, Berlin.

T. Shiga, *Ergodic theorems and exponential decay of sample paths for certain interacting diffusion systems*, Osaka J. Math. **29** (1992), 789–807.

T. Shiga and A. Shimizu, *Infinite-dimensional stochastic differential equations and their applications*, J. Math. Kyoto Univ. **20** (1980), 395–416.

ABSTRACT. We consider the problem of comparing the long term behavior of systems composed of infinitely many interacting components with corresponding systems composed of a large but finite number of interacting components. We formulate and prove an abstract theorem which provides a more systematic approach to some of the ideas discussed in [Cox and Greven 1990], [Cox and Greven 1991], [Dawson and Greven 1993a], and [Cox, Greven and Shiga 1993]. We also consider a new example: interacting Fleming-Viot processes.

MATHEMATICS DEPARTMENT, SYRACUSE UNIVERSITY, SYRACUSE, NY 13244, USA.
E-mail address: jtcox@gumby.syr.edu

INSTITUT FÜR STOCHASTIK, HUMBOLDT UNIVERSITÄT ZU BERLIN, UNTER DEN LINDON 6, D-10099 BERLIN, GERMANY.
E-mail address: greven@namu01.gwdg.de

On Path Properties of Super-2 Processes. I

Donald A. Dawson, Kenneth J. Hochberg, and Vladimir Vinogradov

1. Introduction

In this work, we consider *two-level* $(2, d, \beta_1, \beta_2)$-*measure-valued processes* X_t introduced in [Dawson, Hochberg and Wu 1990]. Such processes generalize so-called *measure-valued branching processes* or *superprocesses*. They can appear as weak limits of certain properly rescaled *two-level branching particle systems* (BPS-2). For convenience of the reader, we will briefly describe such systems as well as characterize their limits. We refer to [Wu 1992, 1993] for a more complete setting and to [Dawson and Hochberg 1991] and [Hochberg 1993] for a detailed review of possible applications of two-level (or multi-level) models in population genetics, computer science, environmental science, and physics.

It is natural to suppose that individual particles within a colony (superparticle) may reproduce or disappear as a result of branching at the individual level, and that whole colonies or groups of individuals (superparticles) may disappear or replicate according to another such process, with splitting rates perhaps depending upon the size of the colony. Between splits, each particle is allowed to undergo some spatial motion (which can be interpreted, e.g., as *change of type (mutation)*) independently of the other particles. Such multilevel processes arise naturally in population biology as models for mitochondrial DNA, where sampling replacement takes place at both the individual and organelle levels. They also arise in models describing the spread of species in competitive environments, where each species attempts to create copies of itself so as to become established in neighboring territory, while the possibility of simultaneous extinction by natural calamity or nearby competitors exists for all members present in a single colony. Other examples include, e.g., the spread of viruses in computerized data collections; random reflection (with replication) of files or directories when transferred through computer networks; and environmental damage caused by propellants in the atmosphere, which can effect individuals (e.g., asthmatic individuals) as well as whole colonies (say, via destruction of the ozone layer over an entire land-mass). It should be mentioned that the absence of independence in branching as well as the presence of a hierarchical structure influence the behavior of such systems. The fact that higher-level branching

---

1991 *Mathematics Subject Classification.* Primary: 60J80, 60J60; Secondary: 60G17, 60F15.

The first author is supported by an NSERC-Canada Operating Grant.

The second author is supported by the U.S. Army Research Office through the Mathematical Sciences Institute of Cornell University and by an NSERC-Canada International Scientific Exchange Award.

The third author is supported by an NSERC-Canada International Research Award hosted at Carleton University.

This is the final form of the paper.

affects groups of particles simultaneously necessitates a finer analysis of the path behavior than is needed for super-1 processes. Note that a different model which undergoes hierarchical branching was considered in [Greven 1991].

Now, consider η independent families of particles (hereinafter referred to as *level-2 particles* or *superparticles*), each of which consists of η independent particles (*level-1 particles*). The parameter η will be assumed to tend to infinity. Let us assume for simplicity that all particles are located at the origin at time $t = 0$, and that each level-1 particle has constant mass $1/\eta^2$ and performs independent Brownian motion in $\mathbb{R}^d$. It is assumed that any level-1 particle splits into a random number of offspring at an exponentially distributed instant of time with mean $\eta^{-\beta_1}$, and that each newly born particle is a copy of its parent and immediately starts to perform d-dimensional Brownian motion. The motions, lifetimes and branchings of all particles are independent of each other. The branching mechanism is assumed to be governed by the particle-production generating function

$$\vartheta_{\beta_1}(s) := s + \frac{1}{1+\beta_1} \cdot (1-s)^{1+\beta_1}, \tag{1.1}$$

where $\beta_1 \in (0, 1]$.

Each superparticle is also assumed to split (independent of everything else) into a random number of superparticles, each of which copies its parent superparticle. The superparticle-lifetime distribution function is exponential with mean $\eta^{-\beta_2}$ and the number of newly born superparticles is governed by the superparticle-production generating function

$$\vartheta_{\beta_2}(s) := s + \frac{1}{1+\beta_2} \cdot (1-s)^{1+\beta_2}, \tag{1.1'}$$

where $\beta_2 \in (0, 1]$.

REMARK. Note that in fact the constants $\frac{1}{1+\beta_1}$ and $\frac{1}{1+\beta_2}$ from (1.1) and (1.1') can be replaced by any arbitrary positive constants. The case of arbitrary positive constants is reduced to our special case by scaling in space and time.

It is obvious that such a BPS-2 can be viewed as an $M_F(M_F(\mathbb{R}^d))$-valued process. In fact, such a system possesses a weak limit as $\eta \to \infty$ that is a *two-level $(2, d, \beta_1, \beta_2)$-measure-valued process* starting with the measure δ_{δ_0} at $t = 0$. We refer to [Wu 1992, Chapter 4], and [Wu 1993] for a detailed study of two-level $(2, d, 1, 1)$-measure-valued processes, which are related to binary branching on both levels and generalize the real-valued birth-and-death process. Following [Hochberg 1993], we will refer to stochastic processes taking values in $M_F(M_F(\mathbb{R}^d))$ as "*super-2 processes*". These $(2, d, \beta_1, \beta_2)$-super-2 processes X_t can be characterized via the Laplace functional $L_{t,\nu}(J)$ given by

$$L_{t,\nu}(J) := \mathbb{E}\left\{ \exp\left(-\int_{M_F(\mathbb{R}^d)} J(\mu) \cdot X(t, d\mu)\right) \Big| X(0) = \nu \right\}$$

$$= \exp\left(-\int u(t,\mu) \cdot \nu(d\mu)\right),$$

where $u(t, \mu)$ satisfies the integral equation

$$u(t,\mu) = T_t u(0,\mu) - \frac{1}{1+\beta_2} \cdot \int_0^t \left[T_{t-s} u^{1+\beta_2}(s, \cdot)\right](\mu) \cdot ds,$$

$$u(0,\mu) = J(\mu)$$

for $J(\mu) = f(\langle\phi,\mu\rangle)$, where the function ϕ is $\mathbb{R}^d$-valued continuous and the function f is real-valued bounded continuous. Here, $\langle\phi,\mu\rangle := \int \phi \cdot d\mu$ and $\{T_t : t \geq 0\}$, $T_t : \mathbb{C}(M_F(\mathbb{R}^d)) \to \mathbb{C}(M_F(\mathbb{R}^d))$ is a semigroup which is associated with the semigroup $\{V_t : t \geq 0\}$ of the single-level $M_F(\mathbb{R}^d)$-valued branching process via the relationship

$$T_t \exp\{-\langle\phi,\cdot\rangle\} := \exp\{-\langle V_t\phi,\cdot\rangle\},$$

where

(1.2) $$V_t\phi(x) := S_t\phi(x) - \frac{1}{1+\beta_1} \cdot \int_0^t S_u\left[(V_{t-u}\phi)^{1+\beta_1}\right] \cdot du$$

and $\{S_t : t \geq 0\}$ is the semigroup associated with the infinitesimal generator $\Delta/2$. For a comprehensive review of the characterization of measure-valued processes, see [Dawson 1993].

Let $\mathbb{P}_\nu$ denote the probability measure of X_t in $\mathbb{D}\big([0,\infty), M_F\big(M_F(\mathbb{R}^d)\big)\big)$, where $\nu \in M_F\big(M_F(\mathbb{R}^d)\big)$ denotes the initial measure. Let

$$Z_t := \int \mu \cdot X_t(d\mu).$$

This $M_F(\mathbb{R}^d)$-valued process is called the *aggregated process* associated with the super-2 process X_t. We will show that the aggregated process Z_t possesses properties analogous to Lévy's modulus of continuity (cf. Theorem 1.1 below). Namely, let

$$r(t) := \inf\{R : S(Z_t) \subseteq \overline{\mathbb{B}(0,R)}\},$$

where $S(Y)$ denotes the closed support of Y in $\mathbb{R}^d$, and $\overline{\mathbb{B}(0,R)}$ is the closed ball centered at the origin with radius R.

Our main result, Theorem 1.1 gives the exact asymptotics of $\sup_{0\leq u\leq t} r(u)$ as $t \to 0$; i.e., it describes the exact almost-sure propagation of mass from a point source.

THEOREM 1.1. *For any positive ε,*

(1.3) $$\mathbb{P}_{\delta_{\delta_0}}\left\{1 + \left(\frac{d-2}{4/\beta_1 + 4/\beta_2} - \varepsilon\right) \cdot \frac{\log\log 1/t}{\log 1/t} \leq \frac{\sup_{0\leq u\leq t} r(u)}{\sqrt{(2/\beta_1 + 2/\beta_2) \cdot t \cdot \log(1/t)}}\right.$$
$$\left. \leq 1 + \left(\frac{d}{4/\beta_1 + 4/\beta_2} + \varepsilon\right) \cdot \frac{\log\log 1/t}{\log 1/t} \text{ for all sufficiently small positive } t\right\} = 1.$$

REMARKS. (i) Obviously, (1.3) implies that

(1.3') $$\mathbb{P}_{\delta_{\delta_0}}\left\{\lim_{t\downarrow 0} \frac{\sup_{0\leq u\leq t} r(u)}{\sqrt{(2/\beta_1 + 2/\beta_2) \cdot t \cdot \log(1/t)}} = 1\right\} = 1.$$

In addition, (1.3) implies that the ratio under the probability sign in (1.3') almost surely converges to one at the rate $O\big((\log\log 1/t)/(\log 1/t)\big)$ as t approaches zero. Moreover, the estimate $O\big((\log\log 1/t)/(\log 1/t)\big)$ is exact at least for $d \geq 3$, since the constant $(d-2)/(4/\beta_1 + 4/\beta_2) - \varepsilon$ can be chosen to be positive for such d. The question of whether or not this constant can be chosen to be positive for $d=1$ or 2 remains open (compare to [Dawson and Vinogradov 1992, Corollary 1.2]).

(ii) Note that the constants $(d-2)/(4/\beta_1 + 4/\beta_2)$ and $d/(4/\beta_1 + 4/\beta_2)$ from (1.3) are consistent with the constants $(d-2)/(4/\beta)$ and $d/(4/\beta)$ which arise in the analogous result for super-1 processes (cf., e.g., Theorem 1.4 of [Dawson and Vinogradov 1992]).

(iii) Note that (1.3′) can be viewed as a local-modulus-of-continuity-type result for super-2 processes. However, the local modulus for the Wiener process has a different form. Heuristically this difference reflects the fact that even though we consider here propagation of the closed support of the aggregated process on *short time intervals*, because of branching we must take into account the possibility of large increments of *many* individual Wiener processes. Thus, our situation is more similar in character to that of the global modulus of continuity for the Wiener process, where the possibility of large increments on a *large* number of short intervals is considered, rather than to that of the local modulus of continuity, where the consideration centers about the possibility of large increments on a *single* short time interval. In this respect, let us emphasize here that our result on the almost-sure rate of convergence in the *local* modulus for super-2 processes is analogous to the result on the almost-sure rate of convergence in the *global* modulus for the d-dimensional Wiener process (cf., e.g., Theorem 2 of [Chung, Erdös, and Sirao 1959]).

The proof of Theorem 1.1 is carried out in Section 3 and is based on a lower estimate (cf. Theorem 2.1 of Section 2) and an upper estimate (cf. Theorem 3.1 of Section 3). In order to get the lower estimate of Theorem 2.1, we apply purely probabilistic methods based on the branching particle system approximation. Arguments of that section are similar to those used in the proofs of Theorem 2.1 of [Tribe 1989] and Theorem 3.1 of [Dawson and Vinogradov 1992]. In particular, Proposition 2.2 of this work easily follows from Lemmas 4.2-4.4 of [Dawson and Vinogradov 1992].

In contrast, the proof of Theorem 3.1 of Section 3 involves the technique of historical processes developed in [Dawson and Perkins 1991], [Mueller and Perkins 1992], and [Dawson 1993], and will be published separately.

2. A Lower estimate for local propagation of mass for super-2 processes

We first proceed with arguments that are similar to those used in the proofs of Theorem 2.1 of [Tribe 1989] and Theorem 3.1 of [Dawson and Vinogradov 1992] in the setting of single-level models. Let $K^{(2)}_{\eta,\beta_2}(t)$ denote the number of initial superparticles of BPS-2 (described in Section 1) which have living descendants at instant t. It is natural to refer to $K^{(2)}_{\eta,\beta_2}(t)$ as the *number of level-2 clusters of age t*. Each individual level-2 cluster is generated by a surviving superparticle from the initial set. Note that some of the superparticles might contain no level-1 particles; hereinafter, we refer to such superparticles as null superparticles. Obviously, $K^{(2)}_{\eta,\beta_2}(t)$ is in fact the sum of η i.i.d. 0/1-valued Bernoulli random variables, and the probability of success $Q^{(2)}_{\eta,\beta_2}(t)$ in a single trial (i.e., survival of descendants of an individual superparticle from the initial set at instant t) is given by

$$(2.1) \qquad Q^{(2)}_{\eta,\beta_2}(t) = \left(1 + \frac{\beta_2}{\beta_2+1} \cdot t \cdot \eta^{\beta_2}\right)^{-1/\beta_2}$$

(cf., e.g., [Zolotarev 1957, Section 5], or [Dawson and Vinogradov 1992, Formula

(1.12)]). In addition,

$$\mathbb{P}\{K^{(2)}_{\eta,\beta_2}(t) > 0\} = 1 - \left(1 - Q^{(2)}_{\eta,\beta_2}(t)\right)^{\eta} \tag{2.2}$$

and

$$\mathbb{P}\{K^{(2)}_{\eta,\beta_2}(t) = \ell\} = \binom{\eta}{\ell} \cdot \left(Q^{(2)}_{\eta,\beta_2}(t)\right)^{\ell} \cdot \left(1 - Q^{(2)}_{\eta,\beta_2}(t)\right)^{\eta-\ell}, \tag{2.3}$$

where $0 \le \ell \le \eta$.

Now, fix the n-th superparticle alive at time t (which might be a null superparticle). Clearly, it is comprised of a random number $K^{(1)}_{\eta,\beta_1}(t;n)$ of level-1 clusters. Each level-1 cluster is comprised of the surviving descendants of an individual particle from the initial set. Since each newly born superparticle is a copy of its parent superparticle, by analogy with the arguments behind (2.1)-(2.3) it follows that $K^{(1)}_{\eta,\beta_1}(t;n)$ can be viewed as the sum of η i.i.d. 0/1-valued Bernoulli random variables, where the probability of success $Q^{(1)}_{\eta,\beta_1}(t)$ in a single trial (i.e., survival of descendants of an individual particle belonging to the n-th superparticle at instant t) is given by

$$Q^{(1)}_{\eta,\beta_1}(t) = \left(1 + \frac{\beta_1}{\beta_1 + 1} \cdot t \cdot \eta^{\beta_1}\right)^{-1/\beta_1}. \tag{2.4}$$

In other words, in order to determine the distribution of the number of level-1 clusters within an *individual superparticle* at time t, we can suppress the second-level branching and regard this superparticle as a single-level branching particle system.

By analogy with (2.2)-(2.3), we obtain that

$$\mathbb{P}\{K^{(1)}_{\eta,\beta_1}(t;n) > 0\} = 1 - \left(1 - Q^{(1)}_{\eta,\beta_1}(t)\right)^{\eta}, \tag{2.5}$$

and

$$\mathbb{P}\{K^{(1)}_{\eta,\beta_1}(t;n) = \ell\} = \binom{\eta}{\ell} \cdot \left(Q^{(1)}_{\eta,\beta_1}(t)\right)^{\ell} \cdot \left(1 - Q^{(1)}_{\eta,\beta_1}(t)\right)^{\eta-\ell}, \tag{2.6}$$

where $0 \le \ell \le \eta$.

Recall that by our choice of the initial distribution, any initial superparticle $SP_n(0)$ ($1 \le n \le \eta$) consists of η particles. On the other hand, superparticles $SP_n(t)$ alive at time t consist of random numbers $K^{(1)}_{\eta,\beta_1}(t;n)$ of level-1 clusters, and the random numbers $K^{(1)}_{\eta,\beta_1}(t;1)$, $K^{(1)}_{\eta,\beta_1}(t;2)$, ... are identically distributed, though some of them – those related to superparticles that belong to the same level-2 clusters – are dependent. However, the random numbers $K^{(1)}_{\eta,\beta_1}(t;n_1)$ and $K^{(1)}_{\eta,\beta_1}(t;n_2)$ of level-1 clusters related to superparticles belonging to *different* level-2 clusters are independent. The main idea behind the derivation of the lower bound for BPS-2 consists in choosing one and only one representative (hereinafter referred to as the *tagged superparticle*) from each surviving level-2 cluster, and then one and only one representative (hereinafter referred to as the *tagged particle*) from each surviving level-1 cluster (which belongs to a tagged superparticle). Note that any tagged particle belongs to a particular tagged superparticle. It is clear that the exclusion from our consideration of certain superparticles and particles will give us a rougher lower bound. On the other hand, such simplification will enable us to employ the independence of tagged superparticles as well as the independence of tagged particles within a superparticle. This independence property and the above representations in terms of sums of i.i.d. Bernoulli variables enable us to use the compound Poisson approximation for BPS-2 (cf. Proposition 2.1(iv) below). Recall that a similar idea was used by [Tribe 1989] (see Theorem 2.1 therein) in the single-level setting.

We now proceed with the following auxiliary result, which easily follows from the classical Poisson theorem for the triangular array of 0/1-valued Bernoulli variables and the characterization of weak convergence of non-negative random variables in terms of pointwise convergence of their Laplace transforms (compare to Proposition 1.2 of [Dawson and Vinogradov 1992]):

PROPOSITION 2.1. *Consider the BPS-2 described in Section 1 with branching mechanisms governed by* (1.1) *and* (1.1'). *Then for any fixed real* $t > 0$:

(i)
$$K^{(2)}_{\eta,\beta_2}(t) \xrightarrow{d} \Pi\left(\left(\frac{\beta_2}{\beta_2+1}\cdot t\right)^{-1/\beta_2}\right),$$
$$K^{(1)}_{\eta,\beta_1}(t;1) \xrightarrow{d} \Pi\left(\left(\frac{\beta_1}{\beta_1+1}\cdot t\right)^{-1/\beta_1}\right)$$

as $\eta \to \infty$, *where* $\Pi(\kappa)$ *denotes a random variable having the Poisson distribution with parameter* κ.

(ii) *The Laplace transforms* $f_t^{(\beta_i)}(\cdot)$ *of* $\Pi\left((\beta_i \cdot t/(\beta_i+1))^{-1/\beta_i}\right)$, *(i = 1, 2) are given by*

(2.7) $$f_t^{(\beta_i)}(s) = \exp\{(\beta_i \cdot t/(\beta_i+1))^{-1/\beta_i} \cdot (e^{-s} - 1)\}.$$

(iii) *The Laplace transforms* $\psi^{(\beta_2)}_{\eta,t}(\cdot)$ *and* $\psi^{(\beta_1)}_{\eta,t}(\cdot)$ *of* $K^{(2)}_{\eta,\beta_2}(t)$ *and* $K^{(1)}_{\eta,\beta_1}(t;1)$ *are given by*

$$\psi^{(\beta_i)}_{\eta,t}(s) = \left(1 - Q^{(i)}_{\eta,\beta_i}(t) \cdot (1-e^{-s})\right)^\eta.$$

(iv) *The random sum* $\mathcal{L}_{\eta,\beta_1,\beta_2}(t) := \sum_{n=1}^{K^{(2)}_{\eta,\beta_2}(t)} K^{(1)}_{\eta,\beta_1}(t;n)$ *has Laplace transform*

$$\psi^{(\beta_1,\beta_2)}_{\eta,t}(s) = \psi^{(\beta_2)}_{\eta,t}\left(\log \frac{1}{\psi^{(\beta_1)}_{\eta,t}(s)}\right).$$

The random sum $\mathcal{L}_{\eta,\beta_1,\beta_2}(t)$ *converges weakly as* $\eta \to \infty$ *to a random variable* $\mathcal{L}_{\infty,\beta_1,\beta_2}(t)$ *having Laplace transform*

(2.8) $$f_t^{(\beta_1,\beta_2)}(s) := \mathbb{E}\left\{\exp\left\{-s\mathcal{L}_{\infty,\beta_1,\beta_2}(t)\right\}\right\} = f_t^{(\beta_2)}\left(\log \frac{1}{f_t^{(\beta_1)}(s)}\right).$$

Note that the random sum $\mathcal{L}_{\eta,\beta_1,\beta_2}(t)$ and the random variable $\mathcal{L}_{\infty,\beta_1,\beta_2}(t)$ can be viewed as numbers of *tagged level-1 clusters of age t* of the BPS-2 and of the super-2 process, respectively.

Now, we state the following auxiliary result, which easily follows from Lemmas 4.2-4.4 in [Dawson and Vinogradov 1992] :

PROPOSITION 2.2. *For any $i = 1, 2$ there exists a positive constant $C(\beta_i)$, such that for any positive t there is a value η_i $(= \eta_i(t))$ with the property that for any $\eta \geq \eta_i$,*

$$\psi_{\eta,t}^{(\beta_i)}(s) \leq C(\beta_i) \cdot f_t^{(\beta_i)}(s) \tag{2.9}$$

for any $s \in \mathbb{R}_+^1$.

It is clear that almost-sure lower estimates for the propagation of support can be derived from lower estimates for

$$\mathbb{P}_{\delta_{b_0}} \left\{ \int \mu(\overline{\mathbb{B}(0,R)^c}) \cdot X_s(d\mu) > 0 \text{ for some } 0 \leq s \leq t \right\}$$

by the use of Borel-Cantelli arguments. In turn, estimating the above probability is equivalent to the derivation of upper estimates for the probability of the event that the $(2, d, \beta_1, \beta_2)$-super-2 process X_t will remain inside the closed ball $\overline{\mathbb{B}(0,R)}$ during the time period $[0, t]$. This approach is developed in Theorem 2.1 below by the use of BPS-2 approximation, Propositions 2.1-2.2 and Borel-Cantelli arguments. Our main idea lies in the simultaneous consideration of the $(2, d, \beta_1, \beta_2)$-super-2 process and the BPS-2. Specifically, Propositions 2.1-2.2 will enable us to apply the Borel-Cantelli arguments for establishing convergence of a certain series made up of probabilities of certain events related to the $(2, d, \beta_1, \beta_2)$-super-2 process X_t. Indeed, such probabilities can be obtained as limits (as $\eta \to \infty$) of probabilities of certain events related to the corresponding BPS-2. We then estimate the latter probabilities from above in terms of $\psi_{\eta,t}^{(\beta_1,\beta_2)}(\cdot)$, and this function can be properly estimated by the use of Propositions 2.1 and 2.2. In other words, we can establish almost-sure lower bounds (i.e., find proper lower functions) for the asymptotic behavior of $\sup_{0 \leq u \leq t} r(u)$ as $t \to 0$ (cf. Theorem 2.1 below). Apart from this, only purely probabilistic arguments and some well-known properties of d-dimensional Wiener process are used.

We next introduce the following family of continuous increasing functions on the interval $[0, e^{-1}]$:

$$\nu_{\beta_1,\beta_2,\varepsilon}(s) = \sqrt{\left(\frac{2}{\beta_1} + \frac{2}{\beta_2}\right) \cdot s \cdot \left(\log \frac{1}{s} + \left(\frac{d-2}{2/\beta_1 + 2/\beta_2} - \varepsilon\right) \cdot \log\log \frac{1}{s}\right)}$$

for $s \in (0, e^{-1}]$,

$\nu_{\beta_1,\beta_2,\varepsilon}(0) = 0$.

In what follows, the parameter ε is assumed to take any fixed (sufficiently small) positive value less than one if $d \leq 2$ and less than $(d-2)/(2/\beta_1 + 2/\beta_2)$ if $d \geq 3$.

THEOREM 2.1. *For any $\varepsilon > 0$,*

$$\mathbb{P}_{\delta_{b_0}} \left\{ \sup_{0 \leq u \leq t} r(u) \geq \nu_{\beta_1,\beta_2,\varepsilon}(t) \text{ for all sufficiently small positive } t \right\} = 1. \tag{2.10}$$

PROOF OF THEOREM 2.1. The cornerstone of the proof is the following lemma related to the behavior of $r(\cdot)$ at specific time instants.

LEMMA 2.1 (compare to [Ugbebor 1980, pp. 42–43]). *Let $\gamma_n := \exp(-n^\rho)$, where $0 < \rho < \frac{1}{2}$ is fixed. Then for any $\varepsilon > 0$,*

(2.11) $\quad \mathbb{P}_{\delta_{\delta_0}}\{r(\gamma_{n+1}) \geq \nu_{\beta_1,\beta_2,\varepsilon}(\gamma_n) \text{ for all sufficiently large } n\} = 1.$

Indeed, (2.11) easily implies (2.10), since for $\gamma_{n+1} \leq t \leq \gamma_n$,

$$\sup_{0 \leq u \leq t} r(u) \geq \sup_{0 \leq u \leq \gamma_{n+1}} r(u) \geq r(\gamma_{n+1})$$
$$\geq \nu_{\beta_1,\beta_2,\varepsilon}(\gamma_n) \geq \nu_{\beta_1,\beta_2,\varepsilon}(t), \quad \mathbb{P}_{\delta_{\delta_0}}\text{-a.s.}$$

REMARK. Note that [Tribe 1989, Theorem 2.1] used a slightly different, more classical, technique for the derivation of a lower bound in the single-level setting. Namely, his lower bound followed from the lower bound for specific time instants and subsequent application of almost-sure upper estimates. However, that approach is not sufficient to isolate the principal error term $(\log \log 1/t)/(\log 1/t)$ in (1.3).

PROOF OF LEMMA 2.1. Note that the first Borel-Cantelli lemma implies that in order to establish (2.11), it suffices to prove that

(2.12) $\quad \sum_{n=1}^{\infty} \mathbb{P}_{\delta_{\delta_0}}\{r(\gamma_{n+1}) \leq \nu_{\beta_1,\beta_2,\varepsilon}(\gamma_n)\} < \infty.$

Let $Y_t^{(\eta)}$ denote the corresponding BPS-2. It can be shown by applying arguments similar to those of Lemma 4.8 of [Dawson, Iscoe and Perkins 1989] that

(2.13) $\mathbb{P}_{\delta_{\delta_0}}\{r(\gamma_{n+1}) \leq \nu_{\beta_1,\beta_2,\varepsilon}(\gamma_n)\} \leq \liminf_{\eta \to \infty} \mathbb{P}_{\delta_{\delta_0}}\left\{S(Y_{\gamma_{n+1}}^{(\eta)}) \in \overline{\mathbb{B}(0, \nu_{\beta_1,\beta_2,\varepsilon}(\gamma_n))}\right\}$

for any fixed integer $n \geq 1$, where $\overline{\mathbb{B}(0, \nu_{\beta_1,\beta_2,\varepsilon}(\gamma_n))}$ is the closed ball centered at the origin with radius $\nu_{\beta_1,\beta_2,\varepsilon}(\gamma_n)$. Therefore, in order to obtain (2.12), we will first derive auxiliary estimates for the support process of $Y_t^{(\eta)}$, and then obtain analogous estimates for the terms of (2.12) by letting $\eta \to \infty$ and applying Propositions 2.1-2.2 and formula (2.13).

Let $\{w_i(\cdot)\}$ denote independent copies of the standard d-dimensional Wiener process $w(\cdot)$. Now, take a tagged superparticle from each surviving level-2 cluster and then take a tagged particle from each surviving level-1 cluster belonging to a specific tagged superparticle. Successive application of the independence of tagged superparticles and of tagged particles within a superparticle yields that

$$\mathbb{P}_{\delta_{\delta_0}}\{S(Y_{\gamma_{n+1}}^{(\eta)}) \in \overline{\mathbb{B}(0, \nu_{\beta_1,\beta_2,\varepsilon}(\gamma_n))}\},$$

the probability of the event that all the particles of BPS-2 $Y_{\gamma_{n+1}}^{(\eta)}$ remain inside the closed ball $\overline{\mathbb{B}(0, \nu_{\beta_1,\beta_2,\varepsilon}(\gamma_n))}$, does not exceed

(2.14) $\quad \mathbb{P}\left\{\sup_{1 \leq i \leq \mathcal{L}_{\eta,\beta_1,\beta_2}(\gamma_{n+1})} |w_i(\gamma_{n+1})| \leq \nu_{\beta_1,\beta_2,\varepsilon}(\gamma_n)\right\}$

$$\leq \sum_{\ell=0}^{\eta^2} \mathbb{P}\{\mathcal{L}_{\eta,\beta_1,\beta_2}(\gamma_{n+1}) = \ell\} \cdot \{1 - \mathbb{P}\{|w(\gamma_{n+1})| > \nu_{\beta_1,\beta_2,\varepsilon}(\gamma_n)\}\}^\ell$$

$$\leq \sum_{\ell=0}^{\eta^2} \mathbb{P}\{\mathcal{L}_{\eta,\beta_1,\beta_2}(\gamma_{n+1}) = \ell\} \cdot \exp\{-\ell \cdot \mathbb{P}\{|w(\gamma_{n+1})| > \nu_{\beta_1,\beta_2,\varepsilon}(\gamma_n)\}\}.$$

Recall the self-similarity property $w(t) \stackrel{d}{=} w(kt)/\sqrt{k}$ with $k = 1/t$ and the well-known bound
$$\mathbb{P}\{|w(1)| > x\} \geq C(d) \cdot e^{-x^2/2} \cdot x^{d-2}$$
for the tail probability of the d-dimensional Wiener process. It then follows that
$$\mathbb{P}\{|w(\gamma_{n+1})| > \nu_{\beta_1,\beta_2,\varepsilon}(\gamma_n)\} = \mathbb{P}\{|w(1)| > \nu_{\beta_1,\beta_2,\varepsilon}(\gamma_n)/\gamma_{n+1}^{1/2}\}.$$

Now, let us show that the latter probability is greater than or equal to
$$\tau_{n+1} := C(d,\beta_1,\beta_2,\rho,\varepsilon) \cdot \gamma_{n+1}^{1/\beta_1+1/\beta_2} \cdot \left(\log \frac{1}{\gamma_{n+1}}\right)^{\varepsilon \cdot (1/\beta_1+1/\beta_2)},$$
where $C(d,\beta_1,\beta_2,\rho,\varepsilon)$ is some positive constant depending on the specified parameters.

Indeed,
$$\mathbb{P}\{|w(1)| > \nu_{\beta_1,\beta_2,\varepsilon}(\gamma_n)/\gamma_{n+1}^{1/2}\}$$
$$\geq C(d) \exp\{-\nu_{\beta_1,\beta_2,\varepsilon}^2(\gamma_n)/2\gamma_{n+1}\} \cdot (\nu_{\beta_1,\beta_2,\varepsilon}^2(\gamma_n)/\gamma_{n+1})^{(d-2)/2}.$$

After some relatively simple computations, we obtain the result that the product of the second and the third factors on the right-hand side of the above inequality is greater than or equal to
(2.15)
$$C(d,\beta_1,\beta_2,\varepsilon) \cdot \exp\left\{-\frac{\gamma_n}{\gamma_{n+1}} \cdot \left(\frac{1}{\beta_1} + \frac{1}{\beta_2}\right) \cdot \log \frac{1}{\gamma_n}\right\}$$
$$\cdot \exp\left\{-\frac{\gamma_n}{\gamma_{n+1}} \cdot \left(\frac{1}{\beta_1} + \frac{1}{\beta_2}\right) \cdot \left(\frac{d-2}{2/\beta_1 + 2/\beta_2} - \varepsilon\right) \log\log \frac{1}{\gamma_n}\right\}$$
$$\cdot \left(\frac{\gamma_n}{\gamma_{n+1}} \log \frac{1}{\gamma_n}\right)^{(d-2)/2}$$
$$\geq C(d,\beta_1,\beta_2,\rho,\varepsilon)\left(\gamma_n^{\frac{\gamma_n}{\gamma_{n+1}}}\right)^{\frac{1}{\beta_1}+\frac{1}{\beta_2}}$$
$$\cdot \left(\log \frac{1}{\gamma_n}\right)^{(d-2)/2} \cdot \left(\log \frac{1}{\gamma_n}\right)^{-\left((d-2)/2 - \varepsilon(1/\beta_1+1/\beta_2)\right)\gamma_n/\gamma_{n+1}}.$$

(Note that
$$\frac{\gamma_n}{\gamma_{n+1}} = \exp\left\{n^\rho\left(\left(1+\frac{1}{n}\right)^\rho - 1\right)\right\} = \exp\left\{\frac{\rho}{n^{1-\rho}} + o\left(\frac{1}{n^{1-\rho}}\right)\right\} = 1 + \frac{\rho}{n^{1-\rho}} + o\left(\frac{1}{n^{1-\rho}}\right)$$
as $n \to \infty$, and hence is greater than or equal to some positive constant $c(\rho)$ for all sufficiently large n, by our choice of $\{\gamma_n\}$.) In addition,
$$\frac{\gamma_n^{(\gamma_n/\gamma_{n+1})(\frac{1}{\beta_1}+\frac{1}{\beta_2})}}{\gamma_{n+1}^{\frac{1}{\beta_1}+\frac{1}{\beta_2}}} = \left(\frac{\gamma_n}{\gamma_{n+1}} \cdot \gamma_n^{\frac{\gamma_n}{\gamma_{n+1}}-1}\right)^{\frac{1}{\beta_1}+\frac{1}{\beta_2}}$$
$$= \left(\left(1 + \frac{\rho}{n^{1-\rho}} + o\left(\frac{1}{n^{1-\rho}}\right)\right)\left(1 + \frac{\rho}{n^{1-2\rho}} + o\left(\frac{1}{n^{1-\rho}}\right)\right)\right)^{1/\beta_1+1/\beta_2}$$

as $n \to \infty$, and hence is greater than or equal to some positive constant $C(\beta_1, \beta_2, \rho)$ for all sufficiently large n (recall that $0 < \rho < \frac{1}{2}$). Therefore, the expression on the right-hand side of (2.15) is greater than or equal to

$$C(d, \beta_1, \beta_2, \rho, \varepsilon)(\gamma_{n+1})^{1/\beta_1 + 1/\beta_2} \left(\log \frac{1}{\gamma_n}\right)^{-[(d-2)/2](\gamma_n/\gamma_{n+1} - 1)}$$

$$\times \left(\log \frac{1}{\gamma_n}\right)^{\varepsilon(1/\beta_1 + 1/\beta_2)\gamma_n/\gamma_{n+1}}$$

Now, it is easily seen that

$$\left(\log \frac{1}{\gamma_n}\right)^{-[(d-2)/2](\gamma_n/\gamma_{n+1} - 1)} = (n^\rho)^{-[(d-2)/2][\rho/n^{1-\rho} + o(1/n^{1-\rho})]} \quad (\text{as } n \to \infty)$$

$$= \exp\left\{-\rho^2 \cdot \frac{d-2}{2} \cdot \frac{\log n}{n^{1-\rho}} + o\left(\frac{\log n}{n^{1-\rho}}\right)\right\},$$

and hence is greater than or equal to some positive constant $C(\rho)$ for all sufficiently large n (recall that $0 < \rho < \frac{1}{2}$). On the other hand, as n increases,

$$\frac{(\log 1/\gamma_n)^{\varepsilon(1/\beta_1 + 1/\beta_2)(\gamma_n/\gamma_{n+1})}}{(\log 1/\gamma_{n+1})^{\varepsilon(1/\beta_1 + 1/\beta_2)}}$$

$$= (\log 1/\gamma_n)^{\varepsilon(1/\beta_1 + 1/\beta_2)((\gamma_n/\gamma_{n+1}) - 1)} \cdot \left(\frac{n^\rho}{(n+1)^\rho}\right)^{\varepsilon(1/\beta_1 + 1/\beta_2)}$$

$$= n^{\varepsilon(1/\beta_1 + 1/\beta_2)\rho\left(\rho/n^{1-\rho} + o(1/n^{(1-\rho)})\right)} \cdot \left(1 - \frac{1}{n+1}\right)^{\rho\varepsilon(1/\beta_1 + 1/\beta_2)}$$

as $n \to infty$, and hence is greater than or equal to some positive constant $C(\beta_1, \beta_2, \rho, \varepsilon)$ for all sufficiently large n. Therefore, the expression on the right-hand side of (2.15) is greater than or equal to τ_{n+1} for all sufficiently large n.

Thus, the probability on the left-hand side of (2.14) does not exceed

$$(2.16) \qquad \sum_{\ell=0}^{\eta^2} \mathbb{P}\{\mathcal{L}_{\eta,\beta_1,\beta_2}(\gamma_{n+1}) = \ell\} \cdot \exp\{-\ell \cdot \tau_{n+1}\},$$

which equals $\psi_{\eta,\gamma_{n+1}}^{(\beta_1,\beta_2)}(\tau_{n+1})$ by definition, where $\psi_{\eta,t}^{(\beta_1,\beta_2)}(\cdot)$ denotes the Laplace transform defined in Proposition 2.1.iv. Hence, an application of Proposition 2.2 with $i = 2$, $t = \gamma_{n+1}$ and $s = \log(1/\psi_{\eta,\gamma_{n+1}}^{(\beta_1)}(\tau_{n+1}))$ (cf. (2.9)), along with the explicit expressions for $\psi_{\eta,\gamma_{n+1}}^{(\beta_1)}(\cdot)$ and $f_{\gamma_{n+1}}^{(\beta_2)}(\cdot)$ given in Proposition 2.1, enables one to obtain that the expression (2.16) does not exceed

$$C(\beta_2) \cdot \exp\left\{(\beta_2 \cdot \gamma_{n+1}/(\beta_2 + 1))^{-1/\beta_2} \cdot (e^{-s} - 1)\right\}$$

$$= C(\beta_2) \cdot \exp\left\{(\beta_2 \gamma_{n+1}/(\beta_2 + 1))^{-1/\beta_2} \cdot \left(\psi_{\eta,\gamma_{n+1}}^{(\beta_1)}(\tau_{n+1}) - 1\right)\right\}.$$

The first factor on the right-hand side above is a certain positive constant depending only on β_2, which entered from inequality (2.9). To estimate the second factor, we apply Proposition 2.1.iii:

$$\exp\{(\beta_2 \cdot \gamma_{n+1}/(\beta_2+1))^{-1/\beta_2} \cdot \left(\psi^{(\beta_1)}_{\eta,\gamma_{n+1}}(\tau_{n+1}) - 1\right)\}$$

$$= \exp\left\{\left(\frac{\beta_2 \cdot \gamma_{n+1}}{\beta_2+1}\right)^{-1/\beta_2} \cdot \left(\left(1 - Q^{(1)}_{\eta,\beta_1}(\gamma_{n+1}) \cdot (1 - \exp\{-\tau_{n+1}\})\right)^\eta - 1\right)\right\}.$$

Now, our choice of $\{\gamma_n\}$ and $\{\tau_n\}$ and the representation (2.4) for $Q^{(1)}_{\eta,\beta_1}(\gamma_{n+1})$ imply that for all sufficiently large n and any $\eta \geq \eta(n)$,

$$\left(1 - Q^{(1)}_{\eta,\beta_1}(\gamma_{n+1}) \cdot (1 - \exp\{-\tau_{n+1}\})\right)^\eta - 1$$
$$\leq -\eta \cdot Q^{(1)}_{\eta,\beta_1}(\gamma_{n+1}) \cdot (1 - \exp\{-\tau_{n+1}\})$$
$$\quad + \frac{\eta \cdot (\eta - 1)}{2} \cdot \left(Q^{(1)}_{\eta,\beta_1}(\gamma_{n+1})\right)^2 \cdot (1 - \exp\{-\tau_{n+1}\})^2$$
$$\leq -\frac{1}{2} \cdot \eta \cdot Q^{(1)}_{\eta,\beta_1}(\gamma_{n+1}) \cdot (1 - \exp\{-\tau_{n+1}\})$$
$$\leq -\frac{1}{2} \cdot \eta \cdot Q^{(1)}_{\eta,\beta_1}(\gamma_{n+1}) \cdot (\tau_{n+1} - \tau_{n+1}^2/2) \leq -\frac{1}{4} \cdot \eta \cdot Q^{(1)}_{\eta,\beta_1}(\gamma_{n+1}) \cdot \tau_{n+1}.$$

Replacing τ_{n+1} by $C(d,\beta_1,\beta_2,\rho,\varepsilon) \cdot \gamma_{n+1}^{1/\beta_1 + 1/\beta_2} \cdot \left(\log \frac{1}{\gamma_{n+1}}\right)^{\varepsilon \cdot (1/\beta_1 + 1/\beta_2)}$ in the rightmost expression and then making some relatively simple transformations, we obtain the result that

$$\left(1 - Q^{(1)}_{\eta,\beta_1}(\gamma_{n+1}) \cdot (1 - \exp\{-\tau_{n+1}\})\right)^\eta - 1$$
$$\leq -\frac{C(d,\beta_1,\beta_2,\rho,\varepsilon)/4}{(1/\gamma_{n+1} \cdot \eta^{\beta_1} + \beta_1/(\beta_1+1))^{1/\beta_1}} \cdot \gamma_{n+1}^{1/\beta_2} \cdot \left(\log \frac{1}{\gamma_{n+1}}\right)^{\varepsilon \cdot (1/\beta_1 + 1/\beta_2)}.$$

Hence, the expression on the right-hand side of (2.16) does not exceed

$$C(\beta_2) \cdot \exp\{(\beta_2 \cdot \gamma_{n+1}/(\beta_2+1))^{-1/\beta_2} \cdot (\psi^{(\beta_1)}_{\eta,\gamma_{n+1}}(\tau_{n+1}) - 1)\}$$
$$\leq C(\beta_2) \cdot \exp\left\{-C_1(d,\beta_1,\beta_2,\rho,\varepsilon) \cdot \left(\log \frac{1}{\gamma_{n+1}}\right)^{\varepsilon \cdot (1/\beta_1 + 1/\beta_2)}\right\}$$

for all sufficiently large n and any $\eta \geq \eta(n)$, where $C(\beta_2)$ and $C_1(d,\beta_1,\beta_2,\rho,\varepsilon)$ are some positive constants depending on the specified parameters. Together with (2.14) and (2.16), this implies that for all sufficiently large n and any $\eta \geq \eta(n)$,

$$(2.17) \quad \mathbb{P}_{\delta_{\delta_0}}\left\{S(Y^{(\eta)}_{\gamma_{n+1}}) \in \overline{\mathbb{B}(0, \nu_{\beta_1,\beta_2,\varepsilon}(\gamma_n))}\right\}$$
$$\leq C(\beta_2) \cdot \exp\left\{-C_1(d,\beta_1,\beta_2,\rho,\varepsilon) \cdot \left(\log \frac{1}{\gamma_{n+1}}\right)^{\varepsilon \cdot (1/\beta_1 + 1/\beta_2)}\right\}.$$

Note that the expression on the right-hand side of (2.17) does not depend on η. Hence, an application of (2.13) implies the following estimate for the probability on the left-hand side of (2.13) for all sufficiently large n:

$$(2.18) \quad \mathbb{P}_{\delta_{\delta_0}}\{r(\gamma_{n+1}) \leq \nu_{\beta_1,\beta_2,\varepsilon}(\gamma_n)\}$$
$$\leq C(\beta_2) \cdot \exp\left\{-C_1(d,\beta_1,\beta_2,\rho,\varepsilon) \cdot \left(\log \frac{1}{\gamma_{n+1}}\right)^{\varepsilon \cdot (1/\beta_1 + 1/\beta_2)}\right\}.$$

Now, by our choice of $\{\gamma_n\}$, the expression on the right-hand side in (2.18) is majorized by

(2.19) $$\text{Const} \cdot \exp\{-C(d, \beta_1, \beta_2, \rho, \varepsilon) \cdot n^{\varepsilon \cdot \rho \cdot (1/\beta_1 + 1/\beta_2)}\}.$$

Obviously, for any positive ε, expression (2.19) is the general term of a convergent series. Hence, the expression on the left-hand side of (2.18) is also the general term of a convergent series, i.e., the series (2.12) is convergent. □

3. Proof of Theorem 1.1

We first formulate the local-modulus-of-continuity-type result for super-2 processes.

Let

$$h_\kappa(t) := \sqrt{\left(\frac{2}{\beta_1} + \frac{2}{\beta_2}\right) \cdot t \cdot \left(\log \frac{1}{t} + \kappa \cdot \log\log \frac{1}{t}\right)} \quad \text{for } t \in (0, e^{-1}],$$

$$h_\kappa(0) := 0.$$

THEOREM 3.1. *Let $m \in M_F(M_F(\mathbb{R}^d))$ and $\kappa > d/(2/\beta_1 + 2/\beta_2)$. Let $T > 0$ be fixed. Then for each fixed $0 \leq t \leq T$ and for $\mathbb{P}_m$-a.e. ω, there exists a $\delta(\omega, \kappa) > 0$ such that if $0 < s < \delta$ and $t + s \leq T$, then*

$$S(Z_{t+s}) \subseteq S(Z_t)^{h_\kappa(s)}.$$

Recall that the proof of Theorem 3.1 involves the technique of historical processes and will be published separately.

Now, we proceed with the proof of our main result.

PROOF OF THEOREM 1.1. In order to get an upper estimate, we apply the assertion of Theorem 3.1 with $t = 0$ and $m = \delta_{\delta_0}$. In particular, we get that $S(Z_0) = \{0\}$, $\mathbb{P}_m$-a.e., and hence $S(Z_0)^{h_\kappa(s)} = \overline{\mathbb{B}(0, h_\kappa(s))}$. A subsequent application of Theorem 3.1 yields that for each $\kappa > d/(2/\beta_1 + 2/\beta_2)$ and for each $0 < s < \delta \wedge T$,

(3.1) $$\mathbb{P}_{\delta_{\delta_0}}\{S(Z_u) \subseteq \overline{\mathbb{B}(0, h_\kappa(s))} \text{ for } 0 \leq u \leq s\} = 1.$$

It follows from the explicit form of $h_\kappa(\cdot)$ that (3.1) implies that for any positive $\kappa > d/(2/\beta_1 + 2/\beta_2)$,

(3.2) $$\mathbb{P}_{\delta_{\delta_0}}\left\{\sup_{0 \leq u \leq t} r(u) \leq \sqrt{\left(\frac{2}{\beta_1} + \frac{2}{\beta_2}\right) \cdot t \cdot \left(\log \frac{1}{t} + \kappa \cdot \log\log \frac{1}{t}\right)}\right.$$
$$\left. \text{for all sufficiently small positive } t\right\} = 1.$$

Combining (3.2) and (2.10), we obtain the result that for any sufficiently small positive ε,

$$\mathbb{P}_{\delta_{\delta_0}}\left\{\sqrt{\left(\frac{2}{\beta_1}+\frac{2}{\beta_2}\right)\cdot t\cdot\left(\log\frac{1}{t}+\left(\frac{d-2}{2/\beta_1+2/\beta_2}-\varepsilon\right)\cdot\log\log\frac{1}{t}\right)}\leq\sup_{0\leq u\leq t}r(u)\right.$$
$$\left.\leq\sqrt{\left(\frac{2}{\beta_1}+\frac{2}{\beta_2}\right)\cdot t\cdot\left(\log\frac{1}{t}+\left(\frac{d}{2/\beta_1+2/\beta_2}+\varepsilon\right)\cdot\log\log\frac{1}{t}\right)}\right.$$
$$\text{for all sufficiently small positive }t\right\}=1.$$

Dividing by $\sqrt{(2/\beta_1+2/\beta_2)\cdot t\cdot\log 1/t}$ all the three terms of the inequality under the probability sign and keeping in mind that $\sqrt{1+\theta}=1+\theta/2+O(\theta^2)$ as $\theta\to 0$, we immediately obtain (1.3). $\square$

ACKNOWLEDGEMENT. We would like to thank Ed Perkins for many helpful discussions.

References

K. L. Chung, P. Erdös, and T. Sirao, *On the Lipschitz's condition for Brownian motions*, J. Math. Soc. Japan **11** (1959), 263-274.

D. A. Dawson, *Measure-valued Markov processes*, École d'Eté de Probabilités de Saint-Flour XXI (P. L. Hennequin, ed.), Lecture Notes in Math., vol. 1451, Springer, 1993, pp. 1–260.

D. A. Dawson and K. J. Hochberg, *A multilevel branching model*, Adv. in Appl. Probab. **23** (1991), 701-715.

D. A. Dawson, K. J. Hochberg, and Y. Wu, *Multilevel branching systems*, White Noise Analysis: Mathematics and Applications (T. Hida, H.H. Kuo, J. Potthoff and L. Streit, ed.), World Scientific, 1990, pp. 93-107.

D. A. Dawson, I. Iscoe, and E. A. Perkins, *Super-Brownian motion: path properties and hitting probabilities*, Probab. Theor. Relat. Fields **83** (1989), 135-205.

D. A. Dawson and E. A. Perkins, *Historical processes*, vol. 93, Mem. Amer. Math. Soc., 1991.

D. A. Dawson and V. Vinogradov, *Almost-sure path properties of $(2,d,\beta)$-superprocesses*, LRSP Tech. Report 195 (1992).

A. Greven, *A phase transition for the coupled branching process* I, Probab. Theor. Relat. Fields **87** (1991), 417-458.

K. J. Hochberg, *Hierarchically structured branching populations with spatial motion*, Rocky Mountain J. Math. (1993) (to appear).

C. Mueller and E. A. Perkins, *The compact support property for solutions to the heat equation with noise*, Probab. Theor. Relat. Fields **93** (1992), 325-358.

R. Tribe, *Path properties of superprocesses*, Ph.D. Thesis, U.B.C., 1989.

O. O. Ugbebor, *Uniform variation results for Brownian motion*, Z. Wahrsch. Verw. Gebiete **51** (1980), 39-48.

Y. Wu, *Dynamic particle systems and multilevel measure branching processes*, Ph.D. Thesis, Carleton University, 1992.

Y. Wu, *Multilevel birth and death particle system and its continuous diffusion*, Adv. in Appl. Probab. **25** (1993) (to appear).

V. M. Zolotarev, *More exact statements of several theorems in the theory of branching processes*, Theory Probab. Appl. **2** (1957), 245-253.

ABSTRACT. We consider two-level measure-valued branching processes introduced in [Dawson, Hochberg and Wu 1990]. Such processes generalize so-called *super-Brownian motion*. They can appear as high-density limits of certain properly rescaled hierarchically structured branching particle systems. Such systems can be viewed as populations of individuals or particles which not only undergo both diffusion and branching, but which are also affected by an additional branching process that acts upon groups of particles concurrently. The possibility of simultaneous branching at the upper level leads to the absence of independence of particle behavior. Nonetheless, we show that such processes possess path properties which are analogous to the Lévy modulus of continuity for Brownian motion.

DEPARTMENT OF MATHEMATICS AND STATISTICS, CARLETON UNIVERSITY, OTTAWA, ONTARIO, CANADA K1S 5B6

E-mail address: ddawson@math.carleton.ca

DEPARTMENT OF MATHEMATICS AND COMPUTER SCIENCES, BAR-ILAN UNIVERSITY, 52900 RAMAT-GAN, ISRAEL

E-mail address: hochberg@bimacs.cs.biu.ac.il

DEPARTMENT OF MATHEMATICS AND STATISTICS, CONCORDIA UNIVERSITY, SIR GEORGE WILLIAMS CAMPUS, MONTRÉAL, QUÉBEC, CANADA H3G 1M8

E-mail address: vvinogra@poincare.concordia.ca

A Type of Interaction between Superprocesses and Branching Particle Systems

Eugene B. Dynkin

1. In [Evans and O'Connell 1993], Evans and O'Connell suggested a way to construct a Dawson-Watanabe superprocess $\widehat{X}$ with the branching parameter

$$\widehat{\psi}(w) = \frac{1}{2} cw^2 - bw \tag{1.1}$$

(b and c are positive constants) from an analogous superprocess X with the parameter

$$\psi(v) = \frac{1}{2} cv^2 + bv \tag{1.2}$$

and a branching particle system Y with the offspring generating function

$$\varphi(u) = u^2 \tag{1.3}$$

(Y provides the mass creation mechanism which transforms a subcritical process X to a supercritical process $\widehat{X}$).

We investigate a class of interactions of superprocesses X with branching particle systems Y which produce again superprocesses $\widehat{X}$. The result of Evans and O'Connell is extended to superprocesses X with parameters (ξ, K, ψ) where K is a continuous additive functional of ξ and ψ has the form (1.2). Proofs are different and seem to be simpler than in [Evans and O'Connell 1993].

2. The set $\mathcal{M}(E)$ of all finite measures on a measurable Luzin space can be considered again as a measurable Luzin space. A superprocess $X = (X_t, P_{r,\mu})$ with parameters (ξ, K, ψ) is a Markov process in $\mathcal{M}(E)$ with the property: for every measure σ on $\mathbb{R}_+ = [0, \infty)$ with compact support and for every positive measurable function f on $S = \mathbb{R}_+ \times E$,

$$P_{r,\mu} \exp \int_r^\infty \langle -f^s, X_s \rangle \sigma(ds) = e^{-\langle v^r, \mu \rangle} \tag{2.1}$$

where v satisfies the equation

$$v^r(x) + \Pi_{r,x} \int_r^\infty \psi(v)(s, \xi_s) dK_s = \Pi_{r,x} \int_r^\infty f^s(\xi_s) \sigma(ds). \tag{2.2}$$

1991 *Mathematics Subject Classification.* Primary: 60J80; Secondary: 60J55.

Partially supported by National Science Foundation Grant DMS-9146347 and by The U.S. Army Research Office through the Mathematical Sciences Institute at Cornell University.

This is the final form of the paper.

(usually (2.1), (2.2) are required only for Dirac's measures $\sigma = \delta_t$, $t \in \mathbb{R}_+$; however this implies that (2.1), (2.2) hold for all σ with compact supports, see Theorem I.1.8 in [Dynkin 1993] or Theorem 1.2 in [Dynkin 1991a]). A wide class of admissible values of the branching parameter ψ is described in Theorem 1.1 in [Dynkin 1991b] and Theorem I.1.1 in [Dynkin 1993]. It includes

$$(2.3) \qquad \psi(v)(s,x) = v(s,x)^2 - b(s,x)v(s,x)$$

where b is a measurable function on S which is bounded on every set $[0,t] \times E$.

3. For every interval Δ, we set $S(\Delta) = \Delta \times E$. Denote by $\mathfrak{M}$ the class of measures J on S whose restriction to every set $S[0,t]$ is finite. To every $J \in \mathfrak{M}$ there corresponds a probability measure P_J on S such that

$$(3.1) \qquad P_J \exp \int_r^\infty \langle -f^s, X_s \rangle \sigma(ds) = \exp\left\{ -\int_r^\infty \int_E v^s(x) J(ds, dx) \right\}$$

for the same class of f and σ as in formula (2.1) (see [Dynkin 1991a], [Dynkin 1991b] or [Dynkin 1993]). Denote by $J_{r,\mu}$ the measure on S which coincides on $S(r,\infty)$ with J and which is defined on $S[0,r]$ by the formula

$$J_{r,\mu}(C) = \mu(C_r)$$

where C_r is the r-section of C. Put $P^J_{r,\mu} = P_{J_{r,\mu}}$. It is easy to prove (cf. proof of Theorem I.1.3 in [Dynkin 1993]) that $(X_t, P^J_{r,\mu})$ is a Markov process. We call it *the process with immigration rule* J. (Heuristically, $J(\Delta, B)$ can be interpreted as the amount of immigrants admitted during time interval Δ to the region B.)

An immigration rule can be random with the probability distribution depending on r, μ. Such a rule is given by a kernel $J(\omega'; ds, dx)$ from a measurable space $(\Omega', \mathcal{F}')$ to S and by a family of probability measures $\Gamma_{r,\mu}$ on Ω'. We call $\widetilde{X} = (X_t, \widetilde{P}_{r,\mu})$ where

$$(3.2) \qquad \widetilde{P}_{r,\mu} = \int_{\Omega'} \Gamma_{r,y}(d\omega') P^{J(\omega')}_{r,\mu}$$

the process with immigration rule $(J, \Gamma_{r,\mu})$. By (3.1),

$$(3.3) \quad \widetilde{P}_{r,\mu} \exp \int_r^\infty \langle -f^s, X_s \rangle \sigma(ds) = e^{-\langle v^r, \mu \rangle} P'_{r,y} \exp\left\{ -\int_r^\infty \int_E v^s(x) J(ds, dx) \right\}.$$

Our objective is to find immigration rules for which $\widetilde{X}$ is a superprocess.

We start from a measure-valued additive functional $J(\omega'; ds, dx)$ of a Markov process $X' = (X'_t, P'_{r,y})$ and we put

$$(3.4) \qquad \Gamma_{r,\mu} = \int_{E'} N_r(\mu, dy) P'_{r,y}$$

where $N_r(\mu, dy)$ is a kernel from $\mathcal{M}(E)$ to the state space E' of X'. The immigration rule $(J, \Gamma_{r,\mu})$ and the corresponding process $\widetilde{X}$ can be considered as the output of an interaction between X and X'.

We concentrate on the case when E' is the space $\mathcal{N}(E)$ of all integer-val measures on E, X' is a branching particle system and $N_r(\mu, \cdot)$ is the Poisson random measure on $\mathcal{N}(E)$ with intensity proportional to μ.

4. Let $(Y, Q_{r,\nu})$ be the branching particle system with parameters (ξ, L, φ) (ξ describes the motion of each particle, L is a continuous additive functional of ξ characterizing intensity of branching and φ is the offspring generating function (see, e.g., [Dynkin 1991a]). To every additive functional A of ξ there corresponds a random measure J_A on $S = \mathbb{R}_+ \times E$ given by the formula

$$J_A(C) = \sum_a \int_{\Delta_a} 1_C(s, \xi_s) A(ds)$$

where the sum is taken over all particles and ξ_s^a, $s \in \Delta_a$ is the path of the particle with the label a. $J_A(ds, dx)$ is a measure-valued additive functional of Y. In particular, if $A(ds) = \rho^s(\xi_s)\sigma(ds)$ where σ is an arbitrary measure on $\mathbb{R}_+$, then $J_A(ds, dx) = \rho^s(x) Y_s(dx) \sigma(ds)$. By formulae (2.6), (2.7) in [Dynkin 1991a],

(4.1) $$u^r(x) = Q_{r,\delta_x} \exp[-J_A(r, \infty)]$$

satisfies the equation

(4.2) $$u^r(x) = \Pi_{r,x} \bigg\{ \exp[-A(r, \infty) - L(r, \infty)] + \int_r^t \exp[-A(r, s) - L(r, s)] L(ds) \varphi^s(u^s(\xi_s)) \bigg\}.$$

LEMMA. *Let A and L be continuous additive functionals of ξ and be a positive measurable function on S. Put*

$$H(r, s) = e^{-A(r,s)}.$$

The equation

(4.3) $$g^r(x) = \Pi_{r,x} \left[\int_r^t H(r, s) h^s(\xi_s) L(ds) + H(r, t) \right] \quad \text{for } r < t$$

implies

(4.4) $$g^r(x) + \Pi_{r,x} \int_r^t g^s(\xi_s) A(ds) = \Pi_{r,x} \left[1 + \int_r^t h^s(\xi_s) L(ds) \right] \quad \text{for } r < t.$$

This is a slight modification of Lemma 2.3 in [Dynkin 1991b] and it can be proved in the same way.

COROLLARY. *Function u given by (4.1) satisfies the equation*

(4.5) $$u^r(x) + \Pi_{r,x} \int_r^\infty u^s(\xi_s) A(ds) = 1 + \Pi_{r,x} \int_r^\infty \Phi^s(u^s(\xi_s)) L(ds)$$

where $\Phi(u) = \varphi(u) - u$.

This follows from (4.2) if we apply Lemma to, $g = u$, $h = \varphi(u)$ and A replaced by $A + L$.

5. Now we return to the situation described in Section 3 taking $X' = Y$, $J = J_A$ and $N_r(\mu, d\nu) = N_{\rho\mu}(d\nu)$ where ρ is a positive constant and N_μ is the Poisson random measure on $\mathcal{N}(E)$ described by Laplace functional

$$\int N_\mu(d\nu) e^{-\langle h, \nu \rangle} = \exp \langle e^{-h} - 1, \mu \rangle. \tag{5.1}$$

By (3.3), (3.4) and (5.1),

$$\widetilde{P}_{r,\mu} \exp \int_r^\infty \langle -f^s, X_s \rangle \sigma(ds) = e^{-\langle w^r, \mu \rangle}$$

where $w^r = \rho(1 - u^r) + v^r$. By (2.2) and (4.5),

$$w^r(x) + \Pi_{r,x} \int_r^\infty [\psi(v) dK + \rho \Phi(u) dL - \rho uv dA] = \Pi_{r,x} \int_r^\infty f^s(\xi_s) \sigma(ds).$$

Suppose that

$$\psi(v) dK + \rho \Phi(u) dL - \rho uv dA = \widehat{\psi}(\rho - \rho u + v) d\widehat{K}. \tag{5.2}$$

Then $(X_t, \widetilde{P}_{r,\mu})$ is a superprocess with parameters $(\xi, \widehat{K}, \widehat{\psi})$.

Put $B = K + L + A + \widehat{K}$. If ξ is a right process, then $K(ds) = k(s, \xi_s) B(ds)$, $L(ds) = \ell(s, \xi_s) B(ds)$, $A(ds) = a(s, \xi_s) B(ds)$, $\widehat{K}(ds) = \hat{k}(s, \xi_s) B(ds)$. The equation (5.2) holds if

$$\psi(v) k + \rho \Phi(u) \ell - \rho uv a = \widehat{\psi}(\rho - \rho u + v) \hat{k}. \tag{5.3}$$

Suppose

$$\widehat{\psi}(w)(s, x) = w(s, x)^2 - b(s, x) w(s, x). \tag{5.4}$$

Then (5.3) is equivalent to

$$a = 2\hat{k}, \quad k\psi(v) = \hat{k} \left[v^2 + (2\rho - b) v \right],$$
$$\varphi(u) = (\rho q) u^2 + [1 + (b - 2\rho) q] u + (\rho - b) q / 2 \tag{5.5}$$

where $q = \hat{k}/\ell$. The condition $\varphi(u) = 1$ is satisfied if and only if $\rho = b$. In this case $\varphi(u) = bq u^2 + (1 - bq) u$ is a generating function if $0 \leq bq \leq 1$. We get $\varphi(u) = u^2$ by taking $q = 1/b$.

References

E. B. Dynkin, *Path processes and historical superprocesses*, Probab. Theor. Relat. Fields **90** (1991), 1–36.

———, *Branching particle systems and superprocesses*, Ann. Probab. **19** (1991), 1157–1194.

———, *Superprocesses and partial differential equations*, Ann. Probab. **21** (1993), 1185–1262.

S. N. Evans and N. O'Connell, *Weighted occupation times for branching particle systems and a representation for the supercritical superprocesses* (1993) (preprint).

DEPARTMENT OF MATHEMATICS, WHITE HALL, CORNELL UNIVERSITY, ITHACA, NY 14853-7901, USA.

E-mail address: ebd1@cornell.edu

Neutral Allelic Genealogy

S. N. Ethier and Tokuzo Shiga

1. Introduction

In a recent article discussing the different but related genealogies that describe the ancestral relationships among genes, Takahata [Takahata 1991] wrote the following (see also [Takahata and Nei 1990]):

> In allelic genealogy, any ancestral relationship among genes that belong to the same line [of descent] is ignored .... An allelic divergence occurs only when a gene in an older line mutates to form a new one, and may or may not accompany a gene divergence. In principle ... allelic genealogy can be studied by the coalescent on which mutational events are superimposed.

The aim of the present paper is to formulate a model of allelic genealogy and use it to prove Takahata's assertion. We restrict our attention to the neutral case (i.e., the case of no selection) throughout. The selective case is of considerable interest but is substantially more difficult.

The basic model is a Fleming-Viot [Fleming and Viot 1979] probability-measure-valued diffusion process, which is described by its type space E and its mutation operator A. In general, E is a compact metric space, and A is a (possibly unbounded) linear operator on $C(E)$ whose closure generates a Feller semigroup on $C(E)$. The Fleming-Viot process with type space E and mutation operator A is a diffusion process in $\mathcal{P}(E)$, the set of Borel probability measures on E with the topology of weak convergence, with generator

$$(1.1) \quad (\mathcal{L}\varphi)(\mu) = \frac{1}{2} \int_E \int_E \mu(dx)(\delta_x(dy) - \mu(dy)) \frac{\delta^2 \varphi(\mu)}{\delta\mu(x)\,\delta\mu(y)}$$
$$+ \int_E \mu(dx) A\left(\frac{\delta\varphi(\mu)}{\delta\mu(\cdot)}\right)(x),$$

where $\delta\varphi(\mu)/\delta\mu(x) = \lim_{\varepsilon \to 0+} \varepsilon^{-1}\{\varphi(\mu + \varepsilon\delta_x) - \varphi(\mu)\}$ and $\delta_x \in \mathcal{P}(E)$ denotes the unit mass at $x \in E$. The domain of $\mathcal{L}$ can be taken to be the set of all functions $\varphi \in C(\mathcal{P}(E))$ of the form $\varphi(\mu) = \langle f_1, \mu \rangle \ldots \langle f_m, \mu \rangle$, where $m \geq 1$ and $f_1, \ldots, f_m \in \mathcal{D}(A)$. Here $\langle f, \mu \rangle = \int_E f\,d\mu$ for $f \in B(E)$ and $\mu \in \mathcal{P}(E)$.

The Fleming-Viot process is the limit in distribution of a sequence of probability-measure-valued Wright-Fisher models (see, e.g., [Ethier and Kurtz 1993]). When E is finite, it reduces to the usual finite-dimensional diffusion process sometimes referred to as the Wright-Fisher diffusion.

1991 *Mathematics Subject Classification.* Primary: 92D15; Secondary 60G57, 60J60.
Research supported in part by NSF grants DMS-8902991 and DMS-9102925.
This is the final form of the paper.

Three examples will help to clarify the process and motivate the model proposed here. Each depends on three "parameters," a compact metric space S (the set of alleles), a constant $\theta > 0$ (twice the mutation intensity), and a one-step Feller transition function $P(x, d\xi)$ for a Markov chain in S (the type distribution of a mutant offspring of a type x parent), which is nonatomic for each $x \in S$, i.e.,

$$(1.2) \qquad P(x, \{\xi\}) = 0, \quad x, \xi \in S.$$

EXAMPLE 1.1. The infinitely-many-neutral-alleles model. Here $E = S$ and

$$(1.3) \qquad (Af)(x) = \tfrac{1}{2}\theta \int_S \bigl(f(\xi) - f(x)\bigr) P(x, d\xi).$$

This is the generator of a pure-jump Markov process in S that jumps at rate $\theta/2$ according to P. The nonatomic assumption on P models the basic requirement of the model that every mutant be of a type that has never before appeared.

EXAMPLE 1.2. The infinitely-many-neutral-alleles model with ages [Ethier 1990]. Here $E = S \times [0, \infty]$ and

$$(1.4) \qquad (Af)(x, a) = \left(\frac{\partial}{\partial a} f\right)(x, a) + \tfrac{1}{2}\theta \int_S \bigl(f(\xi, 0) - f(x, a)\bigr) P(x, d\xi).$$

The type space is the set of ordered pairs of alleles and ages. The age of an allele increases at rate 1, and it starts over at 0 for a new mutant.

EXAMPLE 1.3. The infinitely-many-sites model [Ethier and Griffiths 1987]. Here $E = S^{\mathbb{Z}_+}$ and

$$(1.5) \qquad (Af)(\mathbf{x}) = \tfrac{1}{2}\theta \int_S \bigl(f(\xi, \mathbf{x}) - f(\mathbf{x})\bigr) P(x_0, d\xi).$$

Actually, for this model $P(x_0, d\xi)$ is assumed independent of x_0. However, the formulation (1.5) also describes a version of the infinitely-many-neutral-alleles model in which the "type" of an individual includes information about its ancestors.

Notice that Examples 1.2 and 1.3 can be regarded as elaborations of Example 1.1. The idea is to keep track of certain aspects of the history of the process by augmenting the type space. This is also the key to the model introduced below.

2. The model

Let S, θ, and P be as in the examples above. We define $E = S^{\mathbb{Z}_+} \times [0, \infty]^{\mathbb{Z}_+}$. An individual is of type $(\mathbf{x}, \mathbf{a}) = \bigl((x_0, x_1, \ldots), (a_0, a_1, \ldots)\bigr) \in E$ if
  $x_0 = $ its allelic type,
  $a_0 = $ the (current) age of x_0,
  $x_i = $ the allelic type of the individual that produced x_{i-1} by mutation, $i \geq 1$,
  $a_i = $ the age of x_i at the time x_{i-1} first appeared, $i \geq 1$.
With this interpretation, it is reasonable to define A by

$$(2.1) \quad (Af)(\mathbf{x}, \mathbf{a}) = \left(\frac{\partial}{\partial a_0} f\right)(\mathbf{x}, \mathbf{a}) + \tfrac{1}{2}\theta \int_S \bigl(f((\xi, \mathbf{x}), (0, \mathbf{a})) - f(\mathbf{x}, \mathbf{a})\bigr) P(x_0, d\xi)$$

with $\mathcal{D}(A) = \{f \in C(E) : (\partial/\partial a_0) f \text{ exists and belongs to } C(E)\}$. E and A will have the above meanings hereafter.

The resulting Fleming-Viot process can be regarded as the infinitely-many-neutral-alleles diffusion model with allelic genealogies. It includes the processes of Examples 1.1–1.3 above as marginal processes. Although sample-path properties are of some interest, we will focus our attention on the stationary distribution of the process.

We will need the weak ergodicity assumption that there exists $\nu_0 \in \mathcal{P}(S)$ such that

$$(2.2) \qquad \lim_{n \to \infty} \int_S f(\xi) P^n(x, d\xi) = \langle f, \nu_0 \rangle, \quad f \in C(S), \; x \in S,$$

in what follows.

LEMMA 2.1. *Assume (2.2). Then the Fleming-Viot process with type space E and mutation operator A has a unique stationary distribution $\Pi \in \mathcal{P}(\mathcal{P}(E))$ and is weakly ergodic.*

PROOF. By a well-known result (see, e.g., [Ethier and Kurtz 1993], Theorem 5.2), it is enough to show that there exists $\eta \in \mathcal{P}(E)$ such that the Feller semigroup $\{T(t)\}$ on $C(E)$ generated by the closure of A satisfies

$$(2.3) \qquad \lim_{t \to \infty} T(t) f(\mathbf{x}, \mathbf{a}) = \langle f, \eta \rangle, \quad f \in C(E), \; (\mathbf{x}, \mathbf{a}) \in E.$$

In fact, we need only verify this for f of the form

$$(2.4) \qquad f(\mathbf{x}, \mathbf{a}) = g_0(x_0) \ldots g_k(x_k) h_0(a_0) \ldots h_k(a_k),$$

where $k \geq 0$, $g_0, \ldots, g_k \in C(S)$, and $h_0, \ldots, h_k \in C([0, \infty])$.

Given $(\mathbf{x}, \mathbf{a}) \in E$, let $\{X_n\}$ be the Markov chain in S with one-step transition function $P(x, d\xi)$ and initial state $X_0 = x_0$. Let $\tau_0, \tau_1, \ldots$ be i.i.d. exponential $(\theta/2)$ random variables, independent of $\{X_n\}$. Then

$$(2.5)$$
$$T(t)f(\mathbf{x}, \mathbf{a}) = e^{-\theta t/2} f(\mathbf{x}, (t + a_0, a_1, a_2, \ldots))$$
$$+ \sum_{n=1}^{\infty} \frac{(\theta t/2)^n}{n!} e^{-\theta t/2} E\big[f\big((X_n, \ldots, X_1, \mathbf{x}),$$
$$(t - (\tau_0 + \cdots + \tau_{n-1}), \tau_{n-1}, \ldots, \tau_1, \tau_0 + a_0, a_1, a_2, \ldots)\big)$$
$$\big| \tau_0 + \cdots + \tau_{n-1} \leq t < \tau_0 + \cdots + \tau_n \big]$$

for all $f \in C(E)$ and $t \geq 0$. Assuming (2.4), the zeroeth term on the right side of (2.5) as well as the first k terms of the sum tend to 0 as $t \to \infty$. The remaining terms amount to

$$(2.6) \qquad \sum_{n=k+1}^{\infty} \frac{(\theta t/2)^n}{n!} e^{-\theta t/2} E[g_0(X_n) \ldots g_k(X_{n-k})]$$
$$\cdot E\big[h_0\big(t - (\tau_0 + \cdots + \tau_{n-1})\big) h_1(\tau_{n-1}) \ldots h_k(\tau_{n-k})$$
$$\big| \tau_0 + \cdots + \tau_{n-1} \leq t < \tau_0 + \cdots + \tau_n \big].$$

By (2.2), the first expectation in the sum in (2.6) converges as $n \to \infty$ to $E[g_0(Y_k) \ldots g_k(Y_0)]$, where $\{Y_n\}$ is the stationary Markov chain in S with one-step transition function $P(x, d\xi)$ and initial distribution ν_0. Adding and subtracting this limit, we find that it is enough to examine

$$(2.7) \quad \sum_{n=k+1}^{\infty} E\big[h_0(t-(\tau_0+\cdots+\tau_{n-1}))h_1(\tau_{n-1})\ldots h_k(\tau_{n-k});$$
$$\tau_0+\cdots+\tau_{n-1} \leq t < \tau_0+\cdots+\tau_n\big]$$
$$= \sum_{n=k+1}^{\infty} E[h_0(\tau_0)h_1(\tau_1)\ldots h_k(\tau_k); \tau_0+\cdots+\tau_{n-1} \leq t < \tau_0+\cdots+\tau_n]$$
$$= E[h_0(\tau_0)h_1(\tau_1)\ldots h_k(\tau_k); \tau_0+\cdots+\tau_k \leq t];$$

here the first equality follows from the fact that, given that a Poisson process has exactly n arrivals during $[0, t]$ and $n \geq k+1$, the conditional joint distribution of the last $k+1$ interarrival times in $[0, t]$ is the same as that of the first $k+1$. Since (2.7) tends to $E[h_0(\tau_0)\ldots h_k(\tau_k)]$ as $t \to \infty$, the proof is complete.

Next, we need the following technical lemma, which extends $\mathcal{L}$ to a more useful class of functions. For $n \geq 1$, $g \in B((S^{\mathbb{Z}_+})^n)$, and $h \in C^1(([0, \infty]^{\mathbb{Z}_+})^n)$, we denote by $g \times h$ the function in $B(E^n)$ given by

$$(2.8) \quad (g \times h)\big((\mathbf{x}_1, \mathbf{a}_1), \ldots, (\mathbf{x}_n, \mathbf{a}_n)\big) = g(\mathbf{x}_1, \ldots, \mathbf{x}_n)h(\mathbf{a}_1, \ldots, \mathbf{a}_n).$$

For $n \geq 1$ and $f \in B(E^n)$, we define $\varphi_f \in B(\mathcal{P}(E))$ by $\varphi_f(\mu) = \langle f, \mu^n \rangle$.

LEMMA 2.2. *Assume* (2.2), *and define the operator* $\mathcal{L}^+$ *by*

$$(2.9) \quad (\mathcal{L}^+ \varphi_f)(\mu) = \sum_{1 \leq i < j \leq n} (\langle \Phi_{ij}^{(n)} f, \mu^{n-1}\rangle - \langle f, \mu^n \rangle) + \sum_{i=1}^{n} \langle A_i^{(n)} f, \mu^n \rangle,$$
$$\mathcal{D}(\mathcal{L}^+) = \{\varphi_f : f = g \times h,\ g \in B((S^{\mathbb{Z}_+})^n),\ h \in C^1(([0, \infty]^{\mathbb{Z}_+})^n),\ n \geq 1\},$$

where $\Phi_{ij}^{(n)} f$ is the function in $B(E^{n-1})$ obtained from f by replacing $(\mathbf{x}_i, \mathbf{a}_i)$ by $(\mathbf{x}_j, \mathbf{a}_j)$ and renumbering the variables, and $A_i^{(n)} f$ is A acting on f as a function of its ith coordinate $(\mathbf{x}_i, \mathbf{a}_i)$. (Clearly, A can be extended using (2.1) to functions $f \in B(E)$ of the form $f(\mathbf{x}, \mathbf{a}) = g(\mathbf{x})h(\mathbf{a})$, where $g \in B(S^{\mathbb{Z}_+})$ and $h \in C^1([0, \infty]^{\mathbb{Z}_+})$.) Then $\int_{\mathcal{P}(E)} (\mathcal{L}^+ \varphi)(\mu)\, \Pi(d\mu) = 0$ for all $\varphi \in \mathcal{D}(\mathcal{L}^+)$.

PROOF. The proof is analogous to that of Lemma 2.6 of [Ethier 1990].

Our aim in this paper is to describe the unique stationary distribution Π. We begin by showing that if μ is a $\mathcal{P}(E)$-valued random variable with distribution Π, then with probability 1 a random sample of size n from μ forms what we call a genealogical n-tree.

If $n \geq 1$, we say that $((\mathbf{x}_1, \mathbf{a}_1), \ldots, (\mathbf{x}_n, \mathbf{a}_n)) \in E^n$ forms a *genealogical n-tree* if
  (a) the coordinates x_{ij}, $j \in \mathbb{Z}_+$, of $\mathbf{x}_i$ are distinct for fixed $i \in \{1, \ldots, n\}$ and the coordinates a_{ij}, $j \in \mathbb{Z}_+$, of $\mathbf{a}_i$ are finite for $i = 1, \ldots, n$,
  (b) if $i, i' \in \{1, \ldots, n\}$, $j, j' \in \mathbb{Z}_+$, and $x_{ij} = x_{i'j'}$, we have $(x_{i,j+l}, a_{i,j+l}) = (x_{i',j'+l}, a_{i',j'+l})$ for all $l \geq 1$ and $\sum_{l=0}^{j} a_{il} = \sum_{l=0}^{j'} a_{i'l}$,
  (c) there exist $j_1, \ldots, j_n \in \mathbb{Z}_+$ such that $x_{1j_1} = \cdots = x_{nj_n}$.

Condition (a) implies that no allele appears more than once in the sequence of new mutants in a given line of descent. This will be a consequence of (1.2). Condition (b) is the tree property and condition (c) asserts that there is a common ancestor. (Cf. [Ethier and Griffiths 1987].) It is instructive to consider an example with $n = 3$ and to sketch the resulting genealogical 3-tree.

For each $n \geq 1$, let $\mathcal{T}_n \subset E^n$ denote the set of all genealogical n-trees. We denote by $\mathcal{P}_a(E)$ the set of purely atomic Borel probability measures on E, and we put

$$(2.10) \qquad \mathcal{P}_a^0(E) = \{\mu \in \mathcal{P}_a(E) : \mu^n(\mathcal{T}_n) = 1 \text{ for every } n \geq 1\}.$$

LEMMA 2.3. *Assume* (1.2) *and* (2.2). *Then* $\Pi\big(\mathcal{P}_a^0(E)\big) = 1$.

PROOF. The proof that $\Pi(\mathcal{P}_a(E)) = 1$ is analogous to the proof of Proposition 2.5 of [Ethier 1990]. It is therefore enough to show that

$$(2.11) \qquad \Pi\{\mu \in \mathcal{P}(E) : \mu^n(\mathcal{T}_n) = 1\} = 1$$

for every $n \geq 1$. Suppose we can verify (2.11) for $n = 1$ and $n = 2$. Then, noting that

$$(2.12) \quad \mathcal{T}_n = \bigcap_{1 \leq i < j \leq n} \big\{\big((\mathbf{x}_1, \mathbf{a}_1), \ldots, (\mathbf{x}_n, \mathbf{a}_n)\big) \in E^n : \big((\mathbf{x}_i, \mathbf{a}_i), (\mathbf{x}_j, \mathbf{a}_j)\big) \in \mathcal{T}_2\big\}$$

for each $n \geq 3$, the general case will follow.

We begin with the case $n = 1$. For each $m \geq 0$, define $D_m = \{\mathbf{x} \in S^{\mathbb{Z}_+} : x_0, \ldots, x_m \text{ are distinct}\}$ (note that $D_0 = S^{\mathbb{Z}_+}$), and put

$$(2.13) \qquad f_m^\lambda(\mathbf{x}, \mathbf{a}) = 1_{D_m}(\mathbf{x}) e^{-\lambda(a_0 + \cdots + a_m)}$$

for each $\lambda > 0$. Then, using (1.2), we have $A f_m^\lambda = -\lambda f_m^\lambda + \frac{1}{2}\theta(f_{m-1}^\lambda - f_m^\lambda)$, where $f_{-1}^\lambda \equiv 1$, so since

$$(2.14) \qquad 0 = \int_{\mathcal{P}(E)} (\mathcal{L}^+ \varphi_{f_m^\lambda})(\mu) \, \Pi(d\mu) = \int_{\mathcal{P}(E)} \langle A f_m^\lambda, \mu \rangle \, \Pi(d\mu)$$

by Lemma 2.2, we have

$$(2.15) \qquad \int_{\mathcal{P}(E)} \langle f_m^\lambda, \mu \rangle \, \Pi(d\mu) = \frac{\frac{1}{2}\theta}{\lambda + \frac{1}{2}\theta} \int_{\mathcal{P}(E)} \langle f_{m-1}^\lambda, \mu \rangle \, \Pi(d\mu) = \left(\frac{\frac{1}{2}\theta}{\lambda + \frac{1}{2}\theta}\right)^{m+1}.$$

Letting $\lambda \to 0+$, we conclude that $\mu\{(\mathbf{x}, \mathbf{a}) \in E : x_0, \ldots, x_m \text{ are distinct}, a_0, \ldots, a_m \in [0, \infty)\} = 1$, $\Pi(d\mu)$-a.s. Letting $m \to \infty$ gives $\mu(\mathcal{T}_1) = 1$, $\Pi(d\mu)$-a.s.

We turn next to the case $n = 2$. Observe that we can partition $\mathcal{T}_2$ as the union over $i, j \geq 0$ of the sets

$$(2.16) \quad T_{i,j} = \big\{\big((\mathbf{x}, \mathbf{a}), (\mathbf{y}, \mathbf{b})\big) \in E^2 : (\mathbf{x}, \mathbf{y}) \in \Lambda_{i,j}, \, a_0, a_1, \ldots, b_0, b_1, \ldots \in [0, \infty),$$
$$a_{i+l} = b_{j+l} \text{ for all } l \geq 1, \, a_0 + \cdots + a_i = b_0 + \cdots + b_j\big\},$$

where

$$(2.17) \quad \Lambda_{i,j} = \big\{(\mathbf{x}, \mathbf{y}) \in \big(S^{\mathbb{Z}_+}\big)^2 : x_0, \ldots, x_{i-1}, y_0, y_1, \ldots \text{ are distinct},$$
$$y_0, \ldots, y_{j-1}, x_0, x_1, \ldots \text{ are distinct}, \, x_{i+l} = y_{j+l} \text{ for all } l \geq 0\big\}.$$

It will suffice to show that

$$(2.18) \qquad \sum_{i,j \geq 0} \int_{\mathcal{P}(E)} \mu^2(T_{i,j}) \, \Pi(d\mu) = 1.$$

With this in mind, we define for each $i, j \geq 0$ and $m, r \geq 1$

$$(2.19) \quad f_{i,j}^{m,r}((\mathbf{x}, \mathbf{a}), (\mathbf{y}, \mathbf{b})) = 1_{\Lambda_{i,j}}(\mathbf{x}, \mathbf{y}) \prod_{l=1}^{m} h(r(a_{i+l} - b_{j+l}))$$
$$\cdot h(r(a_0 + \cdots + a_i - b_0 - \cdots - b_j))$$
$$\cdot h\left(\frac{a_0 + \cdots + a_{i+m} + b_0 + \cdots + b_{j+m}}{r}\right)$$

on E^2, where $h \in C^1([-\infty, \infty])$ satisfies $1_{[-1,1]} \leq h \leq 1_{[-2,2]}$ and $\infty - \infty = 0$, and we observe using (1.2) that

$$(2.20) \quad (\mathcal{L}^+ \varphi_{f_{i,j}^{m,r}})(\mu) = \delta_{i0}\delta_{j0} \int_E 1_D(\mathbf{x}) h\left(\frac{2a_0 + \cdots + 2a_m}{r}\right) \mu(d(\mathbf{x}, \mathbf{a}))$$
$$- (1+\theta)\varphi_{f_{i,j}^{m,r}}(\mu) + \zeta_{i,j}^{m,r}(\mu) + \tfrac{1}{2}\theta\varphi_{f_{i-1,j}^{m,r}}(\mu) + \tfrac{1}{2}\theta\varphi_{f_{i,j-1}^{m,r}}(\mu)$$

for all $i, j \geq 0$, $m, r \geq 1$, and $\mu \in \mathcal{P}(E)$, where $D = \{\mathbf{x} \in S^{\mathbb{Z}_+} : x_0, x_1, \ldots$ are distinct$\}$ and $|\zeta_{i,j}^{m,r}(\mu)| \leq 2\|h'\|/r$ for all i, j, m, r, μ. Letting $r \to \infty$ and then $m \to \infty$ and using the $n = 1$ case already established as well as Lemma 2.2, we find that the probabilities

$$(2.21) \quad p_{i,j} \equiv \int_{\mathcal{P}(E)} \mu^2(T_{i,j}) \Pi(d\mu)$$

satisfy

$$(2.22) \quad (1+\theta)p_{i,j} = \delta_{i0}\delta_{j0} + \tfrac{1}{2}\theta p_{i-1,j} + \tfrac{1}{2}\theta p_{i,j-1}$$

for all $i, j \geq 0$, where $p_{i,-1} = p_{-1,j} = 0$. As noted in [Ethier and Griffiths 1987], this implies that

$$(2.23) \quad p_{i,j} = \binom{i+j}{i}\left(\frac{\theta}{2(1+\theta)}\right)^{i+j} \frac{1}{1+\theta},$$

and hence that $\sum_{i,j \geq 0} p_{i,j} = 1$, as required.

3. Description of the stationary distribution

In this section we assume the existence of $\nu_0 \in \mathcal{P}(S)$ such that

$$(3.1) \quad P(x, d\xi) = \nu_0(d\xi), \quad x \in S.$$

We assume further that ν_0 is nonatomic, so that (1.2) holds. Note that (2.2) is trivially satisfied.

Fix $n \geq 1$. Our main theorem expresses the nth moment measure of Π, i.e., $\int_{\mathcal{P}(E)} \mu^n(\cdot) \Pi(d\mu)$, in terms of the n-coalescent with mutation and an independent i.i.d. ν_0 sequence.

For $1 \leq k \leq n$, let $\pi(n, k)$ denote the set of partitions β of $\{1, \ldots, n\}$ into k nonempty subsets $\beta_1, \ldots, \beta_k$, labeled so that $\min \beta_1 < \cdots < \min \beta_k$. The n-coalescent with mutation [Kingman 1982a] is a pure-jump Markov process with state space

$$(3.2) \quad \mathcal{E}_n(\mathbb{Z}_+) \equiv \bigcup_{k=1}^{n} \{\pi(n, k) \times (\mathbb{Z}_+)^k\},$$

initial state $(\{1\},\ldots,\{n\};0,\ldots,0)$, and transitions from state

(3.3) $$(\beta_1,\ldots,\beta_k; m_1,\ldots,m_k)$$

to state

(3.4) $$(\beta_1,\ldots,\beta_{i-1},\beta_i\cup\beta_j,\beta_{i+1},\ldots,\beta_{j-1},\beta_{j+1},\ldots,\beta_k; 0,\ldots,0)$$

with rate 1, $1\leq i<j\leq k$, and to state

(3.5) $$(\beta_1,\ldots,\beta_k; m_1,\ldots,m_{i-1},m_i+1,m_{i+1},\ldots,m_k)$$

with rate $\theta/2$, $1\leq i\leq k$. Transitions from (3.3) to (3.4) are called *coalescences*, and those from (3.3) to (3.5) are called *mutations*. Note that from (3.3) the overall jump rate is $k(k-1+\theta)/2$.

Let

(3.6) $$\{(\beta_1(t),\ldots,\beta_{k(t)}(t); m_1(t),\ldots,m_{k(t)}(t)),\ t\geq 0\}$$

denote the n-coalescent with mutation. Observe that $\{(\beta_1(t),\ldots,\beta_{k(t)}(t)),\ t\geq 0\}$ is the usual n-coalescent (without mutation). For $i=1,\ldots,n$, let

(3.7) $$\begin{aligned}M_i(t) &= m_j(t) \quad \text{if}\ \ i\in\beta_j(t),\\ \tau_{i0} &= \inf\{t>0: M_i(t)-M_i(t-)=1\},\\ \tau_{il} &= \inf\{t>\tau_{i,l-1}: M_i(t)-M_i(t-)=1\},\quad l\geq 1;\end{aligned}$$

$M_i(t)$ is the number of mutations in the ith line before time t and after the last coalescence prior to time t. $\tau_{i0},\tau_{i1},\ldots$ is the sequence of times at which mutations occur in the ith line.

Arrange the distinct elements of $\{\tau_{il}: i=1,\ldots,n,\ l\in\mathbb{Z}_+\}$ in ascending order $0<\tau_{(0)}<\tau_{(1)}<\tau_{(2)}<\cdots$. These are the times at which mutations occur. Label them with an i.i.d. ν_0 sequence $\xi_0,\xi_1,\ldots$ (independent of (3.6)). More precisely, for $i=1,\ldots,n$ and $l\in\mathbb{Z}_+$, let $x_{il}=\xi_j$ if $\tau_{il}=\tau_{(j)}$.

THEOREM 3.1. *Assume* (1.2) *and* (3.1), *and fix* $n\geq 1$. *Then the random vectors*

(3.8) $$\big((x_{i0},x_{i1},\ldots),(\tau_{i0},\tau_{i1}-\tau_{i0},\tau_{i2}-\tau_{i1},\ldots)\big),\quad i=1,\ldots,n,$$

defined in terms of the n-coalescent with mutation (3.6) *and an independent i.i.d. ν_0 sequence* $\xi_0,\xi_1,\ldots$ *as above, have joint distribution* $\int_{\mathcal{P}(E)}\mu^n(\cdot)\,\Pi(d\mu)$.

PROOF. We first treat the case $n=1$. Given $m\geq 0$, $g_0,\ldots,g_m\in C(S)$, and $\kappa_0,\ldots,\kappa_m>0$, define

(3.9) $$f^{g_0,\ldots,g_m}_{\kappa_0,\ldots,\kappa_m}(\mathbf{x},\mathbf{a})=g_0(x_0)\ldots g_m(x_m)e^{-(\kappa_0 a_0+\cdots+\kappa_m a_m)}$$

in $\mathcal{D}(A)$, and observe that

(3.10) $$Af^{g_0,\ldots,g_m}_{\kappa_0,\ldots,\kappa_m}=-(\kappa_0+\tfrac{1}{2}\theta)f^{g_0,\ldots,g_m}_{\kappa_0,\ldots,\kappa_m}+\tfrac{1}{2}\theta\langle g_0,\nu_0\rangle f^{g_1,\ldots,g_m}_{\kappa_1,\ldots,\kappa_m}.$$

Since $0 = \int_{\mathcal{P}(E)} \mathcal{L}\langle f, \mu\rangle \Pi(d\mu) = \int_{\mathcal{P}(E)} \langle Af, \mu\rangle \Pi(d\mu)$ for all $f \in \mathcal{D}(A)$ (note that Lemma 2.2 is not needed here),

$$(3.11) \quad \int_{\mathcal{P}(E)} \langle f_{\kappa_0,\ldots,\kappa_m}^{g_0,\ldots,g_m}, \mu\rangle \Pi(d\mu) = \frac{\frac{1}{2}\theta}{\kappa_0 + \frac{1}{2}\theta} \langle g_0, \nu_0\rangle \int_{\mathcal{P}(E)} \langle f_{\kappa_1,\ldots,\kappa_m}^{g_1,\ldots,g_m}, \mu\rangle \Pi(d\mu)$$

$$= \prod_{l=0}^{m} \frac{\frac{1}{2}\theta}{\kappa_l + \frac{1}{2}\theta} \prod_{l=0}^{m} \langle g_l, \nu_0\rangle.$$

It follows that, under $\int_{\mathcal{P}(E)} \mu(\cdot) \Pi(d\mu)$, $x_0, x_1, \ldots$ are i.i.d. ν_0, and $a_0, a_1, \ldots$ are i.i.d. exponential $(\theta/2)$ random variables, independent of $x_0, x_1, \ldots$. But the 1-coalescent with mutation is in effect a Poisson process with rate $\theta/2$, and so the conclusion of the theorem certainly holds in this case.

The general case $(n \geq 2)$ is conceptually no more difficult than the case $n = 2$, but the notation is awkward and may obscure the main idea. So, to complete the proof, we treat only the case $n = 2$.

Fix $i, j \geq 0$ and recall the set $T_{i,j}$ of (2.16). We partition $T_{i,j}$ into $\binom{i+j}{i}$ subsets as follows. For each $I \subset \{1, 2, \ldots, i+j\}$ with $|I| = i$, let $J = \{1, 2, \ldots, i+j\} - I$ (so $|J| = j$) and define

$$(3.12) \quad T_{I,J} = \big\{\big((\mathbf{x}, \mathbf{a}), (\mathbf{y}, \mathbf{b})\big) \in T_{i,j} : k \in I \text{ if and only if the } k\text{th largest of}$$
$$a_0, a_0 + a_1, \ldots, a_0 + \cdots + a_{i-1}, b_0, b_0 + b_1, \ldots, b_0 + \cdots + b_{j-1}$$
$$\text{is among } a_0, a_0 + a_1, \ldots, a_0 + \cdots + a_{i-1}\big\}.$$

(In particular, denoting the empty set by $\varnothing$, we have $T_{\varnothing,\varnothing} = T_{0,0}$.) If $\big((\mathbf{x}, \mathbf{a}), (\mathbf{y}, \mathbf{b})\big) \in T_{i,j}$ and $i + j \geq 1$, then $x_0, \ldots, x_{i-1}, y_0, \ldots, y_{j-1}$ are called *segregating sites* because they appear on one but not both branches of the genealogical 2-tree. The quantities $a_0, a_0 + a_1, \ldots, a_0 + \cdots + a_{i-1}, b_0, b_0 + b_1, \ldots, b_0 + \cdots + b_{j-1}$ are their (current) ages. If these ages are distinct (they will be with probability 1) and if $z_1, \ldots, z_{i+j}$ denote the $i + j$ segregating sites ordered from oldest to youngest, then $\{z_k : k \in I\}$ and $\{z_k : k \in J\}$ denote respectively the sets of segregating sites on the left and right branches of the tree. We denote by $c_1, \ldots, c_{i+j}$ the (current) ages of $z_1, \ldots, z_{i+j}$. Specifically,

$$(3.13) \quad c_k = \begin{cases} a_0 + \cdots + a_l & \text{if } k \in I \text{ and } |\{k+1, \ldots, i+j\} \cap I| = l \\ b_0 + \cdots + b_l & \text{if } k \in J \text{ and } |\{k+1, \ldots, i+j\} \cap J| = l \end{cases}$$

for $k = 1, \ldots, i + j$.

Given $m \geq 0$, $g_0, \ldots, g_m, h_0, \ldots, h_m \in C(S)$, $\kappa_0, \ldots, \kappa_m, \lambda_0, \ldots, \lambda_m \geq 0$, and I, J as above, define

$$(3.14) \quad f_{\kappa_0,\ldots,\kappa_m;\lambda_0,\ldots,\lambda_m}^{g_0,\ldots,g_m;h_0,\ldots,h_m;I,J}\big((\mathbf{x}, \mathbf{a}), (\mathbf{y}, \mathbf{b})\big)$$
$$= g_0(x_0) \ldots g_m(x_m) h_0(y_0) \ldots h_m(y_m)$$
$$\cdot e^{-(\kappa_0 a_0 + \cdots + \kappa_m a_m)} e^{-(\lambda_0 b_0 + \cdots + \lambda_m b_m)} 1_{T_{I,J}}\big((\mathbf{x}, \mathbf{a}), (\mathbf{y}, \mathbf{b})\big)$$

on E^2. This choice of f does not fit into the framework of Lemma 2.2, so we replace it temporarily by

$$
\begin{aligned}
(3.15)\quad & f^{g_0,\ldots,g_m;h_0,\ldots,h_m;I,J}_{\kappa_0,\ldots,\kappa_m;\lambda_0,\ldots,\lambda_m;r}\big((\mathbf{x},\mathbf{a}),(\mathbf{y},\mathbf{b})\big) \\
&= g_0(x_0)\ldots g_m(x_m)h_0(y_0)\ldots h_m(y_m)1_{\Lambda_{i,j}}(\mathbf{x},\mathbf{y})e^{-(\kappa_0 a_0+\cdots+\kappa_m a_m)} \\
&\quad \cdot e^{-(\lambda_0 b_0+\cdots+\lambda_m b_m)}u\big(r(c_1-c_2)\big)\ldots u\big(r(c_{i+j-1}-c_{i+j})\big),
\end{aligned}
$$

where $r \geq 1$, $\Lambda_{i,j}$ is as in (2.17), and $u \in C^1([-\infty,\infty])$ satisfies $1_{[1,\infty]} \leq u \leq 1_{[0,\infty]}$. Applying Lemmas 2.2 and 2.3, using the $n=1$ case already established, and letting $r \to \infty$, we find that

$$
(3.16)\quad (\kappa_0+\lambda_0+1+\theta)\int_{\mathcal{P}(E)}\langle f^{g_0,\ldots,g_m;h_0,\ldots,h_m;I,J}_{\kappa_0,\ldots,\kappa_m;\lambda_0,\ldots,\lambda_m},\mu^2\rangle\Pi(d\mu)
$$

$$
=\begin{cases}
\prod_{l=0}^{m}\langle g_l h_l,\nu_0\rangle \prod_{l=0}^{m}\dfrac{\frac{1}{2}\theta}{\kappa_l+\lambda_l+\frac{1}{2}\theta} & \text{if } i=j=0, \\
\frac{1}{2}\theta\langle g_0,\nu_0\rangle \int_{\mathcal{P}(E)}\langle f^{g_1,\ldots,g_m,1;h_0,\ldots,h_m;I-\{i+j\},J}_{\kappa_1,\ldots,\kappa_m,0;\lambda_0,\ldots,\lambda_m},\mu^2\rangle\Pi(d\mu) & \text{if } i+j\in I, \\
\frac{1}{2}\theta\langle h_0,\nu_0\rangle \int_{\mathcal{P}(E)}\langle f^{g_0,\ldots,g_m;h_1,\ldots,h_m,1;I,J-\{i+j\}}_{\kappa_0,\ldots,\kappa_m;\lambda_1,\ldots,\lambda_m,0},\mu^2\rangle\Pi(d\mu) & \text{if } i+j\in J.
\end{cases}
$$

This is a recursive system that uniquely determines the integrals

$$
(3.17)\quad \int_{\mathcal{P}(E)}\langle f^{g_0,\ldots,g_m;h_0,\ldots,h_m;I,J}_{\kappa_0,\ldots,\kappa_m;\lambda_0,\ldots,\lambda_m},\mu^2\rangle\Pi(d\mu)
$$

and hence the second moment measure $\int_{\mathcal{P}(E)}\mu^2(\cdot)\,\Pi(d\mu)$.

Now let $((\mathbf{x},\mathbf{a}),(\mathbf{y},\mathbf{b}))$ be the pair of random vectors in the statement of the theorem ($n=2$), which are specified in terms of the 2-coalescent with mutation and an independent i.i.d. ν_0 sequence. We claim that

$$
(3.18)\quad E\big[f^{g_0,\ldots,g_m;h_0,\ldots,h_m;I,J}_{\kappa_0,\ldots,\kappa_m;\lambda_0,\ldots,\lambda_m}\big((\mathbf{x},\mathbf{a}),(\mathbf{y},\mathbf{b})\big)\big]
$$

satisfies a recursive system identical to (3.16).

To see this, note that the indicator function in (3.14) completely specifies the behavior of the embedded chain of the 2-coalescent with mutation. Given this event, the waiting times in each state are conditionally independent exponential random variables with parameters as in the unconditional case. (Cf. [Ethier and Kurtz 1986], Eq. (2.3) of Chapter 4; in particular, the waiting time in state $(\{1\},\{2\};k,l)$ is exponential $(1+\theta)$ if $k\leq i$ and $l<j$, whereas it is exponential $(\theta/2)$ in state $(\{1,2\};k)$ for $k\geq 0$. Thus, if $i=j=0$, (3.18) equals

$$
\begin{aligned}
(3.19)\quad & P\{((\mathbf{x},\mathbf{a}),(\mathbf{y},\mathbf{b}))\in T_{\varnothing,\varnothing}\}E\big[g_0(x_0)\ldots g_m(x_m)h_0(y_0)\ldots h_m(y_m) \\
&\quad \cdot e^{-(\kappa_0 a_0+\cdots+\kappa_m a_m)}e^{-(\lambda_0 b_0+\cdots+\lambda_m b_m)}\,\big|\,((\mathbf{x},\mathbf{a}),(\mathbf{y},\mathbf{b}))\in T_{\varnothing,\varnothing}\big] \\
&= \frac{1}{1+\theta}E\big[g_0(x_0)\ldots g_m(x_m)h_0(x_0)\ldots h_m(x_m)e^{-(\kappa_0+\lambda_0)a_0}\ldots e^{-(\kappa_m+\lambda_m)a_m} \\
&\qquad\qquad\qquad\qquad\qquad\qquad\qquad\qquad\big|\,((\mathbf{x},\mathbf{a}),(\mathbf{y},\mathbf{b}))\in T_{\varnothing,\varnothing}\big] \\
&= \frac{1}{1+\theta}\prod_{l=0}^{m}\langle g_l h_l,\nu_0\rangle \frac{1+\theta}{\kappa_0+\lambda_0+1+\theta}\prod_{l=0}^{m}\frac{\frac{1}{2}\theta}{\kappa_l+\lambda_l+\frac{1}{2}\theta},
\end{aligned}
$$

since a_0 is conditionally the sum of two independent exponential random variables.

If $i+j\in I$, we take a different approach. Let H be the event that the first jump of the 2-coalescent with mutation is to state $(\{1\},\{2\};1,0)$ (from the initial state $(\{1\},\{2\};0,0))$. Let τ be the time of that first jump. Given that H occurs, let $((\mathbf{x}',\mathbf{a}'),(\mathbf{y}',\mathbf{b}'))$ be the pair of random vectors in the statement of the theorem, but defined in terms of the 2-coalescent with mutation restarted at time τ from $(\{1\},\{2\};0,0)$. Observe that then

$$(3.20) \quad \begin{aligned} & x'_0 = x_1,\ x'_1 = x_2, \ldots,\quad a'_0 = a_1,\ a'_1 = a_2, \ldots, \\ & y'_0 = y_0,\ y'_1 = y_1, \ldots,\quad b'_0 = b_0 - a_0,\ b'_1 = b_1, \ldots. \end{aligned}$$

It follows that, if $i + j \in I$, (3.18) equals

$$(3.21)\quad P(H) E\left[f_{\kappa_0,\ldots,\kappa_m;\lambda_0,\ldots,\lambda_m}^{g_0,\ldots,g_m;h_0,\ldots,h_m;I,J}((\mathbf{x},\mathbf{a}),(\mathbf{y},\mathbf{b})) \mid H\right]$$

$$= P(H)\langle g_0,\nu_0\rangle E[e^{-(\kappa_0+\lambda_0)a_0}] E[g_1(x'_0)\cdots g_m(x'_{m-1}) h_0(y'_0)\cdots h_m(y'_m)$$

$$\cdot e^{-(\kappa_1 a'_0+\cdots+\kappa_m a'_{m-1})} e^{-(\lambda_0 b'_0+\cdots+\lambda_m b'_m)} 1_{T_{I-\{i+j\},J}}((\mathbf{x}',\mathbf{a}'),(\mathbf{y}',\mathbf{b}')) \mid H]$$

$$= \frac{\tfrac{1}{2}\theta}{1+\theta}\langle g_0,\nu_0\rangle \frac{1+\theta}{\kappa_0+\lambda_0+1+\theta} E\left[f_{\kappa_1,\ldots,\kappa_m,0;\lambda_0,\ldots,\lambda_m}^{g_1,\ldots,g_m,1;h_0,\ldots,h_m;I-\{i+j\},J}((\mathbf{x},\mathbf{a}),(\mathbf{y},\mathbf{b}))\right].$$

The third case $(i+j \in J)$ is similar to this one, so the proof is complete.

We conclude this section with a corollary that extends Theorem 3.1 to the case $n = \infty$.

For $k \geq 1$, let $\pi(\mathbb{N},k)$ denote the set of partitions β of $\mathbb{N} = \{1,2,\ldots\}$ into k nonempty subsets $\beta_1,\ldots,\beta_k$, labeled so that $\min \beta_1 < \cdots < \min \beta_k$. Given $\beta \in \pi(\mathbb{N},k)$, $\mathbf{m} \in (\mathbb{Z}_+)^k$, and $n \geq 1$, we denote by $\rho_n \beta$ the restriction of β to $\{1,\ldots,n\}$ and by $\rho_n \mathbf{m}$ the corresponding restriction of $\mathbf{m}$ (i.e., if $1 \leq l \leq k$ and $(\beta_1 \cap \{1,\ldots,n\},\ldots,\beta_l \cap \{1,\ldots,n\})$ defines an element of $\pi(n,l)$, then $\rho_n \mathbf{m} = (m_1,\ldots,m_l)$).

There is a process

$$(3.22)\qquad \{(\beta(t),\mathbf{m}(t)),\ t > 0\}$$

in

$$(3.23)\qquad \mathcal{E}_\infty(\mathbb{Z}_+) \equiv \bigcup_{k=1}^{\infty}\{\pi(\mathbb{N},k) \times (\mathbb{Z}_+)^k\},$$

called the *coalescent with mutation*, with the property that $\{(\rho_n\beta(t),\rho_n\mathbf{m}(t)),\ t \geq 0\}$ (defined at time 0 by right continuity) is an n-coalescent with mutation for each $n \geq 1$. This can be deduced from [Kingman 1982b] or from [Donnelly and Kurtz 1993].

For $i \in \mathbb{N}$, the definitions (3.7) can be made. Arrange the distinct elements of $\{\tau_{il} : i \in \mathbb{N},\ l \in \mathbb{Z}_+\}$ in ascending order $\cdots < \tau_{(-1)} < \tau_{(0)} < \tau_{(1)} < \cdots$ and label them with an i.i.d. ν_0 sequence $\ldots,\xi_{-1},\xi_0,\xi_1,\ldots$ (independent of (3.22)). More precisely, for $i \in \mathbb{N}$ and $l \in \mathbb{Z}_+$, let $x_{il} = \xi_j$ if $\tau_{il} = \tau_{(j)}$.

COROLLARY 3.2. *Under the assumptions of Theorem 3.1, the random vectors*

$$(3.24)\qquad (\mathbf{x}_i,\mathbf{a}_i) \equiv ((x_{i0},x_{i1},\ldots),(\tau_{i0},\tau_{i1}-\tau_{i0},\tau_{i2}-\tau_{i1},\ldots)),\quad i \in \mathbb{N},$$

defined in terms of the coalescent with mutation (3.22) and an independent i.i.d. ν_0 sequence $\ldots,\xi_{-1},\xi_0,\xi_1,\ldots$ as above, have joint distribution $\int_{\mathcal{P}(E)}\mu^\infty(\cdot)\,\Pi(d\mu)$. In particular, the sequence of empirical measures $n^{-1}\sum_{i=1}^n \delta_{(\mathbf{x}_i,\mathbf{a}_i)}$ converges almost surely in the topology of weak convergence to a $\mathcal{P}(E)$-valued random variable with distribution Π.

PROOF. For each $n \geq 1$, Theorem 3.1 implies that $(\mathbf{x}_1, \mathbf{a}_1), \ldots, (\mathbf{x}_n, \mathbf{a}_n)$ have joint distribution $\int_{\mathcal{P}(E)} \mu^n(\cdot) \Pi(d\mu)$, so the first conclusion follows easily. That the second conclusion follows from the first is a standard result (see Lemma 6.1 of [Dawson and Hochberg 1982] or Lemma 3.2 of [Ethier and Kurtz 1992]).

This shows that a $\mathcal{P}(E)$-valued random variable with distribution Π can be constructed from the coalescent with mutation and an i.i.d. ν_0 sequence. But the coalescent with mutation cannot be constructed from a $\mathcal{P}(E)$-valued random variable with distribution Π. In particular, the coalescence times do not appear in the allelic genealogy.

ACKNOWLEDGMENT. The research for this paper was begun during the first author's visit to the Department of Applied Physics at the Tokyo Institute of Technology. He would like to say *domo arigato gozaimashita* to his hosts for their hospitality.

References

D. A. Dawson and K. J. Hochberg, *Wandering random measures in the Fleming-Viot model*, Ann. Probab. **10** (1982), 554–580.

P. Donnelly and T. G. Kurtz, *A particle representation of infinite population genetic models* (1993) (in preparation).

S. N. Ethier, *The infinitely-many-neutral-alleles diffusion model with ages*, Adv. Appl. Probab. **22** (1990), 1–24.

S. N. Ethier and R. C. Griffiths, *The infinitely-many-sites model as a measure-valued diffusion*, Ann. Probab. **15** (1987), 515–545.

S. N. Ethier and T. G. Kurtz, *Markov processes: Characterization and convergence* (1986), Wiley, New York.

____, *On the stationary distribution of the neutral diffusion model in population genetics*, Ann. Appl. Probab. **2** (1992), 24–35.

____, *Fleming-Viot processes in population genetics*, SIAM J. Control Opt. **31** (1993), 345–386.

W. H. Fleming and M. Viot, *Some measure-valued Markov processes in population genetics theory*, Indiana Univ. Math. J. **28** (1979), 817–843.

J. F. C. Kingman, *On the genealogy of large populations*, J. Appl. Probab. **19A** (1982a), 27–43.

____, *The coalescent*, Stochastic Processes Appl. **13** (1982b), 235–248.

N. Takahata, *A trend in population genetics theory*, New Aspects of the Genetics of Molecular Evolution (M. Kimura and N. Takahata, eds.), Japan Sci. Soc. Press, Tokyo. Springer-Verlag, Berlin, 1991, pp. 27–47.

N. Takahata and M. Nei, *Allelic genealogy under overdominant and frequency-dependent selection and polymorphism of major histocompatibility complex loci*, Genetics **124** (1990), 967–978.

ABSTRACT. A Fleming-Viot probability-measure-valued diffusion process is introduced that can be regarded as the infinitely-many-neutral-alleles model with allelic genealogies. Its stationary distribution is described in terms of the coalescent with mutation.

DEPARTMENT OF MATHEMATICS, UNIVERSITY OF UTAH, SALT LAKE CITY, UTAH 84112, USA.

E-mail address: ethier@math.utah.edu

DEPARTMENT OF APPLIED PHYSICS, TOKYO INSTITUTE OF TECHNOLOGY, OH-OKAYAMA, MEGURO-KU, TOKYO 152, JAPAN.

E-mail address: shiga@ap.titech.ac.jp

Superprocesses in Catalytic Media

Klaus Fleischmann

1. Introduction

In the last 25 years, *spatial branching processes* have extensively been investigated. Concerning branching particle systems in discrete time, we refer to Chapter 10 of [Kerstan et al. 1982], to [Liemant et al. 1988], and to references in these books. *Superprocesses* are a more continuous version, where most of all the population mass changes in a "continuous" way; see, for instance, [Dawson and Fleischmann 1992] for a recent survey.

In a series of papers ([Dawson and Fleischmann 1992, Dawson and Fleischmann 1991, Dawson et al. 1991]), Dawson, Fleischmann, and partly Roelly, started the investigation of a particular class of superprocesses $\mathfrak{X} = \{\mathfrak{X}_t\,;\ t \geq 0\}$, in fact, *one-dimensional super-α-stable processes* $(1 < \alpha \leq 2)$ *with branching index* $1+\beta \in (1,2]$ where *the branching rate ϱ may vary in time and space* in a rather general way. Roughly speaking, if $\varrho(t, dz)$ is a tempered kernel of $\mathbb{R}_+$ into $\mathbb{R}$, then its *generalized* Radon-Nikodym derivatives with respect to the Lebesgue measure, by an abuse of notation we simply write $\varrho(t, \cdot) = \int \varrho(t, dy)\, \delta_y$ for them, serve as *branching rate*.

This, in particular, covers cases as $\varrho(t, \cdot) = \sum_i a_i \delta_{z(i,t)}$ where again δ_z denotes the Dirac δ-function at $z \in \mathbb{R}$. Here a term $a_i \delta_{z(i,t)}$ is interpreted as a *point catalyst* situated at time t at "site" $z(i,t)$ and which *acts with weight* $a_i \geq 0$. In this case, branching is allowed only at the positions of these moving catalysts, but there even with an *unbounded* rate (just in the sense of δ-functions). Consequently, such branching rate ϱ describes a *moving collection of weighted catalytic particles* which initiate the continuous state branching. [For a discussion of *fractal catalysts* from a physical, physicochemical, or biological point of view, we refer to §8.8 in [Nicolis and Prigogine 1977], and [Sapoval 1991].]

In our case, the question of existence of branching models in point catalytic media arose in connection with the unsolved problem of existence of nontrivial space-time-mass scaling limit fluctuations in *higher*-dimensional branching particle models in *random media with infinite overall density;* see also the discussion in Subsection 1.3 of [Dawson and Fleischmann 1991]. [For a fluctuation limit theorem in a random medium with a *finite* expectation, we refer to [Dawson et al. 1989].]

Despite such possible singularity of the *varying* (or even *random*) *medium* ϱ, serving as branching rate, under fairly general conditions the states $\mathfrak{X}_t$ of these one-dimensional superprocesses are *absolutely continuous* measures $\mathfrak{X}_t(dz) = \mathfrak{x}_t(z)dz$

1991 *Mathematics Subject Classification.* Primary 60J80; Secondary 60J65, 60G57.

Key words and phrases. Point catalytic medium, critical branching, super-Brownian motion, occupation time, occupation density, local extinction, sample continuity, scaling limit theorem, absolutely continuous states, clumping effects, Hausdorff dimension, measure-valued branching.

This is the final form of the paper.

with probability one (provided the time point t is fixed); see [Dawson et al. 1991]. But even in the case of a branching mechanism with a *finite* second moment (i.e. $\beta = 1$), the variance of the random density function $\mathfrak{x}_t$ *blows up* at a point catalyst's position (see (2.7) below). Since the catalysts may even be *densely* situated in space, it is rather difficult to imagine samples of the density function $\mathfrak{x}_t$, or even of a density field $\mathfrak{x}$.

To get some insight, in the next section we turn to a very particular case of an irregular ϱ. Section 3 is devoted to some analytical tools, whereas in the final section clumping effects are studied and a remark to the more-dimensional case is added.

2. A super-Brownian motion with a single point catalyst

2.1. Intuitive description. The extremely simplified situation $\varrho(t,dz) \equiv \delta_{\mathfrak{c}}(z)dz$ where the medium consists of a single non-random and non-moving point catalyst (situated at $\mathfrak{c} \in \mathbb{R}$, and with weight one) has been investigated in [Dawson and Fleischmann 1993a] in some detail. Moreover, the consideration there is restricted to the heat flow as motion component ($\alpha = 2$), and to the simplest critical continuous state branching mechanism (case $\beta = 1$) which, isolated considered, is a one-dimensional diffusion process ζ according to the stochastic equation

$$(2.1) \qquad d\zeta_t = \sqrt{2\zeta_t}\, dW_t, \quad t \geq 0, \quad \zeta_0 \geq 0,$$

where W is a one-dimensional Wiener process with generator Δ. Putting symbolically all these things together, we arrive at the *model*

$$(2.2) \qquad d\mathfrak{X}_t = \Delta \mathfrak{X}_t\, dt + \sqrt{2\delta_{\mathfrak{c}}\mathfrak{X}_t}\, dW_t$$

(with the same one-dimensional Wiener process W). Consequently, in this *super-Brownian motion $\mathfrak{X}$ with a single point catalyst*, branching takes place only at the point catalyst's position $\mathfrak{c} \in \mathbb{R}$, but there with an unbounded rate, whereas off $\mathfrak{c}$ only the heat flow acts. Of course, the symbolism (2.2) is only a formal expression, which defines $\mathfrak{X}$ on an intuitive base. To the question of a rigorous definition of this critical measure-valued branching process $\mathfrak{X}$ we will come back in Subsection 3.2 below.

In order to get a first impression on properties of this process, at this point we provide some *formal calculations*. Using (2.2), it is easy to compute formally the first two *moments* of $(\mathfrak{X}_t, \varphi) := \int \mathfrak{X}_t(dz)\, \varphi(z)$ where φ runs in a suitable set Φ_+ of non-negative test functions:

$$(2.3) \qquad \mathbb{E}_\mu (\mathfrak{X}_t, \varphi) = (\mu S_t, \varphi),$$

$$(2.4) \qquad \mathrm{Var}_\mu (\mathfrak{X}_t, \varphi) = 2 \int \mu(da) \int_0^t dr\, p(r, \mathfrak{c} - a) \left[S_{t-r}\varphi(\mathfrak{c})\right]^2,$$

$t > 0$, $\varphi \in \Phi_+$, $\mu \in M_f$. Here μ is the (deterministic) *initial state* $\mathfrak{X}_0$ of $\mathfrak{X}$ which runs in the set M_f of all finite measures defined on $\mathbb{R}$, equipped with the topology of *weak convergence*. Moreover, $\{S_t; t \geq 0\}$ denotes the *Brownian semigroup* related to the Laplacian Δ, and p the corresponding *Brownian transition density function*, that is,

$$p(r, y) = (4\pi r)^{-1/2} \exp\left[-y^2/4r\right], \quad r > 0, \quad y \in \mathbb{R}.$$

Replacing now formally the test function φ by the δ-function δ_z to get with $(\mathfrak{X}_t, \delta_z)$ the density $\mathfrak{x}_t(z)$ at z, we arrive at the following *expectation and variance formulas*, for Lebesgue almost all z (provided that the density $\mathfrak{x}_t$ exists almost surely, for t fixed):

$$\mathbb{E}_\mu \mathfrak{x}_t(z) = \int \mu(da) \, p(t, z-a), \tag{2.5}$$

$$\mathrm{Var}_\mu \mathfrak{x}_t(z) = 2 \int \mu(da) \int_0^t dr \, p(r, \mathfrak{c}-a) \, p^2(t-r, z-\mathfrak{c}), \tag{2.6}$$

$t > 0$, $z \neq \mathfrak{c}$, $\mu \in M_f$. This implies, for $a \in \mathbb{R}$ and $t > 0$ fixed,

$$\mathrm{Var}_{\delta_a} \mathfrak{x}_t(z) \approx \Big| \log |z - \mathfrak{c}| \Big| \to \infty \quad \text{as} \quad z \to \mathfrak{c}. \tag{2.7}$$

That is, in the vicinity of the catalyst the (random) density of mass has an *infinite dispersion*. This behavior is of course different from the regular medium case, i.e. if the branching rate ϱ is given by a regular function. On the other hand, such moment computations alone do not provide too much information on the behavior of the *trajectories* of this superprocess $\mathfrak{X}$.

2.2. Sample path properties. In this subsection, our main interest concerns path properties of the single atomic super-Brownian motion $\mathfrak{X}$ of the previous subsection, in particular, its behavior close to the catalyst's position $\mathfrak{c}$. As indicated above, let $\mathbb{P}_\mu$ denote the law of $\mathfrak{X}$ with initial state $\mathfrak{X}_0 = \mu \in M_f$.

THEOREM 2.1 (CONTINUOUS PATHS). *The process $\mathfrak{X}$ can be constructed in the set $\mathbf{C}[\mathbb{R}_+, M_f]$ of continuous measure-valued paths. Moreover, for each $\mu \in M_f$ with $\mathbb{P}_\mu$-probability one, $\{\mathfrak{X}_t, t > 0\}$ has no mass in each given space point, in particular, $\mathfrak{X}_t(\{\mathfrak{c}\}) = 0$, $t > 0$. Finally, the expectation and variance formulas (2.3) and (2.4) hold.*

Consequently, on this level, $\mathfrak{X}$ behaves just as the usual super-Brownian motion in a regular medium. But note that intuitively it is not so obvious that in our case at positive times there is never any mass at the catalyst's position $\mathfrak{c}$. In fact, by the blow-up property as expressed in (2.7) the heat flow will transport (with infinite speed) arbitrarily large amounts of density of mass into c where by the infinite branching rate and criticality of branching also a huge mass production may occasionally occur.

Now we come back to the question of existence of density functions. The next theorem shows that $\{\mathfrak{X}_t; \, t > 0\}$ *entirely* lives on the set of absolutely continuous measures. Moreover, the corresponding density field $\mathfrak{x}$ can be chosen to be *jointly continuous* on $\{t > 0\} \times \{z \neq \mathfrak{c}\}$. (Continuity *outside* the catalyst's position $\mathfrak{c}$ is not so surprising in that there *locally* only the Laplacian acts.)

THEOREM 2.2 (JOINTLY CONTINUOUS DENSITY FIELD). *There exists a version of $\mathfrak{X}$ satisfying*

$$\mathfrak{X}_t(dz) = \mathfrak{x}_t(z) dz \quad \text{for all } t > 0, \quad \mathbb{P}_\mu\text{-a.s.}, \quad \mu \in M_f, \tag{2.8}$$

where the density field $\mathfrak{x} := \{\mathfrak{x}_t(z); \, t > 0, z \neq \mathfrak{c}\}$ is jointly continuous. Moreover, the expectation and variance formulas (2.5) and (2.6) are actually true.

Recall that in the classical case of a constant branching rate, $\varrho = \text{const}$, [Konno and Shiga 1988] as well as [Reimers 1989] showed that the density field $\mathfrak{x}$ even satisfies a stochastic equation.

Coming back to our single atomic model, we now ask for the behavior of the density field $\mathfrak{x}$ *at the catalyst's position* $\mathfrak{c}$ which was excluded in the previous theorem:

THEOREM 2.3 (STOCHASTICALLY VANISHING DENSITY AT THE CATALYST).
For fixed $t > 0$,

$$(2.9) \qquad \mathfrak{x}_t(z) \xrightarrow[z \to \mathfrak{c}]{} 0 \quad \text{in } \mathbb{P}_\mu\text{-probability}, \quad \mu \in M_f.$$

That is, although the expectation of the random density converges to a positive value (formula (2.5)) and the variance even explodes (cf. (2.7)), the density itself degenerates stochastically to zero in approaching the catalyst at a fixed time. In particular, the probability for the density to be large will become very small.

Heuristically this can be explained as follows. By the infinitely large branching rate at $\mathfrak{c}$ which lets run the branching mechanism infinitely fast, density of mass, which is eventually present at $\mathfrak{c}$, will be killed with overwhelming probability leading to density zero at $\mathfrak{c}$, but with an exceptional set depending on the time point t. (Of course, the joint continuity of the density field $\mathfrak{x}$ outside of $\mathfrak{c}$, and its stochastic disappearance approaching $\mathfrak{c}$ at a fixed time does not exclude that $\mathfrak{x}$ could "oscillate" around $\mathfrak{c}$.)

On the other hand, this kind of stochastic disappearance of $\mathfrak{x}_t(\mathfrak{c})$ for a fixed t is only one side of the story. Having in mind the dynamics of the process $\mathfrak{X}$, as long as it is not extinct, there is a permanent and infinitely fast flow of (absolutely continuous) mass into $\mathfrak{c}$ where not only a killing takes place but "occasionally" also a production of mass according to the critical continuous state branching. By the "infinite" branching rate at $\mathfrak{c}$ this should happen only with an extremely small probability.

Heuristically we conclude that the random set of times where a production of population mass will actually occur should be "rather thin" and will not be "met" at a fixed time.

2.3. Occupation densities. To put these ideas on a firm base, we first introduce the so-called *occupation time process* $\mathfrak{Y}$ and its *occupation density field (super-Brownian local time)* $\mathfrak{y}$:

$$(2.10) \quad \mathfrak{Y}_t := \int_0^t ds\, \mathfrak{X}_s(\cdot), \quad \mathfrak{y}_t(z) := \int_0^t ds\, \mathfrak{x}_s(z), \quad t \geq 0, \quad z \neq \mathfrak{c}.$$

(Note that by the path continuities according to the Theorems 2.1 and 2.2, these pathwise integrals exist.) The next theorem constitutes that the occupation density field $\mathfrak{y}$ can *continuously* be extended to the line $z = \mathfrak{c}$, and, opposed to the stochastic disappearance $\mathfrak{x}_t(\mathfrak{c}) = 0$ according to Theorem 2.3, does *not* vanish at $\mathfrak{c}$ (see also the statement (i) in Theorem 2.8 below).

THEOREM 2.4 (EVERYWHERE JOINTLY CONTINUOUS OCCUPATION DENSITY).
There exists a version of $\mathfrak{X}$ such that the occupation density field (super-Brownian local time) $\mathfrak{y}$ defined in (2.10) continuously extends to all of $\mathbb{R}_+ \times \mathbb{R}$. Moreover, the

following expectation and variance formulas hold:

$$\mathbb{E}_\mu \mathfrak{y}_t(z) = \int \mu(da) \int_0^t dr\, p(r, z-a), \tag{2.11}$$

$$\operatorname{Var}_\mu \mathfrak{y}_t(z) = 2\int \mu(da) \int_0^t dr\, p(r, \mathfrak{c}-a) \left[\int_r^t ds\, p(s-r, z-\mathfrak{c})\right]^2 < \infty, \tag{2.12}$$

$\mu \in M_f$, $t \geq 0$, $z \in \mathbb{R}$.

Note that in contrast to the blow-up effect (2.7), here the variance at the catalyst's position remains finite. This can be understood by the smoothing effect of the integration in time which neglects the behavior on time sets of measure 0. Consequently, the occupation density field $\mathfrak{y}$ is relatively "tame", just as in the case of a regular branching rate (cf. [Sugitani 1989]).

What can further be done to enlighten the behavior of the model at the catalyst? For each fixed space point z, the occupation density $\mathfrak{y}_t(z)$ is obviously *non-decreasing* in t. Hence, for $z \in \mathbb{R}$ fixed, it defines a (locally finite) random measure $\lambda^z(dt) := d\mathfrak{y}_t(z)$ on $\mathbb{R}_+$ which we call the *occupation density measure* at z. Of course, if z is different from $\mathfrak{c}$, by (2.10) this occupation density measure is almost surely absolutely continuous, indeed, has density function $\{\mathfrak{x}_s(z);\ s > 0\}$. But $\lambda^\mathfrak{c}$ is also uniquely defined, by Theorem 2.4. Heuristically $\lambda^\mathfrak{c}$ measures the "thin" set of times where a non-vanishing density of mass occurs at $\mathfrak{c}$, as heuristically explained at the end of Subsection 2.2. In fact, in [Dawson *et al.* 1991], the following theorem is proved (we restrict the attention to the case where $\mathfrak{X}$ starts off at time 0 with a unit mass at $\mathfrak{c}$):

THEOREM 2.5 (SINGULARITY OF THE CATALYST'S OCCUPATION DENSITY).
Assume $\mathfrak{X}_0 = \delta_\mathfrak{c}$. The occupation density measure $\lambda^\mathfrak{c}$ at the catalyst's position is with probability one a singularly continuous random measure on $\mathbb{R}_+$.

Despite that singularity, nevertheless, relatively "much" happens at $\mathfrak{c}$, although at a fixed time we do not meet any density of mass there (recall Theorem 2.3):

THEOREM 2.6 (CARRYING DIMENSION ONE). *Given $\mathfrak{X}_0(\mathbb{R}) > 0$, the occupation density measure $\lambda^\mathfrak{c}$ at the catalyst's position has almost surely carrying Hausdorff-Besicovitch dimension one.*

Compare this with the usual Brownian local time at a point which determines a singular random measure on $\mathbb{R}_+$ with carrying dimension $1/2$ only.

2.4. Longtime behavior. Concerning the asymptotic behavior as time t tends to infinity, we mention the following results.

THEOREM 2.7 (LOCAL EXTINCTION). *Even if $\mathfrak{X}$ starts off at time 0 with the Lebesgue measure, for each bounded Borel set $B \subset \mathbb{R}$ we have*

$$\mathfrak{X}_t(B) \xrightarrow[t\to\infty]{} 0 \quad \text{in probability.}$$

This is interesting in that the single catalyst kills off all the mass in any bounded region, despite the infinite reservoir of the Lebesgue initial mass which even will be spread out by the heat flow with infinite speed. This ones more stresses the fact that at the catalyst an overwhelming killing takes place.

Concerning the occupation time $\mathfrak{Y}_t$ and the occupation densities $\lambda^z([0,t]) = \mathfrak{y}_t(z)$ we expose the following longtime behavior:

THEOREM 2.8. *Let $\mathfrak{X}_0 = \delta_{\mathfrak{c}}$ and fix $z \in \mathbb{R}$ (for instance $z = \mathfrak{c}$).*
(i) (STRICT POSITIVITY). $\mathfrak{y}.(z) > 0$ *on $(0, +\infty)$ with probability one.*
(ii) (TOTAL OCCUPATION DENSITY). $\mathfrak{y}_t(z)$ *converges in distribution as $t \to \infty$ to some stable random variable $\mathfrak{y}_\infty(z)$ with index $1/2$, determined by its Laplace transform*

$$\mathbb{E}\exp\left[-\theta\mathfrak{y}_\infty(z)\right] = \exp\left[-\sqrt{\theta}\right], \quad \theta \geq 0.$$

(iii) (TOTAL OCCUPATION TIME). $\mathfrak{Y}_t$ *converges in distribution as $t \to \infty$ to the random multiple $\mathfrak{y}_\infty(\mathfrak{c})l =: \mathfrak{Y}_\infty$ of the Lebesgue measure denoted by l.*

Let us compare this longtime behavior of the occupation time quantities with the corresponding properties resulting from the "individual mechanisms" in the model. Indeed, if we neglect the branching mechanism, then $\mathfrak{Y}_t(dz)$ coincides with the "potential measure" $\left(\int_0^t dr\, p(r, z-c)\right)dz$ which approximates $\sqrt{t}\, l(dz)$ as $t \to \infty$ (except a constant factor). Conversely, if we drop the motion component, or replace the point catalytic medium $\delta_{\mathfrak{c}}$ by the constant $\varrho = 1$, then $\mathfrak{Y}_t(\mathbb{R})$ is nothing else than the occupation time process related to the simplest continuous state Galton-Watson process of (2.1) starting at 1; hence, as $t \to \infty$, it has in distribution a stable limit variable σ, say, with index $1/2$ (cf. [Dawson and Fleischmann 1988] p. 198).

Consequently, our single point catalytic super-Brownian motion $\mathfrak{X}$ combines and reflects features of both mechanisms in the model resulting into $\mathfrak{Y}_\infty \stackrel{\mathcal{D}}{=} \sigma l$. That is, adding a single point catalyst to the pure heat flow leads to a randomization and significant reduction of the "uniform" limiting mass.

2.5. Martingale approach.
Recall that in the present single point catalytic superprocess $\mathfrak{X}$ the occupation density measure $\lambda^{\mathfrak{c}}(dt) = d\mathfrak{y}_t(\mathfrak{c})$ is an important tool to investigate the behavior at the catalyst. Here we want to stress the fact that it even is an essential for the formulation of the model. Indeed, opposed to the regular branching rate case, a martingale approach to $\mathfrak{X}$ is only possible by use of the jointly continuous occupation density field $\mathfrak{y}$ related to $\mathfrak{X}$.

THEOREM 2.9 (MARTINGALE PROBLEM). *For all $\varphi \in \Phi \cap \mathcal{D}(\Delta)$,*

$$(2.13) \qquad M_t(\varphi) := (\mathfrak{X}_t, \varphi) - (\mathfrak{X}_0, \varphi) - \int_0^t dr\, (\mathfrak{X}_r, \Delta\varphi), \quad t \geq 0,$$

is a continuous martingale with quadratic variation process

$$(2.14) \qquad \ll M.(\varphi) \gg_t := 2\varphi^2(\mathfrak{c})\,\mathfrak{y}_t(\mathfrak{c}), \quad t \geq 0,$$

where the sample continuous $\mathfrak{X}$ and $\mathfrak{y}$ are related by

$$(2.15) \qquad \int_0^t dr\, (\mathfrak{X}_r, \varphi) = \int dz\, \mathfrak{y}_t(z)\, \varphi(z), \quad t \geq 0.$$

3. Singular nonlinear equations as a crucial tool

The famous *Laplace transform* links superprocesses (and functionals of them) to a class of *nonlinear* equations. In our case of a superprocess in a catalytic medium $\varrho(t, dz)$ a more or less *singular coefficient* $\varrho(t, \cdot)$ in front of the nonlinear term in the equation is involved.

3.1. Equations related to the single atomic super-Brownian motion.

To keep the things transparent as much as possible, here we restrict the attention to the case of the relatively simple point catalytic model of Section 2, that is, we assume $\varrho(t,dz) \equiv \delta_{\mathfrak{c}}\, dz$ (and $\alpha = 2$, $\beta = 1$), and we want to mention some relations between quantities of this superprocess and corresponding equations.

First of all, the Laplace functional of the random vector $[\mathfrak{X}_t, \mathfrak{Y}_t]$ has the following representation:

$$(3.1) \qquad \mathbb{E}_\mu \exp\left[-(\mathfrak{X}_t, \varphi) - (\mathfrak{Y}_t, \psi)\right] = \exp\left(\mu, -u(t)\right),$$

$t \geq 0$, $\varphi, \psi \in \Phi_+$, $\mu \in M_f$. Here $u \geq 0$ is the solution of the *formal reaction diffusion equation*

$$(3.2) \qquad \begin{cases} \frac{\partial}{\partial t} u(t,a) = \Delta u(t,a) - \delta_{\mathfrak{c}}(a)\, u^2(t,a) + \psi(a), \\ u(0,a) = \varphi(a), \end{cases}$$

$t \geq 0$, $a \in \mathbb{R}$. Actually, we always understand such equations in a *mild* sense; that is, in this case as the *integral equation* (recall p and S introduced after (2.4))

$$(3.3) \qquad u(t,a) = S_t \varphi(a) + \int_0^t dr\, S_r \psi(a) - \int_0^t dr\, p(t-r, \mathfrak{c}-a)\, u^2(r, \mathfrak{c}),$$

$r \geq 0$, $a \in \mathbb{R}$, which arises by a formal integration (or, conversely, formally differentiate the latter integral equation).

In order to give another example, the density function $\mathfrak{x}_t$ has the Laplace function

$$(3.4) \qquad \mathbb{E}_\mu \exp\left[-\sum_{i=1}^k \mathfrak{x}_t(z_i)\theta_i\right] = \exp\left(\mu, -u(t)\right),$$

$\mu \in M_f$, $t > 0$, $z_i \neq \mathfrak{c}$, $\theta_i \geq 0$, $1 \leq i \leq k$, with the function $u \geq 0$ solving

$$(3.5) \qquad \frac{\partial}{\partial t} u = \Delta u - \delta_{\mathfrak{c}}\, u^2, \quad u|_{t=0+} = \sum_{i=1}^k \theta_i \delta_{z_i},$$

(fundamental solutions). Similarly, the equation related to the occupation density function $\mathfrak{y}_t$ arises by setting $\psi = \sum_{i=1}^k \theta_i \delta_{z_i}$ and $\varphi = 0$ in (3.2).

All these (integral) equations can be well-posed by a *regularization procedure*. For instance, approximate a δ-function δ_x by the smooth functions $p(\varepsilon, x - \cdot) = \delta_x * p(\varepsilon)$ as $\varepsilon \to 0+$. (See [Dawson and Fleischmann 1992] and [Dawson et al. 1991].)

3.2. To the construction of superprocesses in catalytic media.

An essential construction step consists already in establishing the related nonlinear equation, say $\frac{\partial}{\partial t} u = \Delta u - \varrho u^2$ if the branching rate is given by a kernel ϱ (recall [Dawson and Fleischmann 1992]). As already indicated, this is done by regularization via $\varrho(t) * p(\varepsilon)$, $\varepsilon \to 0$. Then, similarly, the superprocess $\mathfrak{X}$ can be defined by starting with superprocesses related to the regular medium $\varrho(t) * p(\varepsilon)$ and passing to the limit in law as $\varepsilon \to 0$ (see [Dawson and Fleischmann 1991]).

Of course, some *conditions* on the branching rate kernel ϱ have to be imposed. If the space dimension is 1, as discussed up to now in this note, the assumption

$$\sup_{0\leq t\leq T}\int \varrho(t,dy)\,\exp\left[-ry^2\right]<\infty, \quad r,T>0,$$

is *sufficient*, for example. Of course, this includes finite collections of weighted δ-functions (finitely many moving atoms).

That in the one-dimensional case for instance $\varrho=\delta_{\mathfrak{c}}$ allows a *nontrivial* effect at all can be understood on a heuristic level as follows. Think in terms of an approximating particle branching process, say the critical binary branching Brownian motion with branching rate $\delta_{\mathfrak{c}}$, or $p(\varepsilon,\mathfrak{c}-\cdot)$ for "small" $\varepsilon>0$. Since each one-dimensional Brownian particle has a positive occupation density (*local time*) $L^{\mathfrak{c}}$ at the catalyst's position $\mathfrak{c}$, a *nontrivial interaction* remains in the limit as $\varepsilon\to 0$.

Of course, various asymptotic properties of the solutions of the singular equations are a main tool to establish results as described in Section 2.

By the way, to mention also another approach, for instance the single atomic super-Brownian motion of Section 2 fits into a class of superprocesses studied by [Dynkin 1991]. In fact, we have only to take the *Brownian local time measure* $L^{\mathfrak{c}}(ds)$ at the catalyst's position $\mathfrak{c}$ as the additive functional $K(ds)$ governing the branching in the model of Dynkin (and to specialize the other quantities in the obvious way).

More generally, if a super-Brownian motion $\mathfrak{X}$ is related to the branching rate kernel $\varrho(t,dy)$, then by

$$K(dt)=\left(\int \varrho(t,dy)\,\delta_y(W_t)\right)dt$$

we can formally associate a continuous additive functional of the (one-dimensional) Brownian motion W. Here $K(dt)$ can be interpreted as the rate of branching at time t at W_t, the position of an *"infinitely small particle hidden in the cloud"* $\mathfrak{X}_t$.

4. Some further aspects

4.1. Clumping properties of $\mathfrak{X}$ in the case of a dense system of point catalysts fluctuating both in time and space. Here we will briefly deal with the case if the branching rate kernel ϱ is realized by the following random measure-valued process $\Gamma=\{\Gamma(t);\,t\geq 0\}$.

In fact, let $\Gamma(0)=\sum_i a_i\delta_{z(i,0)}$ denote the *stable random measure on $\mathbb{R}$ of index* $\gamma\in(0,1)$ defined by its Laplace functional

(4.1) $$\mathcal{E}\exp\left(\Gamma(0),-f\right)=\exp\left[-\int dx\,f^{\gamma}(x)\right],\quad f\in\Phi_+,$$

describing a collection of weighted point catalysts at time 0. Note that these point catalysts are *densely* situated in $\mathbb{R}$, and that the expected mass $\mathcal{E}\left(\Gamma(0),1_B\right)$ in a Borel set B of positive Lebesgue measure is *infinite*; in particular, $\Gamma(0)$ has an infinite asymptotic density.

Given $\Gamma(0)$, define the process $\Gamma(t)=\sum_i a_i\,\delta_{z(i,t)}$, $t\geq 0$, by letting all atoms $a_i\delta_{z(i,0)}$ *independently move* in space according to *Brownian motions* with generator Δ, carrying their action weights a_i with them (for more details, see Section 4 in

[Dawson and Fleischmann 1992]). Let $\mathcal{P}$ denote the law of this *moving system* Γ *of γ-stable point catalysts*.

For $\mathcal{P}$-almost all realizations Γ, we can define the (one-dimensional) *time-inhomogeneous super-Brownian* $\mathfrak{X} = \left[\mathfrak{X}, \mathbb{P}^\Gamma_{s,\mu}, s \geq 0, \mu \in M_f\right]$ with branching rate kernel Γ, related to the *backward equation*

$$(4.2) \qquad -\frac{\partial}{\partial s} u(s, z) = \Delta u(s, z) - \Gamma(s, z) u^2(s, z), \quad s \geq 0, \ z \in \mathbb{R}.$$

[Recall our convention concerning the formal writing of generalized Radon-Nikodym derivatives as $\Gamma(s, \cdot)$.] More precisely, given Γ, the time-inhomogeneous Markov process $\mathfrak{X}$ has Laplace transition functional

$$(4.3) \qquad \mathbb{E}^\Gamma_{s,\mu} \exp\left(\mathfrak{X}_t, -\varphi\right) = \exp\left(\mu, -u(s, t, \cdot|\Gamma)\right),$$

$0 \leq s \leq t$, $\varphi \in \Phi_+$, $\mu \in M_f$, where $u(\cdot, t, \cdot|\Gamma)$ solves (4.2) on $[0, t] \times \mathbb{R}$ with *terminal condition* $u(t, t, \cdot|\Gamma) = \varphi$. We mention that we passed to a *backward formulation* since the (fixed) branching rate kernel Γ is not constant in time (opposed to the case of the single atomic model of Section 2 above).

Since the branching rate kernel Γ is actually random, this $\mathfrak{X}$ is a superprocess in a *random medium* Γ (for more precise formulations we refer to [Dawson and Fleischmann 1991]).

From now on in this subsection, let $\mathfrak{X}$ start off at time 0 even with the Lebesgue measure $l(dx) = dx$.

In order to exhibit *large scale clumping effects* of $\mathfrak{X}$, we introduce the following *time-space-mass rescaling*. Set

$$(4.4) \qquad \eta := (\gamma + 1)/2\gamma$$

(recall that $\gamma \in (0, 1)$ is the index of the stable system Γ of catalysts controlling the branching), and consider

$$(4.5) \quad \mathfrak{X}^K_t(B) := K^{-\eta} \mathfrak{X}_{Kt}(K^\eta B), \quad k > 1, \ t \geq 0, \ B \subset \mathbb{R} \text{ a Borel set.}$$

That is, we speed up the time, contract the space, and renormalize the mass in a specific way (which preserves the expectations). Let $\mathbb{P}^{\Gamma, t, K}$ denote the law of the random measure $\mathfrak{X}^K_t$ given the medium Γ. Since Γ is random (with $\mathcal{P}$ as its distribution), $\mathbb{P}^{\Gamma, t, K}$ is actually a random probability law. The following *scaling limit theorem* follows from the main result in [Dawson and Fleischmann 1991].

THEOREM 4.1 (CLUMPING PICTURE). *Recall that the superprocess $\mathfrak{X}$ in the random medium Γ starts off with the Lebesgue measure l. Fix a (macroscopic) time point $t > 0$. Then as $K \to \infty$, the random probability laws $\mathbb{P}^{\Gamma, t, K}$ of $\mathfrak{X}^K_t$ converge in $\mathcal{P}$-probability to some deterministic probability law $\mathbb{P}^{t, \infty}$ of a random measure $\mathfrak{X}^\infty_t$ on $\mathbb{R}$. This $\mathfrak{X}^\infty_t$ has independent increments, is non-degenerate, has again the Lebesgue measure l as its expectation, and is determined by the Laplace functional*

$$(4.6) \qquad \mathbb{E}^{t,\infty} \exp\left(\mathfrak{X}^\infty_t, -\varphi\right) = \exp\left[-\int dy \, \mathcal{E} u_{\varphi(y)}(0, t, 0|\Gamma)\right], \quad \varphi \in \Phi_+,$$

where $u_c(\cdot, t, \cdot|\Gamma)$ refers to the solution of the equation (4.2) *with* constant *terminal condition $u_c(t, t, z|\Gamma) \equiv c$ (given Γ).*

Consequently, to get the r.h.s. of (4.6), besides t first fix a realization of the random medium Γ and also a point y in $\mathbb{R}$, and solve (4.2) with *constant* terminal condition $\varphi(y)$ at t. Then consider this solution $u_{\varphi(y)}(\cdot, t, \cdot | \Gamma)$ at the initial time $s = 0$ and in the origin $z = 0$. Finally, built the expectation concerning the random Γ, and integrate with respect to the Lebesgue measure dy.

$\mathfrak{X}_t^\infty$ describes *clumping effects* of $\mathfrak{X}$ on a macroscopic scale. Some additional properties of $\mathfrak{X}_t^\infty$ are known, see [Dawson and Fleischmann 1991]. However, it would be of interest to show that the atoms of $\mathfrak{X}_t^\infty$ are *not* isolated, and to reveal the nature of $\mathfrak{X}_t^\infty$ as a *stochastic process* in t. (By the way, in the case of a *constant* branching rate, the limiting clumps are isolated, in fact, they are situated in Poissonian points of finite intensity, and their weights change in time according to the continuous state Galton-Watson process as in (2.1); see [Dawson and Fleischmann 1988].)

Note that in the scaling limit the randomness of Γ "disappears", although the medium Γ cannot be averaged finitely (in space) due to the lack of finite first moments.

4.2. Remark on more-dimensional superprocesses in catalytic media with absolutely continuous states. As already mentioned, superprocesses in *point*-catalytic media are meaningful in a *one*-dimensional space based on the fact that the underlying motion process (an α-stable process, $\alpha > 1$) allows positive local times at points. However, in more than one dimension points are *polar sets* for the underlying motion. Hence, the *more*-dimensional case is more delicate.

One way to attack such problem is to impose more restrictive conditions on the branching rate kernel ϱ, that is, to let it to be *less singular* than in the point catalytic case. On the other hand, keeping ϱ *sufficiently irregular* one can even achieve that a super-Brownian motion with such ϱ gets *absolutely continuous* states also in *higher* dimensions.

To be more precise, think of the following *example* (for more general results, we refer to [Dawson and Fleischmann 1993b]). In $\mathbb{R}^{d+1}$, $d \geq 1$, consider the factored kernel

$$\varrho(t, dy) = \varrho_d(t, y_d) dy_d \, \varrho_1(dy_1), \quad t \geq 0, \quad y = [y_d, y_1] \in \mathbb{R}^d \times \mathbb{R},$$

where ϱ_d is a measurable function, bounded on $[0, T] \times \mathbb{R}^d$, $T > 0$, but ϱ_1 is selected from the stable random measure $\Gamma(0)$ with index $\gamma \in (0, 1)$ defined by (4.1). Hence, branching is allowed only at a *dense set of hyperplanes* in $\mathbb{R}^{d+1}$ (a dense set of lines in $\mathbb{R}^2$ if $d = 1$).

Since ϱ is irregular only in a single coordinate, a super-Brownian motion $\mathfrak{X}$ with such branching rate kernel ϱ indeed *exists*; more precisely, it exists for almost all realizations of $\Gamma(0)$. (In fact, think of the Brownian local time at these hyperplanes, or equivalently, of the Brownian local time of the first coordinate at the ensemble of weighted point catalysts described by $\Gamma(0)$.)

Moreover, if we impose the additional condition $\gamma < (2d+1)^{-1}$ which intuitively says that the hyperplanes are "not too densely" situated in $\mathbb{R}^{d+1}$, then the branching rate kernel ϱ is irregular enough that $\mathfrak{X}$ will even have *absolutely continuous states*.

Of course, if in the previous example we replace $\Gamma(0)$ by a *homogeneous point process* on $\mathbb{R}$ of finite intensity, then *all* such defined factored kernels ϱ have an "adequate portion of irregularity" guaranteeing that corresponding super-Brownian motions exist and even have absolutely continuous states.

Note that such an absolute continuity property of the measure states of some super-Brownian motions in *more than one* dimension is in a sharp contrast to the regular branching rate case. Recall the well-known singularity results of [Dawson and Hochberg 1979] for super-Brownian motions in $\mathbb{R}^d$, $d \geq 2$, with a constant branching rate.

Bibliography

Dawson, D. A., *Measure-valued Markov processes*, École d'été de probabilités de Saint Flour 1991, 1992.

Dawson, D. A. and Fleischmann, K., *Strong clumping of critical space-time branching models in subcritical dimensions*, Stochastic Process. Appl. **30** (1988), 193–208.

———, *Critical branching in a highly fluctuating random medium*, Probab. Theor. Relat. Fields **90** (1991), 241–274.

———, *Diffusion and reaction caused by point catalysts*, SIAM J. Appl. Math. **52** (1992), 163–180.

———, *A Super-Brownian motion with a single point catalyst*, Stochastic Process. Appl. (1993a), (to appear).

———, *Super-Brownian motions in higher dimensions with absolutely continuous measure states* (1993b), (in preparation).

Dawson, D. A., Fleischmann, K. and Gorostiza, L. G., *Stable hydrodynamic limit fluctuations of a critical branching particle system in a random medium*, Ann. Probab. **17** (1989), 1083–1117.

Dawson, D. A., Fleischmann, K., Li, Y. and Mueller, C., *Singularity of super-Brownian local time at a point catalyst* (1993), (in preparation).

Dawson, D. A., Fleischmann, K. and Roelly, S., *Absolute continuity for the measure states in a branching model with catalysts*, Stochastic Processes, Proc. Semin., Vancouver/CA (USA) 1990, Prog. Probab. **24** (1991), 117–160.

Dawson, D. A. and Hochberg, K. J., *The carrying dimension of a stochastic measure diffusion*, Ann. Probab. **7** (1979), 693–703.

Dynkin, E. B., *Branching particle systems and superprocesses*, Ann. Probab. **19** (1991), 1157–1194.

Kerstan, J., Matthes, K. and Mecke, J., *Infinitely divisible point processes*, "Nauka", Moscow, 1982. (Russian)

Konno N. and Shiga, T., *Stochastic partial differential equations for some measure-valued diffusions*, Probab. Theor. Relat. Fields **79** (1988), 201–225.

Liemant, A., Matthes, K. and Wakolbinger, A., *Equilibrium distributions of branching processes*, Kluwer Academic Publishers, Dordrecht, 1988.

Nicolis, G. and Prigogine, I., *Self-organization in nonequilibrium systems. From dissipative structures to order through fluctuations*, Wiley, New York, 1977.

Reimers, M., *One dimensional stochastic partial differential equations and the branching measure diffusion*, Probab. Theor. Relat. Fields **81** (1989), 319–340.

Sapoval, B., *Fractal electrodes, fractal membranes, and fractal catalysts*, Fractals and Disordered Systems, (Bunde, A. and Havlin, S. eds.), Springer-Verlag, Berlin, 1991, pp. 207–226.

Sugitani, S., *Some properties for the measure-valued branching diffusion processes*, J. Math. Soc. Japan **41** (1989), 437–462.

ABSTRACT. This is a review of some recent results on critical measure-valued branching processes $\mathfrak{X}$ in which the branching rate ϱ may be a generalized function. Main attention is paid to the example $\varrho(z) = \delta_{\mathfrak{c}}(z)$, that is, branching occurs only at a single point $\mathfrak{c} \in \mathbb{R}$, at a point catalyst, but there with an infinite rate, whereas outside $\mathfrak{c}$ only the heat flow acts. Although this is an extremely simplified case, nevertheless, this $\mathfrak{X}$ has remarkable properties discussed in some detail. Other aspects are discussed as well as the longtime behavior. Also, a scaling limit theorem is quoted exhibiting clumping effects in the case of a densely situated set of weighted point catalysts fluctuating both in time and space. Finally, the problem of construction of more-dimensional super-Brownian motions with absolutely continuous states is touched.

INSTITUTE OF APPLIED ANALYSIS AND STOCHASTICS, MOHRENSTR.39, D-10117 BERLIN, GERMANY

E-mail address: fleischmann@iaas-berlin.d400.de

A Measure Valued Process Arising from a Branching Particle System with Changes of Mass

Luis G. Gorostiza

1. Introduction

A "mass measure branching process" is obtained in the same way as a "measure branching process" (also called "Dawson-Watanabe process", or "superprocess"), i.e., as a high density-short life-small particle limit of a spatial branching particle system, by superimposing a mass structure which consists in assigning to each particle a mass and specifying a mechanism by which the branching changes the mass form the parent to the offspring. In general this mechanism can be time and position dependent, and there can be randomness in addition to the one produced by the branching. In [Fernández and Gorostiza 1990] we introduced such a branching mass model and we proved laws of large numbers and fluctuation limit theorems under several rescalings. Here we will give a simple example of a mass measure branching process and discuss its relationships to superprocesses.

The function and measure spaces used below are defined at the end.

DEFINITION. The $(d, \alpha, \beta, b, \sigma)$-*mass measure branching process* (MMBP) is an inhomogeneous Markov process $X \equiv \{X(t), t \geq 0\}$ with values in $\mathcal{M}_p(\mathbb{R}^d)$ whose conditional Laplace functional is given by

$$(1.1) \quad E\big[\exp\{-\langle X(t), \varphi\rangle\} \mid X(s) = \mu\big] = \exp\{-\langle \mu, u_\varphi(e^{V(\sigma-b)s}, \cdot, s, t)\rangle\},$$

$$\varphi \in C_p(\mathbb{R}^d)_+, \quad t \geq s \geq 0,$$

where $\{u_\varphi(z, x, s, t) : z > 0, x \in \mathbb{R}^d, t \geq s \geq 0\}$ is the unique (mild) solution of the non-linear partial differential equation

$$(1.2) \quad \frac{\partial}{\partial t} u_\varphi = \left[\Delta_\alpha + Vb + V(\sigma-b)z\frac{\partial}{\partial z}\right] u_\varphi - Vce^{\beta V(\sigma-b)s} u_\varphi^{1+\beta}, \quad t \geq s \geq 0,$$

$$u_\varphi(z, x, s, s) = \varphi(x),$$

with $\Delta_\alpha \equiv -(-\Delta)^{\alpha/2}$, $\alpha \in (0, 2]$ (Δ = Laplacian on $\mathbb{R}^d$), $V > 0$, $\beta \in (0, 1]$, $c \in (0, 1/(1+\beta)]$, $b \in \mathbb{R}$ and $\sigma \in \mathbb{R}$.

In the case $\sigma = b$ there are no changes of mass, and X coincides with the well-known (d, α, β, b)-superprocess. For $b = 0$, which corresponds to critical branching, X is the (d, α, β)-superprocess studied in detail in [Dawson 1993].

THEOREM. *The $(d, \alpha, \beta, b, \sigma)$-mass measure branching process exists.*

1991 *Mathematics Subject Classification.* Primary: 60J80, 60G57 ; Secondary: 60J85, 60K35, 35G20.

This is the final form of the paper.

Since the MMBP arises as a limit of a branching mass particle system, we will outline in Section 2 how the limit is obtained. This approach can be turned into a rigorous proof, which therefore implies existence. However, we will prove existence in terms of particular superprocesses (without an approximating particle system). From the point of view of models it is of course the particle system which is of primary interest.

2. Existence of the $(d, \alpha, \beta, b, \sigma)$-mass measure branching process

We begin by describing a branching mass model in $\mathbb{R}^d$. At time $t = 0$ the particles are distributed by some (random) point measure, and they independently migrate according to a spherically symmetric stable process with exponent $\alpha \in (0, 2]$ (whose generator is Δ_α) during V-exponentially distributed lifetimes, at the end of which they produce offspring according to the generating function

$$(2.1) \qquad \mathcal{F}(s) = s + b(s - 1) + c(1 - s)^{1+\beta}, \quad s \in [0, 1],$$

with $\beta \in (0, 1]$, and the new particles evolve in the same way, starting from the death site of their parent. (The fact that $\mathcal{F}(s)$ belongs to a branching law requires $b \in (-1, c]$, $c \in (0, (1+b)/(1+\beta)]$). For $b = 0$ the branching is critical.

Let $N \equiv \{N(t), t \geq 0\}$ denote the counting measure valued process defined by

$$N(t) = \sum_i \delta_{x_i(t)},$$

where $\{x_i(t)\}_i$ are the locations of the particles present at time t. (This is the process that gives rise to the (d, α, β, b)-superprocess). In addition we assume that each particle has a (positive) mass of its own, and when a particle of mass z branches producing k (> 0) particles, the mass of each one of these particles is equal to $za(k)$, where $a(k)$ (> 0) may depend on k. Here we will consider the simple case where $a \equiv a(k)$ does not depend on k, which is an important simplification (see Remark a below). We assume for simplicity that every initial particle has unit mass. Let $M \equiv \{M(t), t \geq 0\}$ denote the mass process defined by

$$M(t) = \sum_i z_i \delta_{x_i(t)} \quad \text{if} \quad N(t) = \sum_i \delta_{x_i(t)},$$

where z_i is the mass of the particle located at $x_i(t)$. This process is not Markovian because looking at a point mass we do not know if it belongs to a single particle or if a branching is occuring at that instant. (However, the process (M, N) is Markovian because N tells if a branching occurs).

We can see easily by a heuristic argument what the limit should be. For each $n \geq 1$, let M_n and N_n denote the processes M and N, respectively, rescaled as follows: $V_n = Vn^\beta$, $b_n = bn^{-\beta}$ (hence $c \in (0, 1/(1+\beta)]$), and the mass of each particle is divided by n. We recall that if $N_n(0)$ converges weakly as $n \to \infty$, then the process N_n converges weakly in $D(\mathbb{R}_+, \mathcal{M}_p(\mathbb{R}^d))$ as $n \to \infty$ to the (d, α, β, b)-superprocess [Dawson 1993, Gorostiza and López-Mimbela 1993]. Let us see what the rescaled change of mass a_n should be. The mass of any particle present at time $t > 0$ is equal to $a_n^{\nu_n(t)}$, where ν_n is the Poisson process with intensity V_n along the ancestry line of the particle which represents the death times of its ancestors. Hence, letting

$$a_n = (1 + \sigma n^{-\beta})/(1 + bn^{-\beta}), \quad \sigma \in \mathbb{R},$$

and using the renewal theorem, we have

(2.2) $$a_n^{\nu_n(t)} \to e^{V(\sigma-b)t} \quad \text{a.s. as } n \to \infty.$$

On the other hand, the "position part" of M_n is just the process N_n (division of the mass by n is incorporated into N_n). Hence, denoting by Y the (d,α,β,b)-superprocess, M_n should converge weakly as $n \to \infty$ to a process X which is given by

(2.3) $$X(t) = e^{V(\sigma-b)t}Y(t), \quad t \geq 0.$$

This implies existence of X and gives a representation of X in terms of the superprocess Y. Since Y is characterized by (1.1)–(1.2) with $\sigma = b$, and one sees that the right-hand side of (2.3) is described by (1.1)–(1.2), then X is the $(d,\alpha,\beta,b,\sigma)$-MMBP. Note that the representation (2.3) implies the Markov property of X.

The previous argument suggests a way of proving convergence of the particle process. However, it is instructive to see, with a view towards other change of mass mechanisms, how the usual convergence approach works in this simple case.

For each $\varphi \in C_p(\mathbb{R}^d)_+$, let

$$w_\varphi(z,x,s,t) = E\big[\exp\{-\langle M(t),\varphi\rangle\} \mid M(s) = z\delta_x\big], \quad z > 0, \quad x \in \mathbb{R}^d, \quad t \geq s \geq 0.$$

By the usual renewal argument (conditioning on the first branching) we obtain the integral equation

(2.4) $$w_\varphi(z,x,s,t) = e^{-V(t-s)}T_{t-s}e^{-z\varphi}(x)$$
$$+ \int_s^t Ve^{-V(r-s)}T_{r-s}\mathcal{F}\big(w_\varphi(az,\cdot,r-s,t-r)\big)(x)\,dr, \quad t \geq s,$$

where $\{T_t\}_t$ denotes the symmetric α-stable semigroup. We see from (2.4) that w_φ is a (mild) solution of the differential equation

$$\frac{\partial}{\partial t}w_\varphi(z,x,s,t) = \Delta_\alpha w_\varphi(z,x,s,t) + V\big[\mathcal{F}(w_\varphi(z,x,s,t)) - w_\varphi(z,x,s,t)\big]$$
$$+ V\big[\mathcal{F}(w_\varphi(az,x,s,t)) - \mathcal{F}(w_\varphi(z,x,s,t))\big], \quad t \geq s,$$
$$w_\varphi(z,x,s,s) = e^{-z\varphi(x)}.$$

Hence $v_\varphi(z,x,s,t) = 1 - w_\varphi(z,x,s,t)$ satisfies

$$\frac{\partial}{\partial t}v_\varphi(z,x,s,t) = \Delta_\alpha v_\varphi(z,x,s,t) - V\big[\mathcal{F}(1-v_\varphi(z,x,s,t)) - (1 - v_\varphi(z,x,s,t))\big]$$
$$- V\big[\mathcal{F}(1-v_\varphi(az,x,s,t)) - \mathcal{F}(1 - v_\varphi(z,x,s,t))\big],$$
$$v_\varphi(z,x,s,s) = 1 - e^{-z\varphi(x)},$$

and by (2.1),

(2.5) $$\frac{\partial}{\partial t}v_\varphi(z,x,s,t) = (\Delta_\alpha + Vb)v_\varphi(z,x,s,t) - Vcv_\varphi(z,x,s,t)^{1+\beta}$$
$$+ G_\varphi(a,z,x,s,t),$$
$$v_\varphi(z,x,s,s) = 1 - e^{-z\varphi(x)},$$

with

(2.6) $G_\varphi(a,z,x,s,t) = V\big[(1+b)\big(v_\varphi(az,x,s,t) - v_\varphi(z,x,s,t)\big)$
$\qquad\qquad\qquad\qquad - c\big(v_\varphi(az,x,s,t)^{1+\beta} - v_\varphi(z,x,s,t)^{1+\beta}\big)\big].$

Therefore, if $\mu = \sum_i z_i \delta_{x_i}$, we have

(2.7) $E\big[\exp\{-\langle M(t), \varphi\rangle\} \mid M(s) = \mu\big] = \exp\Big\{\sum_i \log\big(1 - v_\varphi(z_i, x_i, s, t)\big)\Big\},$

where $v_\varphi(z,x,s,t)$ solves (2.5).

We now rescale. In the n-th rescaling the masses of the particles are divided by n, and the dependence on n of the constants V_n, b_n and a_n is to be determined (forget what we already know). Let M_n denote the rescaled mass process.

From (2.5), (2.6) and (2.7) we have, if $\mu_n = \sum_i n^{-1} z_i \delta_{x_i}$,

(2.8) $E\big[\exp\{-\langle M_n(t), \varphi\rangle\} \mid M_n(s) = \mu_n\big]$
$\qquad\qquad\qquad = \exp\Big\{\sum_i n \log\big(1 - n^{-1} h^n_\varphi(z_i, x_i, s, t)\big)\Big\},$

where $h^n_\varphi(z,x,s,t) = n v^n_\varphi(z,x,s,t)$ satisfies

(2.9) $\dfrac{\partial}{\partial t} h^n_\varphi(z,x,s,t) = (\Delta_\alpha + V_n b_n) h^n_\varphi(z,x,s,t) - V_n c n^{-\beta}\big(h^n_\varphi(z,x,s,t)\big)^{1+\beta}$
$\qquad\qquad\qquad\qquad\qquad + G^n_\varphi(a_n, z, x, s, t),$
$\qquad h^n_\varphi(z,x,s,s) = n(1 - e^{-n^{-1} z \varphi(x)}),$

with

(2.10) $G^n_\varphi(a_n, z, x, s, t) = V_n\big[(1+b_n)\big(h^n_\varphi(a_n z, x, s, t) - h^n_\varphi(z, x, s, t)\big)$
$\qquad\qquad\qquad - c n^{-\beta}\big((h^n_\varphi(a_n z, x, s, t))^{1+\beta} - (h^n_\varphi(z, x, s, t))^{1+\beta}\big)\big].$

It can be shown (as in [Fernández and Gorostiza 1990]) that

(2.11) $\qquad E\langle M_n(t), \varphi\rangle = E\langle M_n(0), T_t \varphi\rangle \exp\{V_n[a_n(1+b_n) - 1]t\}.$

It follows from (2.9), (2.10) and (2.11) that the right rescaling is of the form

(2.12) $V_n = V n^\beta, \quad b_n = b n^{-\beta}, \quad a_n = (1 + \sigma n^{-\beta})/(1 + b n^{-\beta}),$
$\qquad\qquad\qquad\qquad \sigma \in \mathbb{R}, \quad (\text{and } c \in (0, 1/(1+\beta)]),$

and then h^n_φ satisfies

(2.13) $\dfrac{\partial}{\partial t} h^n_\varphi(z,x,s,t) = (\Delta_\alpha + Vb) h^n_\varphi(z,x,s,t) - Vc\big(h^n_\varphi(z,x,s,t)\big)^{1+\beta}$
$\qquad\qquad + V n^\beta \big(h^n_\varphi(z(1+\sigma n^{-\beta})/(1+b n^{-\beta}), x, s, t) - h^n_\varphi(z,x,s,t)\big)$
$\qquad\qquad + H^n_\varphi(a_n, z, x, s, t),$
$\qquad h^n_\varphi(z,x,s,s) = n(1 - e^{-n^{-1} z \varphi(x)}),$

with

(2.14) $H_\varphi^n(a_n, z, x, s, t) = V\big[b\big(h_\varphi^n(a_n z, x, s, t) - h_\varphi^n(z, x, s, t)\big)$
$\qquad\qquad\qquad\qquad - c\big((h_\varphi^n(a_n z, x, s, t))^{1+\beta} - (h_\varphi^n(z, x, s, t))^{1+\beta}\big)\big].$

If $z_i^n \equiv z_i$ is the mass of the particle at x_i at time s in the n-th rescaling, we know from (2.2) that $z_i^n \to e^{V(\sigma-b)s}$ a.s. as $n \to \infty$. Therefore, if $\mu_n \to \mu \in \mathcal{M}_p(\mathbb{R}^d)$ as $n \to \infty$, then from (2.12), (2.13) and (2.14) we have

(2.15) $\sum_i n\log\big(1 - n^{-1} h_\varphi^n(z_i, x_i, s, t)\big) \to \big\langle \mu, e^{-V(\sigma-b)s} h_\varphi(e^{V(\sigma-b)s}, \cdot, s, t))\big\rangle$

$$\text{as } n \to \infty,$$

where $u_\varphi(z, x, s, t) = e^{-V(\sigma-b)s} h(z, x, s, t)$ satisfies

(2.16)
$$\frac{\partial}{\partial t} u_\varphi(z, x, s, t) = (\Delta_\alpha + Vb) u_\varphi(z, x, s, t)$$
$$- Vc e^{\beta V(\sigma-b)s} u_\varphi(z, x, s, t)^{1+\beta} + V(\sigma-b) z \frac{\partial}{\partial z} u_\varphi(z, x, s, t), \quad t \geq s,$$
$$u_\varphi(z, x, s, s) = \varphi(x),$$

and it follows from (2.8) and (2.15) that if $M_n(0)$ converges weakly to μ as $n \to \infty$, then $M_n(t)$ converges weakly to $X(t)$ as $n \to \infty$, and the Laplace functional $X(t)$ is given by

$$E\big[\exp\{-\langle X(t), \varphi\rangle\} \mid X(s) = \mu\big] = \exp\{-\langle \mu, u_\varphi(e^{V(\sigma-b)s}, \cdot, s, t)\rangle\},$$

where $u_\varphi(z, x, s, t)$ solves (2.16); hence $X(t)$ is the $(d, \alpha, \beta, b, \sigma)$-MMBP.

In order to prove rigorously that M_n converges weakly to X in $D(\mathbb{R}_+, \mathcal{M}_p(\dot{\mathbb{R}}^d))$ as $n \to \infty$, we must prove (2.15), (2.16), weak convergence of finite dimensional distributions (using the Markov processes (M_n, N_n) similary as in [Gorostiza and López-Mimbela 1993]), and tightness (similarly as in [Dawson 1993]). This proof of existence of the X involves some technical difficulties, but it is relevant from the point of view of the behavior of the particle system.

Expression (2.3) gives X in terms of the superprocess Y. We will now relate X to a superprocess on $\mathbb{R}_+ \times \mathbb{R}^d$. This provides another existence proof.

Let $\overline{X}$ denote the $\mathcal{M}_p(\mathbb{R}_+ \times \mathbb{R}^d)$-valued $(W, \mathcal{K}, \Phi)$-superprocess where W is the homogeneous Markovian motion in $\mathbb{R}_+ \times \mathbb{R}^d$ with semigroup

$$S_t F(z, x) = T_t F(z e^{V(\sigma-b)t}, \cdot)(x), \quad t \geq 0,$$

whose generator is

$$AF(z, x) = \left[\Delta_\alpha + V(\sigma-b) z \frac{\partial}{\partial z}\right] F(z, x),$$

(Δ_α acting on x), $\mathcal{K}(dt) = V\,dt$ is the branching rate, and $\Phi(u) = bu - cu^{1+\beta}$, $u \geq 0$, is the branching mechanism (see [Dawson 1993, Dynkin 1991] for the terminology).

The Laplace functional of the Markov process $\overline{X}$ is given by

(2.17) $E\big[\exp\{-\langle \overline{X}(t), \overline{\varphi}\rangle\} \mid \overline{X}(s) = \bar{\mu}\big] = \exp\{-\langle \bar{\mu}, \overline{v}_{\overline{\varphi}}(s, t)\rangle\},$

$$\bar{\mu} \in \mathcal{M}_p(\mathbb{R}_+ \times \mathbb{R}^d), \overline{\varphi} \in C_p(\mathbb{R}_+ \times \mathbb{R}^d)_+,$$

where $\overline{v}_{\overline{\varphi}}(z, x, s, t)$ satisfies

(2.18)
$$\frac{\partial}{\partial t}\overline{v}_{\overline{\varphi}} = [A + Vb]\overline{v}_{\overline{\varphi}} - Vc\overline{v}_{\overline{\varphi}}^{1+\beta}, \quad t \geq s,$$
$$\overline{v}_{\overline{\varphi}}(z, x, s, s) = \overline{\varphi}(z, x).$$

Since the evolution of the z component of $\overline{X}$ is the deterministic curve $t \mapsto z(t) = e^{V(\sigma-b)t}z(0)$, putting $z(0) = 1$ we have

$$E\big[\exp\{-\langle\overline{X}(t), \overline{\varphi}\rangle\} \mid \overline{X}(s) = \delta_{e^{V(\sigma-b)s}} \otimes \mu\big] = \exp\{-\langle\delta_{e^{V(\sigma-b)s}} \otimes \mu, \overline{v}_{\overline{\varphi}}(\cdot, \cdot, s, t)\rangle\}$$
$$= \exp\{-\langle\mu, \overline{v}_{\overline{\varphi}}(e^{V(\sigma-b)s}, \cdot, s, t)\rangle\}.$$

We define the mapping $\Pi : \bar{\mu} \to \Pi\bar{\mu}$ of $\mathcal{M}_p(\mathbb{R}_+ \times \mathbb{R}^d)$ into $\mathcal{M}_p(\mathbb{R}^d)$ by

$$\Pi\bar{\mu}(dx) = \int_0^\infty z\bar{\mu}(dz, dx),$$

and observe that $\langle\Pi\bar{\mu}, \varphi\rangle = \langle\bar{\mu}, \overline{\varphi}\rangle$ for $\bar{\mu} \in \mathcal{M}_p(\mathbb{R}_+ \times \mathbb{R}^d)$, $\varphi \in C_p(\mathbb{R}^d)_+$, $\overline{\varphi}(z, x) = z\varphi(x)$, $\overline{\varphi} \in C_p(\mathbb{R}_+ \times \mathbb{R}^d)_+$, and Π is continuous. Since $\overline{X}(s) = \delta_{e^{V(\sigma-b)s}} \otimes \mu$ (on the curve $e^{V(\sigma-b)s}$) if and only if $\Pi X(s) = e^{V(\sigma-b)s}\mu$, then

$$E\big[\exp\{-\langle\Pi X(t), \varphi\rangle\} \mid \Pi X(s) = e^{V(\sigma-b)s}\mu\big]$$
$$= E\big[\exp\{-\langle\overline{X}(t), \overline{\varphi}\rangle\} \mid \overline{X}(s) = \delta_{e^{V(\sigma-b)s}} \otimes \mu\big]$$
$$= \exp\{-\langle\mu, \overline{v}_{\overline{\varphi}}(e^{V(\sigma-b)s}, \cdot, s, t)\rangle\}.$$

Putting $\mu e^{-V(\sigma-b)s}$ instead of μ we have

$$E\big[\exp\{-\langle\Pi X(t), \varphi\rangle\} \mid \Pi X(s) = \mu\big] = \exp\{-\langle\mu, \overline{u}_{\overline{\varphi}}(e^{V(\sigma-b)s}, \cdot, s, t)\rangle\},$$

where $\overline{u}_{\overline{\varphi}}(z, x, s, t) = e^{-V(\sigma-b)s}\overline{v}_{\overline{\varphi}}(z, x, s, t)$ satisfies

(2.19)
$$\frac{\partial}{\partial t}\overline{u}_{\overline{\varphi}} = [A + Vb]\overline{u}_{\overline{\varphi}} - Vce^{\beta V(\sigma-b)s}\overline{u}_{\overline{\varphi}}^{1+\beta}, \quad t \geq s,$$
$$\overline{u}_{\overline{\varphi}}(z, x, s, s) = e^{-V(\sigma-b)s}z\varphi(x).$$

Since equations (1.2) and (2.19) are the same and the initial conditions coincide (on the curve $e^{V(\sigma-b)s}$), then $\overline{u}_{\overline{\varphi}}(z, x, s, t) = u_\varphi(z, x, s, t)$. (Uniqueness of the mild solution follows from Chapter 6 of [Pazy 1983]).

Hence the $(d, \alpha, \beta, b, \sigma)$-MMBP X is identified as

$$X(t, dx) = \int_0^\infty z\overline{X}(t, dz, dx), \quad t \geq 0.$$

Since the $(d, \alpha, \beta, b, \sigma)$-MMBP with $\sigma \neq b$ is a measure $(\mathcal{M}_p(\mathbb{R}^d))$-valued Markov process which has the multiplicative (or branching) property, it should also be a superprocess in the general sense defined by Dynkin in [Dynkin 1993]. Is it?

Remarks. (a) The assumption $a(k) \equiv a$ for the change of mass mechanism has a simple law of large numbers effect in the limit, which leads to the representation (2.3), because the only source of randomness of the particle mass is the Poisson process of branching times along the ancestry line. In the approximation argument above, this assumption allows an explicit use of the branching generating function $\mathcal{F}$. If $a(k)$ depends on k, then in equation (2.4) $\mathcal{F}(w_\varphi(az,x,s,t))$ is replaced by

$$\sum_{k=0}^{\infty} p_k(w_\varphi(a(k)z,x,s,t))^k,$$

where $\{p_k\}_k$ is the underlying branching law, and the generating function cannot be exploited easily. Other change of mass mechanisms (e. g., nuclear cascades, see [Athreya and Ney 1972]) could have more interesting effects in the limit.

(b) Assuming $\beta = 1$ (i. e., finite variance branching) and using the martingales for M in [Fernández and Gorostiza 1990] ((7) and (11)), the rescaling limit yields the following martingale problem characterization of the $(d,\alpha,1,b,\sigma)$-MMBP X: For each $\varphi \in C_p(\mathbb{R}^d)_+$, the process

$$L_\varphi(t) \equiv \langle X(t),\varphi\rangle - \int_0^t \langle X(s),(\Delta_\alpha + V\sigma)\varphi\rangle\, ds, \quad t \geq 0,$$

is a martingale with increasing process

$$\langle L_\varphi\rangle(t) = \int_0^t e^{V(\sigma-b)s}\langle X(s), 2Vc\varphi^2\rangle\, ds, \quad t \geq 0.$$

For $\sigma = b$, this coincides with the known martingale problem for the $(d,\alpha,1,b)$-superprocess [Dawson 1993, El Karoui and Roelly 1991].

(c) From (2.3) it follows that

$$E\big[\langle X(t),\varphi\rangle \mid X(0) = \mu\big] = \langle \mu, T_t\varphi\rangle e^{V\sigma t}.$$

Hence for $\sigma = 0$ and $\mu = \lambda$ (Lebesgue measure on $\mathbb{R}^d$, which is T_t-invariant) the intensity $EX(t) = \lambda$ is constant in time. In the case $\sigma = b = 0$ it is known that $X(t)$ converges weakly to an equilibrium state $X(\infty)$ as $t \to \infty$, $EX(\infty) = \lambda$ if and only if $d > \alpha/\beta$, $X(t)$ goes to (local) extinction as $t \to \infty$ if $d \leq \alpha/\beta$, and this can be used to derive information on the asymptotic behavior of the solution of the nonlinear equation [Gorostiza, Roelly, and Wakolbinger 1992, Gorostiza and Wakolbinger 1992]. What happens for $\sigma = 0$, $b \neq 0$? Note that this is a question on the large time behavior of the (d,α,β,b)-superprocess Y, and the problem is weak convergence of $e^{-Vbt}Y(t)$ as $t \to \infty$. It seems that for $b > 0$ the limit should be λ.

Notation and spaces

- $\langle \mu,\varphi\rangle = \int \varphi\, d\mu$.
- $C(\mathbb{R}^d)$ (resp. $C(\mathbb{R}_+ \times \mathbb{R}^d)$): real continuous functions on $\mathbb{R}^d$ (resp. $\mathbb{R}_+ \times \mathbb{R}^d$).
- $\varphi_p(x) = (1+|x|^2)^{-p}$, $x \in \mathbb{R}^d$, $(p > 0)$.
- $C_p(\mathbb{R}^d) = \{\varphi \in C(\mathbb{R}^d) : \sup_x |\varphi(x)/\varphi_p(x)| < \infty\}$.
- $C_p(\mathbb{R}_+ \times \mathbb{R}^d) = \{\overline{\varphi} \in C(\mathbb{R}_+ \times \mathbb{R}^d) : \overline{\varphi}(z,\cdot) \in C_p(\mathbb{R}^d) \text{ for all } z \in \mathbb{R}_+\}$.
- $C_p(\mathbb{R}^d)_+$ (resp. $C_p(\mathbb{R}_+ \times \mathbb{R}^d)_+$): non-negative elements of $C_p(\mathbb{R}^d)$ (resp. $C_p(\mathbb{R}_+ \times \mathbb{R}^d)$).
- $\overline{\varphi}_p(z,x) = \varphi_p(x)$, $(z,x) \in \mathbb{R}_+ \times \mathbb{R}^d$, $\overline{\varphi}_p \in C_p(\mathbb{R}_+ \times \mathbb{R}^d)$.
- $\mathcal{M}(\mathbb{R}^d)$ (resp. $\mathcal{M}(\mathbb{R}_+ \times \mathbb{R}^d)$): non-negative Radon measures on $\mathbb{R}^d$ (resp. $\mathbb{R}_+ \times \mathbb{R}^d$).
- $\mathcal{M}_p(\mathbb{R}^d) = \{\mu \in \mathcal{M}(\mathbb{R}^d) : \langle \mu, \varphi_p\rangle < \infty\}$.
- $\mathcal{M}_p(\mathbb{R}_+ \times \mathbb{R}^d) = \{\overline{\mu} \in \mathcal{M}(\mathbb{R}_+ \times \mathbb{R}^d) : \langle \overline{\mu}, \overline{\varphi}_p\rangle < \infty\}$.

- $\mathcal{M}_p(\mathbb{R}^d)$ (resp. $\mathcal{M}_p(\mathbb{R}_+ \times \mathbb{R}^d)$) is endowed with the topology generated by $C_p(\mathbb{R}^d)_+$ (resp. $C_p(\mathbb{R}_+ \times \mathbb{R}^d)_+$).
- $\mathcal{M}_p(\dot{\mathbb{R}}^d)$: extension of $\mathcal{M}_p(\mathbb{R}^d)$, where $\dot{\mathbb{R}}^d = \mathbb{R}^d \cup \{\infty\}$ (∞ isolated point).
- $D(\mathbb{R}_+, \mathcal{M}_p(\dot{\mathbb{R}}^d))$: càdlàg functions from $\mathbb{R}_+$ into $\mathcal{M}_p(\dot{\mathbb{R}}^d)$, with a Skorokhod-type topology.
- p, d and α (in Δ_α) are related by $p > d/2$, and in addition $p < (d+\alpha)/2$ in case $\alpha < 2$. See [Dawson 1993] (and references therein) for more information.

Acknowledgement. I thank Don Dawson and Ed Perkins for fruitful discussions, and CIMAT for hospitality. This research is partially supported by CONACyT grant 2059-E9302.

Bibliography

K. B. Athreya and P. E. Ney, *Branching Processes,* Springer-Verlag, Berlin, 1972.

D. A. Dawson, *Measure-Valued Markov Processes,* Ecole d'Été de Probabilités de Saint-Flour (1991). Lectures Notes in Math. **1541** (1993), 1–260, Springer-Verlag.

E. B. Dynkin, *Branching particle systems and superprocesses,* Ann. Probab. **19** (1991), 1157–1194.

_____, *The 1993 Barrett Memorial Lectures,* University of Tennessee, 1993.

N. El Karoui and S. Roelly, *Propriétés de martingales, explosion et représentation de Lévy-Khintchine d'une classe de processus à valeurs mesures,* Stochastic Process. Appl. **38** (1991), 239–266.

B. Fernández and L. G. Gorostiza, *Hydrodynamic and fluctuation limits of branching particle systems with changes of mass,* Bol. Soc. Mat. Mexicana (2) **35** (1990), 25–37.

L. G. Gorostiza and J. A. López-Mimbela, *A convergence criterion for measure-valued processes, and application to continuous superprocesses,* Barcelona Seminar on Stochastic Analysis (D. Nualart and M. Sanz, eds), Progr. Probab. vol. 32, Birkhäuser, Boston (1993), 62–71.

L. G. Gorostiza, S. Roelly, and A. Wakolbinger, *Persistence of critical multitype particle and measure branching processes,* Probab. Theor. Relat. Fields **92** (1992), 313–335.

L. G. Gorostiza and A. Wakolbinger, *Convergence to equilibrium of critical branching particle systems and superprocesses, and related nonlinear partial differential equations,* Acta Appl. Math. **27** (1992), 269–291.

A. Pazy, *Semigroups of linear operators and applications to partial differential equations,* Springer-Verlag, Berlin, 1983.

ABSTRACT. A rescaling limit of a branching particle system with changes of mass yields a measure-valued Markov processes which contains the (d, α, β, b)-superprocess as a special case.

CENTRO DE INVESTIGACIÓN Y DE ESTUDIOS AVANZADOS, 07000 MÉXICO, D. F., MÉXICO

E-mail address: gortega@redvax1.dgsca.unam.mx

Long Time Behavior of Critical Branching Particle Systems and Applications

Luis G. Gorostiza and Anton Wakolbinger

1. Introduction

The well-known description of spatial branching particle systems in terms of their nonlinear (cumulant) semigroups provides a fruitful interplay between probabilistic properties of the particle systems and analytic properties of the semigroups. A similar situation holds for Dawson-Watanabe (DW) processes (also called measure branching processes, or superprocesses) and their related nonlinear semigroups. (DW processes are high density — short life — small particle limits of branching particle systems [Dawson 1993]). This interplay has been effectively exploited recently to study the branching particle systems and the DW processes by means of the nonlinear semigroups and viceversa (see e.g. [Dynkin 1993, Etheridge 1993a, Gorostiza and Wakolbinger 1993, Iscoe and Lee 1993, Le Gall 1993] and references therein).

In this paper we survey some of the recent results on the long time behavior of critical branching particle systems, critical DW processes, and their related nonlinear semigroups. Our main interest is the relationship between persistence versus extinction of the branching systems and nontrivial versus zero limits of the semigroups, and the dependence of the long time behavior on the parameters of the system for a specific multitype branching particle system. Some of the results and methods presented here are new or extended forms of previous ones. As we shall see, in contrast to the monotype case, in the multitype case the particle system has a wider range of applications than the DW process (Remark 5.2).

The paper is organized as follows. In Section 2 we consider a general class of critical branching particle systems in a Polish space and we give criteria for their persistence and convergence towards equilibrium. In Section 3 we show some applications of the previous criteria, including the case of a multitype branching particle system. In Section 4 we discuss the extinction problem for the critical multitype branching particle system above by means of the "backward tree" method. (For a relation of this method to the behavior of a branching particle system conditioned to nonextinction see [Stöckl and Wakolbinger 1993]). In Section 5 we apply these results to derive the asymptotic behavior of the solution of the system of nonlinear partial differential equations related to the previous critical multitype branching particle system. Appendix 1 contains notation, and Appendix 2 gives a short version of the "renewal argument" for the general branching system in Section 2.

1991 *Mathematics Subject Classification.* Primary: 60J80, 60G57; Secondary: 60J85, 60K35, 35G20.

This is the final form of the paper.

2. Criteria for persistence and convergence towards equilibrium for critical branching particle systems in a Polish space

Let (Z, ρ) be a Polish space on which are defined:
- $(\mathcal{T}_t)_{t \geq 0}$, a semigroup of a homogeneous Markov process with càdlàg paths and infinitesimal generator A.
- $J(z, dy)$, a jump distribution such that

$$\mathcal{J}g(z) = \int_Z g(y) J(z, dy), \quad z \in Z,$$

is continuous for every bounded, measurable function g on Z.
- $\mathcal{K}(z)$, $z \in Z$, a nonnegative, bounded, continuous function.
- $\mathcal{F}(z, y, \zeta)$, a continuous function on $Z \times Z \times [0, 1]$ such that for each $(y, z) \in Z \times Z$, $\mathcal{F}(z, y, \cdot)$ is the generating function of a critical branching law (i.e., a probability measure on $\{0, 1, 2, \ldots\}$ with mean 1).

We denote

$$(2.1) \qquad R(z, y, \zeta) = \frac{1}{\zeta}[\mathcal{F}(z, y, 1 - \zeta) - (1 - \zeta)], \quad \zeta \in (0, 1].$$

It is easy to verify that

$$(2.2) \qquad R(z, y, \cdot) \text{ is concave, increasing and } \lim_{\zeta \to 0+} R(z, y, \zeta) = 0.$$

We consider a (countable and locally finite) particle system in Z evolving by the following branching dynamics. Each particle independently migrates according to $(\mathcal{T}_t)_{t \geq 0}$, it dies with state dependent rate $\mathcal{K}(z)$, and if it dies at site z then it gives birth to particles at a site y chosen by $J(z, dy)$, the number of which is distributed according to the critical generating function $\mathcal{F}(z, y, \cdot)$. The evolution of the system is assumed to be independent of the initial particle configuration. We refer to $\mathcal{K}(z)$ as the *branching rate*.

The state of the system at time $t \geq 0$ is described by the counting measure N_t on Z defined by the locations of the particles at time t. The process $N \equiv (N_t)_{t \geq 0}$ is Markovian and has a version with càdlàg paths.

Given $f \in C_{bb}^+$, let

$$(2.3) \qquad u_f(z, t) = 1 - E[e^{-\langle N_t, f \rangle} \mid N_0 = \delta_z], \quad z \in Z, t \geq 0.$$

A renewal argument (Appendix 2) shows that u_f obeys the nonlinear integral equation

$$(2.4) \qquad u_f(z, t) = \mathcal{U}_t(1 - e^{-f})(z) - \int_0^t \mathcal{U}_{t-s} h_f(\cdot, s)(z) \, ds,$$

or, equivalently, the nonlinear initial value problem

$$(2.5) \qquad \begin{aligned} \frac{\partial}{\partial t} u_f(z, t) &= \mathcal{A} u_f(\cdot, t)(z) - h_f(z, t), \quad t > 0, \\ u_f(z, 0) &= 1 - e^{-f(z)}, \end{aligned}$$

where $(\mathcal{U}_t)_{t \geq 0}$ is the semigroup on C_b generated by the operator

$$(2.6) \qquad \mathcal{A}g(z) = Ag(z) + \mathcal{K}(z)\mathcal{J}[g - g(z)](z),$$

and

(2.7) $$h_f(z,t) = \mathcal{K}(z)\mathcal{J}\bigl[\mathcal{F}\bigl(z,\cdot,1-u_f(\cdot,t)\bigr) - \bigl(1-u_f(\cdot,t)\bigr)\bigr](z).$$

We call $(\mathcal{U}_t)_{t\geq 0}$ the semigroup of the *basic process* on Z because it represents the Markovian motion which evolves according to $(\mathcal{T}_t)_{t\geq 0}$ with jumps at rate $\mathcal{K}$ according to J (i.e., the same dynamics as the branching particle system except for the branching mechanism).

Existence of a global (mild) solution of equation (2.5) follows from the existence of the particle system (see e.g. [Ikeda, Nagasawa, and Watanabe 1968]). Uniqueness holds if the semigroup $(\mathcal{U}_t)_{t\geq 0}$ is strongly continuous (see [Pazy 1983], Chapter 6, Theorem 1.2).

Note that $h_f \geq 0$ due to the criticality of the branching. Hence, from (2.4) we have

(2.8) $$u_f(z,t) \leq \mathcal{U}_t(1-e^{-f})(z) \leq (\mathcal{U}_t f)(z).$$

It follows from (2.1), (2.2), (2.7) and (2.8) that

(2.9) $$h_f(z,t) \leq \mathcal{K}(z)\mathcal{J}\bigl[(\mathcal{U}_t f)(\cdot)R\bigl(z,\cdot,(\mathcal{U}_t f)(\cdot)\bigr)\bigr](z)$$

for all $f \in C_{bb}^+$ such that $\|f\|_\infty \leq 1$.

We assume the existence of a locally finite (nonzero) measure Λ on Z which is invariant for the semigroup $(\mathcal{U}_t)_{t\geq 0}$. Then, if N_0 is Poisson random field on Z with intensity measure Λ, it is easy to verify, by (2.4) and $\mathcal{U}_t$-invariance of Λ, that the Laplace functional of N_t is given by

(2.10)
$$Ee^{-\langle N_t, f\rangle} = \exp\{-\langle \Lambda, u_f(t)\rangle\}$$
$$= \exp\left\{-\langle \Lambda, 1-e^{-f}\rangle + \int_0^t \langle \Lambda, h_f(s)\rangle\, ds\right\}, \quad f \in C_{bb}^+,\ t\geq 0.$$

Since $\langle \Lambda, h_f(s)\rangle \geq 0$, equation (2.10) yields the following result.

PROPOSITION 2.1. *N_t converges in distribution as $t \to \infty$ towards a random point field N_∞ with Laplace functional given by*

(2.11) $$Ee^{-\langle N_\infty, f\rangle} = \exp\left\{-\langle \Lambda, 1-e^{-f}\rangle + \int_0^\infty \langle \Lambda, h_f(s)\rangle\, ds\right\}, \quad f \in C_{bb}^+,$$

and N_∞ is an equilibrium state for the system.

The proof that N_∞ is an equilibrium state is standard and we omit it (see [Liemant, Matthes, and Wakolbinger 1988]). Note that (2.11) implies the finiteness of $\int_0^\infty \langle \Lambda, h_f(s)\rangle\, ds$.

It follows from (2.1), (2.2), (2.3), (2.7) and (2.10) that $EN_t = \Lambda$ for all $t \geq 0$, and therefore, by Proposition 2.1, we have $EN_\infty \leq \Lambda$, i.e., the equilibrium state may loose intensity. If $EN_\infty = \Lambda$, the branching particle system is said to be *persistent* for Λ, and N_∞ is called a *Poisson-type equilibrium measure with intensity Λ* for the system.

We can now state a persistence criterion for the branching particle system and two immediate corollaries.

THEOREM 2.1. *Let us assume that the solution u_f of equation (2.5) is unique. If the semigroup $(\mathcal{U}_t)_{t\geq 0}$ satisfies the condition*

(P) $\quad \int_0^\infty \int_Z \mathcal{K}(z)\mathcal{J}[(\mathcal{U}_t g)(\cdot)R(z,\cdot,(\mathcal{U}_t g)(\cdot))](z)\Lambda(dz)\,dt < \infty$

for all $g \in C_{bb}^+$ such that $\|g\|_\infty \leq 1$,

for a $\mathcal{U}_t$-invariant measure Λ, then the system is persistent for Λ.

COROLLARY 2.1. *If Λ is $\mathcal{T}_t$-invariant and $\mathcal{K}\Lambda$ is $\mathcal{J}$-invariant, then the conclusion of Theorem 2.1 holds with condition (P) replaced by*

(P') $\quad \int_0^\infty H(\|\mathcal{U}_t g\|_\infty)\,dt < \infty$ *for all $g \in C_{bb}^+$ such that $\|g\|_\infty \leq 1$,*

where

(2.12) $$H(\zeta) = \sup_{z,y} R(z,y,\zeta).$$

COROLLARY 2.2. *If Λ is $\mathcal{T}_t$-invariant and $\mathcal{K}\Lambda$ is $\mathcal{J}$-invariant, and the branching law has a finite variance which is bounded on $Z \times Z$, the conclusion of Theorem 2.1 holds if condition (P) is replaced by*

(P'') $\quad \int_0^\infty \langle \Lambda, (\mathcal{U}_t g)^2 \rangle\,dt < \infty$ *for all $g \in C_{bb}^+$ such that $\|g\|_\infty \leq 1$.*

Remark 2.1. If Λ is $\mathcal{T}_t$-invariant, and if $\mathcal{K}\Lambda$ is $\mathcal{J}$-invariant (i.e., Λ is invariant for the process of jumps), then Λ is also $\mathcal{U}_t$-invariant. Hence Corollaries 2.1 and 2.2 are consistent with Theorem 2.1.

Remark 2.2. Note that

$$\langle \Lambda, (\mathcal{U}_t g)^2 \rangle = \int_Z \int_Z g(z)g(w) r_t^\Lambda(dy,dw),$$

where

$$r_t^\Lambda(dy,dw) = \int_Z r_t(z,dy) r_t(z,dw) \Lambda(dz),$$

and $r_t(z,dy)$ is the transition probability of the semigroup $\mathcal{U}_t$. Hence, in the case that Z is a countable space (as in Application 3.1 below), condition (P'') is equivalent to the transience of the symmetrized kernel of r_t with respect to the $\mathcal{U}_t$-invariant measure Λ.

PROOF OF THEOREM 2.1. By (2.11) it suffices to prove that

$$\lim_{\varepsilon \to 0} \int_0^\infty \varepsilon^{-1}\langle \Lambda, h_{\varepsilon f}(t)\rangle\,dt = 0 \text{ for any } f \in C_{bb}^+ \text{ such that } \|f\|_\infty \leq 1.$$

By (2.9) we have

$$\lim_{\varepsilon \to 0} \int_0^\infty \varepsilon^{-1}\langle \Lambda, h_{\varepsilon f}(t)\rangle\,dt$$

$$\leq \limsup_{\varepsilon \to 0} \int_0^\infty \varepsilon^{-1} \int_Z \mathcal{K}(z)\mathcal{J}[(\mathcal{U}_t(\varepsilon f))(\cdot)R(z,\cdot,(\mathcal{U}_t(\varepsilon f))(\cdot))](z)\Lambda(dz)\,dt$$

$$= \limsup_{\varepsilon \to 0} \int_0^\infty \int_Z \mathcal{K}(z)\mathcal{J}[(\mathcal{U}_t f)(\cdot)R(z,\cdot,\varepsilon(\mathcal{U}_t f)(\cdot))](z)\Lambda(dz)\,dt = 0$$

due to (2.2), (P) and Fatou's lemma. □

PROOF OF COROLLARY 2.1. By (2.2) and (2.12) we have

$$\int_Z \mathcal{K}(z)\mathcal{J}[(\mathcal{U}_t g)(\cdot)R(z,\cdot,(\mathcal{U}_t g)(\cdot))](z)\Lambda(dz)$$

$$= \int_Z \int_Z \mathcal{K}(z)(\mathcal{U}_t g)(y)R(z,y,(\mathcal{U}_t g)(y))J(z,dy)\Lambda(dz)$$

$$\leq H(\|\mathcal{U}_t g\|_\infty)\int_Z \mathcal{K}(z)\mathcal{J}(\mathcal{U}_t g)(z)\Lambda(dz)$$

$$\leq H(\|\mathcal{U}_t g\|_\infty)\|\mathcal{K}\|_\infty \langle \Lambda, g\rangle \text{ (by Remark 2.1)},$$

and the conclusion follows by Theorem 2.1. □

PROOF OF COROLLARY 2.2. If the variance of the branching law is bounded on $Z \times Z$, then $R(z,y,\zeta) \leq$ const. ζ uniformly in z,y. Starting as in the proof of Corollary 2.1 we have

$$\int_Z \mathcal{K}(z)\mathcal{J}[(\mathcal{U}_t g)(\cdot)R(z,\cdot,(\mathcal{U}_t g)(\cdot))](z)\Lambda(dz) \leq \text{const.} \int_Z \mathcal{K}(z)\mathcal{J}((\mathcal{U}_t g)^2)(z)\Lambda(dz)$$

$$\leq \text{const.} \|\mathcal{K}\|_\infty \langle \Lambda, (\mathcal{U}_t g)^2\rangle,$$

and the conclusion follows by Theorem 2.1. □

The objective of the convergence towards equilibrium result is to show that, under some additional conditions, a Poisson-type equilibrium measure with intensity Λ is attained not only when the system starts from a Poisson field with intensity Λ, but also when the initial state N_0 belongs to a large class of random point measures which are transported into Λ by the semigroup $\mathcal{U}_t$ as $t \to \infty$. (The Poisson field with intensity Λ has this property).

THEOREM 2.2. *Let us assume the conditions of Corollary 2.1 and the following additional conditions:*
- (C1) $\lim_{t\to\infty} \|\mathcal{U}_t g\|_\infty = 0$ *for all* $g \in C_{bb}^+$.
- (C2) $\lim_{z\to\infty} \mathcal{U}_t g(z) = 0$ *for all* $g \in C_{bb}^+$ *and* $t > 0$ *(where* $\lim_{z\to\infty}$ *means* $\lim_{\rho(z,z_0)\to\infty}$ *for some fixed* $z_0 \in Z$*).*
- (C3) *There exists a Banach space* $(\widehat{C}, \|\ \|)$ *and a set* $\widetilde{C}$ *such that* $C_{bb} \subset \widetilde{C} \subset \widehat{C} \subset C_b$, $|fg| \leq |f|\|g\|$ *for all* $f \in C_{bb}$ *and* $g \in \widehat{C}$, *and* $t \mapsto \mathcal{U}_t g$ *is a continuous curve in* $\widehat{C}$ *for all* $g \in \widetilde{C}$.
- (C4) $J(z,dy)$ *is supported on a ball* $B_r(z)$ *(with center* z *and radius* r*) for all* z,

$$F_g(t) \equiv \left(z \mapsto \sup_{y\in B_r(z)} (\mathcal{U}_t g)(y)\right) \in \widetilde{C} \text{ for all } g \in C_{bb}^+ \text{ and } t > 0,$$

and for each $g \in C_{bb}^+$ *there exists* $\widetilde{g} \in C_{bb}^+$ *such that* $F_g(t) \leq \mathcal{U}_t \widetilde{g}$ *for all* $t > 0$.
- (C5) $\langle N_0, \mathcal{U}_t g\rangle \to \langle \Lambda, g\rangle < \infty$ *in probability as* $t \to \infty$ *for all* $g \in \widetilde{C}$.

Then N_t *converges in distribution as* $t \to \infty$ *towards the Poisson-type equilibrium measure with intensity* Λ.

PROOF. The Laplace functional of N_t is given by

$$(2.13) \qquad E e^{-\langle N_t, f\rangle} = E \exp\{\langle N_0, \log(1 - u_f(t))\rangle\}, \quad f \in C_{bb}^+, \ t \geq 0,$$

where $u_f(t)$ is defined by (2.3). We must prove that the right-hand side of (2.13) converges to that of (2.11) for each $f \in C_{bb}^+$ as $t \to \infty$.

To simplify notation we put $u \equiv u_f$, $h \equiv h_f$ and $g \equiv 1 - e^{-f}$ (note that $g \in C_{bb}^+$.).

First we show that

$$(2.14) \quad |E\exp\{\langle N_0, \log(1-u(t))\rangle\} - E\exp\{-\langle N_0, u(t)\rangle\}| \to 0 \text{ as } t \to \infty,$$

(similarly as in [Gorostiza and Wakolbinger 1992]). By dominated convergence it suffices to show that

$$\exp\{\langle N_0, \log(1-u(t))\rangle\} - \exp\{-\langle N_0, u(t)\rangle\} \to 0 \text{ in probability as } t \to \infty,$$

and for this it suffices to prove that

$$(2.15) \quad (0 \leq) \langle N_0, -[\log(1-u(t)) + u(t)]\rangle \to 0 \text{ in probability as } t \to \infty.$$

We have

$$\langle N_0, -[\log(1-u(t)) + u(t)]\rangle \leq \langle N_0, u(t)^2/(1-u(t))\rangle,$$

and by (2.8), $1 - u(t) \geq \mathcal{U}_t e^{-f} \geq e^{-\|f\|_\infty}$. Hence

$$\langle N_0, -[\log(1-u(t)) + u(t)]\rangle \leq e^{\|f\|_\infty} \langle N_0, u(t)^2\rangle$$
$$\leq e^{\|f\|_\infty} \|u(t)\|_\infty \langle N_0, u(t)\rangle$$
$$\leq e^{\|f\|_\infty} \|\mathcal{U}_t f\|_\infty \langle N_0, \mathcal{U}_t f\rangle,$$

and we obtain (2.15) by (C1) and (C5).

Now, by (2.4),

$$E\exp\{-\langle N_0, u(t)\rangle\} = E\exp\left\{-\langle N_0, \mathcal{U}_t g\rangle + \int_0^t \langle N_0, \mathcal{U}_{t-s} h(s)\rangle \, ds\right\}.$$

Hence, by (2.10), (2.11) and (C5) it suffices to prove that

$$(2.16) \quad \int_0^t \langle N_0, \mathcal{U}_{t-s} h(s)\rangle \, ds \to \int_0^\infty \langle \Lambda, h(s)\rangle \, ds \text{ in probability as } t \to \infty.$$

Fix $z_0 \in Z$, let $B_n = \{z \in Z \mid \rho(z, z_0) \leq n\}$ for $n = 1, 2, \ldots$, and for each n let b_n be a continuous $[0,1]$-valued function on Z, equal to 1 on B_n and to 0 on B_{n+1}^c. For each n and $t > T > 0$ we have

$$\left|\int_0^t \langle N_0, \mathcal{U}_{t-s} h(s)\rangle \, ds - \int_0^\infty \langle \Lambda, h(s)\rangle \, ds\right| \leq I_1(t, T, n)$$
$$+ I_2(t, T, n) + I_3(t, T) + I_4(T, n) + I_5(T),$$

where

$$I_1(t,T,n) = \left| \int_0^T \langle N_0, \mathcal{U}_{t-s} b_n h(s) \rangle \, ds - \int_0^T \langle \Lambda, b_n h(s) \rangle \, ds \right|,$$

$$I_2(t,T,n) = \int_0^T \langle N_0, \mathcal{U}_{t-s}(1-b_n)h(s) \rangle \, ds,$$

$$I_3(t,T) = \int_T^t \langle N_0, \mathcal{U}_{t-s} h(s) \rangle \, ds,$$

$$I_4(T,n) = \int_0^T \langle \Lambda, (1-b_n)h(s) \rangle \, ds,$$

$$I_5(T) = \int_T^\infty \langle \Lambda, h(s) \rangle \, ds.$$

We will show that all I_k, $k = 1, 2, \ldots, 5$, tend to 0 suitably as t, T and $n \to \infty$.

I_1: The fact that $\zeta \mapsto [\mathcal{F}(z, y, 1-\zeta) - (1-\zeta)]$ is Lipschitz-continuous uniformly in (z, y) implies that $t \mapsto u(t)$ is continuous in $\widehat{C}$ [Pazy 1983, Chapter 6, Theorem 1.2]. Then it can be verified that $t \mapsto h(t)$ is continuous in $\widehat{C}$. Hence, given $\delta > 0$ there exists a partition $0 = s_0 < s_1 < \cdots < s_m = T$ such that

$$\|h(s) - h(s_j)\| < \delta \text{ for } s \in [s_j, s_{j+1}], \quad j = 0, \ldots, m-1.$$

Then, by (C3) and (C5),

$$\limsup_{t\to\infty} \left| \int_0^T \langle N_0, \mathcal{U}_{t-s} b_n h(s) \rangle \, ds - \sum_{j=0}^{m-1} \int_{s_j}^{s_{j+1}} \langle N_0, \mathcal{U}_{t-s} b_n h(s_j) \rangle \, ds \right|$$
$$\leq \delta \limsup_{t\to\infty} \left\langle N_0, \int_0^T \mathcal{U}_{t-s} b_n \, ds \right\rangle = \delta T \langle \Lambda, b_n \rangle$$

in probability.

A similar argument yields

$$\lim_{t\to\infty} \sum_{j=0}^{m-1} \int_{s_j}^{s_{j+1}} \langle N_0, \mathcal{U}_{t-s} b_n h(s_j) \rangle \, ds = \sum_{j=0}^{m-1} (s_{j+1} - s_j) \langle \Lambda, b_n h(s_j) \rangle$$

in probability.

On the other hand,

$$\left| \sum_{j=0}^{m-1} (s_{j+1} - s_j) \langle \Lambda, b_n h(s_j) \rangle - \int_0^T \langle \Lambda, b_n h(s) \rangle \, ds \right| \leq \delta T \langle \Lambda, b_n \rangle.$$

Since δ is arbitrary, it follows that

$$\lim_{t\to\infty} I_1(t, T, n) = 0 \text{ in probability}$$

for each T and n.

I_2: By (2.2), (2.9) and (2.12) we have

$$(1 - b_n(z))h(z, s) \leq (1 - b_n(z))\mathcal{K}(z)\mathcal{J}[(\mathcal{U}_s f)(\cdot)H(\mathcal{U}_s f)(\cdot)](z)$$
$$\leq \|\mathcal{K}\|_\infty F_f(z, s) H(C_f^n(s)),$$

where
$$F_f(z,s) = \sup_{y \in B_r(z)} (\mathcal{U}_s f)(y) \text{ and } C_f^n(s) = \sup_{z \in B_n^c} \sup_{y \in B_r(z)} \mathcal{U}_s f(y).$$

Hence, by (C4), (C5) and $\mathcal{U}_t$-invariance of Λ,

$$\limsup_{t \to \infty} \int_0^T \langle N_0, \mathcal{U}_{t-s}(1-b_n)h(s) \rangle \, ds$$

$$\leq \|\mathcal{K}\|_\infty \limsup_{t \to \infty} \left\langle N_0, \int_0^T \mathcal{U}_{t-s} F_f(s) H\big(C_f^n(s)\big) \, ds \right\rangle$$

$$= \|\mathcal{K}\|_\infty \int_0^T \langle \Lambda, F_f(s) \rangle H\big(C_f^n(s)\big) \, ds$$

$$\leq \|\mathcal{K}\|_\infty \langle \Lambda, \widetilde{f} \rangle \int_0^T H\big(C_f^n(s)\big) \, ds.$$

By (2.2), (2.12) and (C2), $\lim_{n \to \infty} H\big(C_f^n(s)\big) = 0$ for each s, and therefore, by (P') and dominated convergence,

$$\lim_{n \to \infty} \limsup_{t \to \infty} I_2(t,T,n) = 0 \text{ in probability for each } T.$$

I_3: Proceeding as in the previous step we have

$$\int_T^t \langle N_0, \mathcal{U}_{t-s} h(s) \rangle \, ds \leq \|\mathcal{K}\|_\infty \int_T^t \langle N_0, \mathcal{U}_{t-s} F_f(s) \rangle H(\|\mathcal{U}_s f\|_\infty) \, ds.$$

Then, by (C4), (C5) and (P'),

$$\lim_{T \to \infty} \limsup_{t \to \infty} I_3(t,T) = 0 \text{ in probability}.$$

I_4, I_5: Since $\int_0^\infty \langle \Lambda, h(s) \rangle \, ds < \infty$, it is clear that $\lim_{n \to \infty} I_4(T,n) = 0$ for each T, and $\lim_{T \to \infty} I_5(T) = 0$.

The previous limits imply the desired conclusion. $\square$

3. Applications of the persistence criteria

3.1. A branching random walk. Cox [Cox 1992] considers a branching random walk on a countable set described as follows. The particles move according to a Markov chain with unit jump rate and jump distribution obeying a strong transience assumption, and branch at constant rate according to a state-independent critical branching law with finite variance. Offspring particles appear at the death site of the parent. His Theorem 0 follows from our Proposition 2.1 and Corollary 2.2, and his Theorem 1(a) follows from our Theorem 2.2. Our methods of proof of Theorems 2.1 and 2.2 were inspired to a considerable extent by [Cox 1992].

3.2. A branching symmetric stable process. Consider a branching particle system on $\mathbb{R}^d$ where the particles move according to a spherically symmetric stable process with index $\alpha \in (0,2]$, the branching rate and the branching law are constant on $\mathbb{R}^d$, and the generating function of the branching law is of the form

$$\mathcal{F}(\zeta) = \zeta + c(1-\zeta)^{1+\beta}, \quad \zeta \in [0,1],$$

where $\beta \in (0,1]$ and $c \in (0, 1/(1+\beta)]$ (this law is critical and belongs to the domain of normal attraction of a stable law with exponent $1+\beta$). The offspring are born at the death site of the parent.

Since the function $H(\zeta)$ defined in (2.12) is $H(\zeta) = c\zeta^\beta$ in this case, condition (P') takes the simple form

$$\int_0^\infty \|\mathcal{U}_t g\|_\infty^\beta \, dt < \infty \text{ for all } g \in C_{bb}^+ \text{ such that } \|g\|_\infty \leq 1.$$

Due to the scaling property of the symmetric α-stable process, $\|\mathcal{U}_t g\|_\infty$ is asymptotically equal to $t^{-d/\alpha}$ for all such g (see [Gorostiza and Wakolbinger 1991]), and therefore, if $d > d_c = \alpha/\beta$, the system is persistent for Lebesgue measure λ on $\mathbb{R}^d$ (which is invariant for the symmetric α-stable semigroup). This result is a special case of Theorem 3.2(a) below, and it was proved in [Gorostiza and Wakolbinger 1991] using a "backward tree" method (the same result holds for the corresponding DW process [Gorostiza, Roelly-Coppoletta, and Wakolbinger 1990]).

In [Gorostiza and Wakolbinger 1991] we also proved the converse, namely, if $d \leq d_c$ the system goes to local extinction as $t \to \infty$. (In Section 4 we will comment on related results and give an extinction criterion for a more general class of models). In particular, for binary branching Brownian motion ($\alpha = 2$, $\beta = 1$) the system goes to local extinction for $d = 2$. We will show from condition (P) that for a suitably integrable state-dependent branching rate $\mathcal{K}$ it is possible to have persistence of finite variance branching Brownian motion for λ when $d = 2$.

THEOREM 3.1. *Consider a branching Brownian motion on $\mathbb{R}^2$ with branching rate $\mathcal{K}(z)$, and state-independent branching law with finite variance. Suppose that $\mathcal{K}$ satisfies*

$$\int_{\mathbb{R}^2 \setminus B_1} \mathcal{K}(x) \|x\|^{-2} \lambda(dx) < \infty,$$

where B_1 is the unit ball in $\mathbb{R}^2$ centered at the origin.

Then the system is persistent for λ.

PROOF. We will verify condition (P) of Theorem 2.1. In the present case we have $R(z,\zeta) \leq \text{const.}\ \zeta$. Hence it suffices to show that for any $r > 1$,

$$I \equiv \int_0^\infty \int_{\mathbb{R}^2} \mathcal{K}(x) \big(\mathcal{U}_t 1_{B_r}(x)\big)^2 \lambda(dx) \, dt < \infty,$$

where B_r is the ball with radius r centered at the origin. We have $I = I_1 + I_2$, where

$$I_1 = \frac{1}{2\pi^2} \int_{B_{2r}} \mathcal{K}(x) \int_{B_r} \int_{B_r} \frac{dy\,dz}{\|y-x\|^2 + \|z-x\|^2} \, dx,$$

$$I_2 = \frac{1}{2\pi^2} \int_{\mathbb{R}^2 \setminus B_{2r}} \mathcal{K}(x) \int_{B_r} \int_{B_r} \frac{dy\,dz}{\|y-x\|^2 + \|z-x\|^2} \, dx.$$

It is easy to show that $I_1 < \infty$ (recall that $\mathcal{K}$ is bounded). For I_2, since $\|y-x\| > \|x\| - r$ and $\|z-x\| > \|x\| - r$, then

$$I_2 \leq \text{const.} \int_{\mathbb{R}^2 \setminus B_{2r}} \mathcal{K}(x)\|x\|^{-2}\, dx < \infty. \quad \square$$

Note that this result includes the case where branching occurs only in a bounded set.

We thank Ted Cox for discussion on this result.

3.3. A multitype branching particle system. In [Gorostiza and Wakolbinger 1993] we study the following branching particle system in $\mathbb{R}^d$ consisting of particles of types $i = 1, \ldots, k$. Particles of type i move according to a spherically symmetric stable process with index $\alpha_i \in (0, 2]$, die at rate V_i, and at death time they produce at their own site particles of a type j chosen with probability m_{ij}, the number of which is governed by the generating function

$$\mathcal{F}_{ij}(\zeta) = \begin{cases} \zeta + \frac{c_{ij}}{m_{ij}}(1-\zeta)^{1+\beta_{ij}} & \text{if } m_{ij} > 0, \\ \zeta & \text{if } m_{ij} = 0, \end{cases}$$

where $\beta_{ij} \in (0, 1]$ and $c_{ij} \in [0, m_{ij}/(1+\beta_{ij})]$. If $m_{ij} > 0$, the case $c_{ij} > 0$ corresponds to a critical branching law, and the case $c_{ij} = 0$ represents the choice of type j without branching. Note that the multitype branching law so defined is critical. The matrix (m_{ij}) is assumed to be irreducible.

This is a special case of the model introduced in Section 2, with

- $Z = \mathbb{R}^d \times \{1, \ldots, k\}$.
- $\mathcal{T}_t f(x, i) = \mathcal{T}_t^i f_i(x)$, $x \in \mathbb{R}$, $i = 1, \ldots, k$, $f_i(x) \equiv f(x, i)$, where $(\mathcal{T}_t^i)_{t \geq 0}$ is the semigroup of the spherically symmetric α_i-stable process (whose infinitesimal generator is $\Delta_{\alpha_i} \equiv -(-\Delta)^{\alpha_i/2}$).
-
$$J\big((x, i), (y, j)\big) = \begin{cases} m_{ij} & \text{if } x = y, \, x, y \in \mathbb{R}^d, \\ 0 & \text{if } x \neq y, \, i, j = 1, \ldots, k, \end{cases}$$

  $m_{ij} \geq 0$, $\sum_{j=1}^k m_{ij} = 1$, $i, j = 1, \ldots, k$.
- $\mathcal{K}(x, i) = V_i > 0$, $x \in \mathbb{R}^d$, $i = 1, \ldots, k$.
- $\mathcal{F}\big((x, i), (y, j), \zeta\big) = \mathcal{F}_{ij}(\zeta)$, $x, y \in \mathbb{R}^d$, $i, j = 1, \ldots, k$.

The generator $\mathcal{A}$ of the semigroup $(\mathcal{U}_t)_{t \geq 0}$ of the basic process, given by (2.6), has the form

$$(3.1) \quad (\mathcal{A}f)_i(x) = \Delta_{\alpha_i} f_i(x) + V_i \sum_{j=1}^k (m_{ij} - \delta_{ij}) f_j(x), \quad i = 1, \ldots, k, \quad f = (f_1, \ldots, f_k).$$

Let $\Gamma = (\gamma_1, \ldots, \gamma_k)$, $\gamma_1, \ldots, \gamma_k > 0$, $\sum_{i=1}^k \gamma_i = 1$, be the normalized solution of the system

$$\sum_{j \neq i} \gamma_j V_j m_{ji} = \gamma_i V_i (1 - m_{ii}), \quad i = 1, \ldots, k.$$

Then $\Gamma = (\gamma_1, \ldots, \gamma_k)$ is an invariant measure for the "Markov chain of types" (i.e., the type component of the basic process), and the measure

$$(3.2) \qquad \Lambda_I = \lambda \otimes \Gamma = (\lambda \gamma_1, \ldots, \lambda \gamma_k),$$

on $\mathbb{R}^d \times \{1, \ldots, k\}$ is invariant for $\mathcal{U}_t$.

Let

$$(3.3) \qquad d_c = \min \alpha_i / \min \beta_{ij},$$

where $\min \beta_{ij} \equiv \min\{\beta_{ij} \mid c_{ij} > 0\}$.

Here we denote $N(t) = \big(N_1(t), \ldots, N_k(t)\big)$, where $N_i(t)$ stands for the configuration of type i particles at time t, $i = 1, \ldots, k$, and $\langle \mu, f \rangle = \sum_{i=1}^{k} \langle \mu_i, f_i \rangle$, where $\mu = (\mu_1, \ldots, \mu_k)$ is a vector measure and $f = (f_1, \ldots, f_k)$ is a vector function.

The long time behavior of the system is given by the following result.

THEOREM 3.2. *Assume $d > d_c$. Then*
(a) *The system is persistent for Λ_I.*
(b) *If the initial configuration $N(0)$ satisfies*

$$\langle N(0), \mathcal{U}_t f \rangle \to \langle \Lambda_I, f \rangle \text{ in probability as } t \to \infty$$

for all $f \in \big(Q_p(\mathbb{R}^d)^+\big)^k$, then $N(t)$ converges in distribution as $t \to \infty$ to the Poisson-type equilibrium measure with intensity Λ_I.

PROOF. (a) We will verify condition (P') of Corollary 2.1. From Proposition 4.2 of [Gorostiza, Roelly, and Wakolbinger 1992] we have

(3.4) $\quad \|\mathcal{U}_t g\|_\infty \leq \text{const. } t^{-d/\min \alpha_i}$ for $t \geq 1$, $\quad g \in \big(C_c(\mathbb{R}^d)^+\big)^k$, $\|g\|_\infty \leq 1$,

(this follows from the scaling properties of the α_i-stable densities). Then

$$H(\|\mathcal{U}_t g\|_\infty) \leq \max_{ij} \|\mathcal{U}_t g\|_\infty^{\beta_{ij}}$$
$$\leq \text{const. } t^{-d \min \beta_{ij} / \min \alpha_i} \text{ for } t \geq 1.$$

Therefore condition (P') holds for $d > d_c$.

(b) We will verify the conditions (C1) to (C4) of Theorem 2.2. The main assumption corresponds to condition (C5).

We have $C_{bb} = \big(C_c(\mathbb{R}^d)\big)^k$ and $C_b = \big(C_b(\mathbb{R}^d)\big)^k$, and we choose the spaces $\widetilde{C}$ and $\widehat{C}$ as $\widetilde{C} = \big(Q_p(\mathbb{R}^d)\big)^k$ and $\widehat{C} = \big(C_p(\mathbb{R}^d)\big)^k$, $\|\cdot\| = \|\cdot\|_p$ (Appendix 1).

(C1) follows from (3.4).

(C2) is a consequence of $\big(C_c(\mathbb{R}^d)\big)^k \subset \big(C_p(\mathbb{R}^d)\big)^k$.

(C3) It is easy to see that $|fg| \leq |f| \|g\|_p$ for $f \in \big(C_c(\mathbb{R}^d)\big)^k$ and $g \in \big(C_p(\mathbb{R}^d)\big)^k$. Continuity of $t \mapsto \mathcal{U}_t g$ in $\big(C_p(\mathbb{R}^d)\big)^k$ for $g \in \big(Q_p(\mathbb{R}^d)\big)^k$ is a consequence of a multitype extension of Lemma 3.3 of [Dawson, Fleischmann and Gorostiza 1989] (proved using Proposition 2.3 of [Iscoe 1986]).

(C4) J satisfies the assumption, membership of $F_g(t)$ in $\big(Q_p(\mathbb{R}^d)\big)^k$ for each $g \in \big(C_c(\mathbb{R}^d)^+\big)^k$ also follows from the multitype extension of Lemma 3.3 of [Dawson, Fleischmann and Gorostiza 1989], and for $\widetilde{g}$ we may take an element $\widetilde{g} \in \big(C_c(\mathbb{R}^d)\big)^k$ such that $\widetilde{g}(x, i) \geq \sup_{\|y-x\| \leq r} \max_{j=1,\ldots,k} g(y, j)$. $\square$

Remark 3.1. This multitype branching particle system can be regarded as a kind of general branching particle system (as in [Gorostiza, Roelly, and Wakolbinger 1992]).

Remark 3.2. A large class of initial configurations $N(0)$ for which the condition in assumption (b) holds is identified in [Stöckl and Wakolbinger 1991] (see also Theorem 4 of [Gorostiza and Wakolbinger 1992]). This result is a multitype version of the result of Dobrushin-Stone [Stone 1968].

4. Extinction of critical branching particle systems and the "backward tree" method

The "backward tree" method was developed by Kallenberg [Kallenberg 1977] to establish criteria for persistence of discrete time critical spatial branching particle systems. The method allows to compute the Palm distributions of the population in the n-th generation of the system. Liemant [Liemant 1981] extended the method to spatially inhomogeneous branching mechanisms. Dawson and Fleischmann [Dawson and Fleischmann 1985] employed this technique to obtain persistence and extinction criteria for a class of branching random walks on a lattice, in discrete time and in a random environment.

The backward tree method in discrete time relies on the generation structure of the branching system and it is not directly applicable to continuous time branching systems, where several generations may coexist at any given time. In [Gorostiza and Wakolbinger 1991] we developed the backward tree method for a class of continuous time branching systems in two different ways: the first one by means of an approximation which takes advantage of the discrete time version, and the second one, directly in continuous time, by an analytic approach using the Feynman-Kac formula. We employed the method to obtain persistence and extinction criteria for a critical branching particle system. In [Gorostiza, Roelly-Coppoletta, and Wakolbinger 1990] we used the previous result to derive persistence and extinction criteria for the corresponding DW process by means of the doubly stochastic Poisson relationship between the particle process and the DW process at each time t, including $t = \infty$. Previous results of this type have been obtained by Dawson [Dawson 1977] assuming finite variance branching, and by Dawson, Fleischmann, Foley and Peletier [Dawson et al. 1988] for branching Brownian motion and a class of branching laws having possibly infinite variance, using an analytic approach. For the model we have considered in [Gorostiza, Roelly-Coppoletta, and Wakolbinger 1990], i.e., spherically symmetric stable motion and "$1 + \beta$ branching", Dawson and Perkins [Dawson and Perkins 1991] have also derived the sufficiency part of the persistence criterion from a "continuous backward tree" representation of the Palm distribution of the canonical measure of the DW process (see Proposition 6.1 of [Dawson and Perkins 1991] and Theorem 3.3 of [Gorostiza and Wakolbinger 1991]), and they have extended their result for the corresponding "historical process". The structure of equilibrium measures for DW processes has also been discussed by Dynkin [Dynkin 1989, Dynkin 1993], whereas equilibrium measures of discrete time branching systems are the subject of the monograph of Liemant, Matthes and Wakolbinger [Liemant, Matthes, and Wakolbinger 1988]. Related results have been proved, among others, by Durrett [Durrett 1979], and Bramson, Cox and Greven [Bramson, Cox, and Greven 1992] for particle systems, by Etheridge [Etheridge 1993a, Etheridge 1993b] for DW processes, and by Wu [Wu 1993] for two-level DW processes, using analytical methods. Palm-type distributions on genealogical trees for a class of branching particle systems are computed in [Chauvin, Rouault, and Wakolbinger 1991].

In [Gorostiza, Roelly, and Wakolbinger 1992] we extended the continuous time backward tree method for a class of critical multitype branching particle systems of a special kind: particles can branch into particles of their own type, or they can mutate into a different type without branching. A further extension is given in [Gorostiza and Wakolbinger 1993] for the critical multitype branching particle system described in Section 3.3, which allows branching into particles of a different type. This branching mechanism requires nontrivial modifications of the previous backward tree method. We will describe roughly the construction of the backward tree for the branching model of Section 3.3, and refer the interested reader to [Gorostiza and Wakolbinger 1991, Gorostiza and Wakolbinger 1993, Gorostiza, Roelly, and Wakolbinger 1992] for details.

The objective is to give an explicit representation of the "random cluster of relatives" of a particle belonging to the equilibrium state $N(\infty)$. First we build the trunk of the tree starting at time 0 from $(x,i) \in \mathbb{R}^d \times \{1,\ldots,k\}$. The trunk follows a random Markovian path whose infinitesimal generator is the adjoint of the generator $\mathcal{A}$ of the basic process, given by (3.1), with respect to the invariant measure Λ_I defined by (3.2). The trunk represents the line of ancestors of the particle at (x,i) going backwards in time. Next we mark on the trunk the death times of all the ancestors; let us denote them by $s_1 < s_2 < \ldots$. For each death time s_n we let Z_{s_n} independent copies of the (forward) multitype branching dynamics issue from the trunk and run during a time interval of length s_n. The random number Z_{s_n} is distributed according to the Palm distribution of the branching law corresponding to the death at s_n, which depends on the types of the parent and the offspring (in forward time). Let $\Phi_{s_n,m}$, $m = 1,\ldots, Z_{s_n}$, denote the particle configurations so obtained. Then the random cluster of relatives of the particle at (x,i) has the "backward tree representation"

$$R_{(x,i)} = \sum_{n=1}^{\infty} \sum_{m=1}^{Z_{s_n}} \Phi_{s_n,m}.$$

The persistence of the multitype branching particle system is related to $R_{(x,i)}$ by the fact that it is equivalent to $R_{(x,i)}$ being locally finite a. s. for Λ_I-a. a. (x,i) (i.e., with probability one $R_{(x,i)}$ has only finitely many points in any bounded subset of $\mathbb{R}^d$) (see Proposition 4.1 of [Gorostiza and Wakolbinger 1991] and Section 2.4 of [Liemant, Matthes, and Wakolbinger 1988]). One of the main technical difficulties in verifying the local finiteness of $R_{(x,i)}$ is accounting for all the possible forms of branching (with and without change of type) that can occur. This is done in detail in [Gorostiza and Wakolbinger 1993], where we prove the following result, with d_c defined by (3.3).

THEOREM 4.1. *If the system is persistent, then $d > d_c$.*

A technical point in the proof is a randomization of all the particle positions and types in $R_{(x,i)}$ "at time 0", which allows to make still another use of Palm distributions. This randomization technique has its origin in [Dawson and Fleischmann 1985].

For the present critical multitype branching model it can be shown that all possible limiting intensities are of the form $a\Lambda_I$, where Λ_I is the invariant measure (3.2), and it can be concluded from this that a is either 0 or 1 (as in [Gorostiza and Wakolbinger 1991]). Hence we have the dichotomy:

Either $EN(\infty) = \Lambda_I$ or $N(\infty) = 0$ a.s.,

and therefore the following corollary of Theorem 4.1:

COROLLARY 4.1. *If $d \leq d_c$, the system goes to local extinction as $t \to \infty$.*

5. Asymptotics of a system of nonlinear partial differential equations

In [Gorostiza and Wakolbinger 1993] we prove the following result.

Let $u(f) = \{u_i(f; x, t), x \in \mathbb{R}^d, t \geq 0, i = 1, \ldots, k\}$ be the (mild) solution of the system of nonlinear partial differential equations

$$\text{(5.1)} \quad \frac{\partial}{\partial t} u_i = \Delta_{\alpha_i} u_i + V_i \sum_{j=1}^{k} (m_{ij} - \delta_{ij}) u_j - V_i \sum_{j=1}^{k} c_{ij} u_j^{1+\beta_{ij}},$$

$$u_i(x, 0) = f_i(x), \quad i = 1, \ldots, k,$$

where $f = (f_1, \ldots, f_k) \in \left(C_c(\mathbb{R}^d)^+\right)^k$, with $\alpha_i \in (0, 2]$, $V_i > 0$, $\beta_{ij} \in (0, 1]$, $c_{ij} \in [0, m_{ij}/(1 + \beta_{ij})]$ and (m_{ij}) is an irreducible nonnegative matrix with maximal eigenvalue 1. (This system has a unique global solution which is nonnegative for all t). Let $d_c = \min \alpha_i / \min \beta_{ij}$ (with $\min \beta_{ij} \equiv \min\{\beta_{ij} \mid c_{ij} > 0\}$).

THEOREM 5.1. *If at least one f_i is not 0, then for each $i = 1, \ldots, k$,*

$$\lim_{t \to \infty} \|u_i(f; t)\|_{L^1} \begin{cases} > 0 & \text{if } d > d_c, \\ = 0 & \text{if } d \leq d_c. \end{cases}$$

We will show that this result is a consequence of the long time behavior of the multitype branching particle process $N(t) = (N_1(t), \ldots, N_k(t))$ considered in Section 3.3. Assuming $N(0)$ is Poisson with intensity measure $\Lambda = (\lambda_1, \ldots, \lambda_k)$, the Laplace functional of $N(t)$ is given by

$$\text{(5.2)} \quad E \exp\{-\langle N(t), f \rangle\} = \exp\{-\langle \Lambda, v(f; t) \rangle\}, \quad f \in (K_p(\mathbb{R}^d))^k,$$

where $v(f; x, t)$ is the (mild) solution of the system of nonlinear partial differential equations

$$\text{(5.3)} \quad \frac{\partial}{\partial t} v_i = \Delta_{\alpha_i} v_i + V_i \sum_{j=1}^{k} (m_{ij} - \delta_{ij}) v_j - V_i \sum_{j=1}^{k} c_{ij} v_j^{1+\beta_{ij}},$$

$$v_i(x, 0) = 1 - e^{-f_i(x)}, \quad i = 1, \ldots, k.$$

(This is equation (2.5) for the present model).

The system (5.3) is a special case of the system (5.1) because the initial condition in (5.3) belongs to $(C_c(\mathbb{R}^d)^+)^k$, but in addition it is restricted to take values in $[0, 1]$, and in (5.3) (m_{ij}) is an irreducible probability matrix, which therefore satisfies the assumption of Theorem 5.1. However, it is shown in [Gorostiza and Wakolbinger 1993] that the asymptotic behaviors of the solutions of the systems (5.1) and (5.3) are the same, and therefore it suffices to prove the theorem for the system (5.3).

The proof given in [Gorostiza and Wakolbinger 1993] for the case $d > d_c$ is a simplified version of the proof of Theorem 3.2(b) which does not use probability. We will show that it also follows from Theorem 3.2(b).

Let $\Lambda_i = (0, \ldots, \lambda, \ldots, 0)$, where λ is the i-th component, $i = 1, \ldots, k$. It was shown in the proof of Theorem 3 of [Gorostiza and Wakolbinger 1992] that $\mathcal{U}_t^* \Lambda_i \to \Lambda_I$ in $\left(\mathcal{M}_p(\mathbb{R}^d)\right)^k$ as $t \to \infty$. (This follows from the irreducibility of

(m_{ij}) by elementary Markov chain theory). Then, if $N(0)$ is Poisson with intensity measure Λ_i we have

$$E\exp\{-\langle N(0), \mathcal{U}_t f\rangle\} = \exp\{-\langle\Lambda_i, 1 - e^{-\mathcal{U}_t f}\rangle\}, \quad f \in (K_p(\mathbb{R}^d))^k,$$

but

$$\lim_{t\to\infty} \exp\{-\langle\Lambda_i, 1 - e^{-\mathcal{U}_t f}\rangle\} = \lim_{t\to\infty} \exp\{-\langle\Lambda_i, \mathcal{U}_t f\rangle\}$$

(proved as A9 of [Gorostiza and Wakolbinger 1992]), hence

$$\lim_{t\to\infty} E\exp\{-\langle N(0), \mathcal{U}_t f\rangle\} = \exp\{-\langle\Lambda_I, f\rangle\}.$$

Thus $N(0)$ satisfies the assumption of Theorem 3.2(b). Therefore, by (5.2),

$$\begin{aligned}
E\exp\{-\langle N(\infty), f\rangle\} &= \lim_{t\to\infty} E\exp\{-\langle N(t), f\rangle\} \\
&= \lim_{t\to\infty} \exp\{-\langle\Lambda_i, v(f;t)\rangle\} \\
&= \lim_{t\to\infty} \exp\{-\|v_i(f;t)\|_{L^1}\}.
\end{aligned}$$

Hence

(5.4) $$\lim_{t\to\infty} \|v_i(f;t)\|_{L^1} = -\log E\exp\{-\langle N(\infty), f\rangle\}.$$

But by Theorem 3.2(a), $E\langle N(\infty), f\rangle = \langle\Lambda_I, f\rangle > 0$ if $f \neq 0$. So the right-hand side of (5.4) is positive if $f \neq 0$.

For $d \leq d_c$ we know from Corollary 4.1 that the system goes to extinction. Therefore if $N(0)$ is Poisson with intensity Λ_I, by (5.2) we have

$$\lim_{t\to\infty}\langle\Lambda, v(f;t)\rangle = -\log E\exp\{-\langle N(\infty), f\rangle\} = 0, \quad f \in \left(C_c(\mathbb{R}^d)^+\right)^k.$$

But $\|v_i(f;t)\|_{L^1} \leq \text{const } \langle\Lambda, v(f;t)\rangle$.

Remark 5.1. For $k = 1$, $\alpha_1 = 2$ and $c_{11} > 0$, equation (5.1) is a nonlinear heat equation,

$$\frac{\partial}{\partial t} u = \Delta u - \gamma u^{1+\beta}, \quad (\gamma > 0),$$

$$u(x, 0) = f(x).$$

In this case Theorem 5.1 has been proved by analytical methods [Escobedo and Kavian 1987, Gmira and Veron 1984].

Remark 5.2. A rescaling limit of the multitype branching particle system above yields the multitype DW process. For each $n = 1, 2, \ldots$, let $V_i^n = V_i n^{\beta_{ii}}$, $m_{ij}^n = \delta_{ij} + (m_{ij} - \delta_{ij})n^{-\beta_{ii}}$, $c_{ij}^n \in [0, m_{ij}^n/(1 + \beta_{ij})]$, and give to each particle a mass n^{-1}. Let X^n denote the corresponding mass process. Then the system of nonlinear partial differential equations corresponding to (5.3) which characterizes X^n is given by

(5.5) $$\frac{\partial}{\partial t} v_i^n = \Delta_{\alpha_i} v_i^n + V_i \sum_{j=1}^k (m_{ij}^n - \delta_{ij})v_j^n - V_i \sum_{j=1}^k c_{ij}^n n^{\beta_{ij}}(v_j^n)^{1+\beta_{ij}},$$

$$v_i^n(x, 0) = n(1 - e^{-n^{-1}f_i(x)}), \quad i = 1, \ldots, k.$$

Assuming $c_{ij}^n \to c_{ij}$ as $n \to \infty$, we observe that the solution of the system (5.5) converges as $n \to \infty$ to the solution of the system (5.1), where necessarily $c_{ij} = 0$ for

$i \neq j$. The multitype DW process is a (vector) measure valued Markov process X which is the limit in distribution as $n \to \infty$ of the process X^n, and it is characterized by

$$E[e^{-\langle X(t), f \rangle} \mid X(0) = \mu] = e^{-\langle \mu, u(f;t) \rangle},$$

where $u(f; x, t)$ is the solution of the system (5.1) with $c_{ij} = 0$ for $i \neq j$. Hence, in this case the asymptotic result for the system (5.1) can be derived from the convergence to equilibrium result for the process X. This was done in [Gorostiza and Wakolbinger 1992]. So the use of the DW process for this purpose is restricted to the special case $c_{ij} = 0$ for $i \neq j$. This is a consequence of the fact that the rescaling which yields the multitype DW process forces each particle to produce in the limit only particles of its own type. From another viewpoint, this restriction on the branching mechanism of the multitype DW process is a consequence of the "multiplicative" (or "branching") property (see [Rhyzhov and Skorokhod 1970, Watanabe 1969], where the general form of continuous state multitype branching processes is discussed). Since a DW process has the multiplicative property, there does not exist a multitype DW process whose cumulant semigroup satisfies the system (5.1), unless $c_{ij} = 0$ for $i \neq j$. (Multitype DW processes can of course be viewed as monotype DW processes with state-dependent characteristics [Gorostiza, Roelly, and Wakolbinger 1992, Li 1992].)

Remark 5.3. In connection with the previous remark it is worthwhile to observe that in some cases, such as the present one, the desired nonlinear equations are related with the particle system but are lost in the DW limit; in other cases, such as in the monotype case, the desired equations appear in both the particle system and the DW process; and in other cases, such as the one in [Gorostiza 1993], the desired equations appear only in the DW limit. Thus one can sometimes take advantage of the choice between particle system and DW limit.

Appendix 1. Notation

- Z: Banach space.
- C_b: bounded (real) continuous functions on Z.
- C_{bb}: subset of C_b of functions with bounded support.
- F^+: nonnegative elements of a function space F.
- $\langle \mu, f \rangle = \int_Z f d\mu$, f: function on Z, μ: measure on Z.
- $C_b(\mathbb{R}^d)$: bounded continuous functions on $\mathbb{R}^d$.
- $C_c(\mathbb{R}^d)$: subset of $C_b(\mathbb{R}^d)$ of functions with compact support.
- $f_p(x) = (1 + |x|^2)^{-p}$, $x \in \mathbb{R}^d$, $(p > 0)$.
- $C_p(\mathbb{R}^d) = \{f \in C_b(\mathbb{R}^d) \mid \|f\|_p < \infty\}$, where $\|f\|_p = \sup_{x \in \mathbb{R}^d} |f(x)/f_p(x)|$.
- $K_p(\mathbb{R}^d) = C_c(\mathbb{R}^d)^+ \cup \{f_p\}$.
- $Q_p(\mathbb{R}^d) = \{f \in C_p(\mathbb{R}^d) \mid \lim_{|x| \to \infty} f(x)/f_p(x) \text{ exists}\}$.
- $\mathcal{M}_p(\mathbb{R}^d)$: space of (nonnegative) Radon measures μ on $\mathbb{R}^d$ such that $\langle \mu, f_p \rangle < \infty$.
- $\mathcal{M}_p(\mathbb{R}^d)$ is equipped with the p-vague topology (i.e., the topology defined by the elements of $K_p(\mathbb{R}^d)$).
- Obvious extensions are defined for Cartesian products of these spaces with themselves.
- Given $\alpha_1, \ldots, \alpha_k \in (0, 2]$, p is chosen so that $p > d/2$, and in addition $p < \frac{1}{2}(d + \min \alpha_i)$ if $\min \alpha_i < 2$.

- More information on these spaces and their use is contained in [Dawson, Fleischmann and Gorostiza 1989, Gorostiza and Wakolbinger 1992, Iscoe 1986].

Appendix 2. Derivation of equations (2.4) and (2.5)

Let
$$v_f(z,t) = E[e^{\langle N_t, f \rangle} \mid N_0 = \delta_z].$$
Conditioning on the time τ of the first branching and separating the cases $t < \tau$ and $t \geq \tau$ we have

(1) $$v_f(z,t) = v^{(1)}(z,t) + \int_0^t v^{(2)}(z,s,t)\,ds$$

with
$$v^{(1)}(z,t) = E\left[\exp\left\{-\int_0^t \mathcal{K}(X_s)\,ds - f(X_t)\right\} \bigg| X_0 = z\right],$$

and
$$v^{(2)}(z,s,t) = E\left[\exp\left\{-\int_0^{t-s} \mathcal{K}(X_r)\,dr\right\}\mathcal{K}(X_{t-s})\mathcal{JF}(X_{t-s},\cdot,v_f(\cdot,s))(X_{t-s}) \bigg| X_0 = z\right],$$

where X is the Markov process with semigroup $(\mathcal{T}_t)_{t \geq 0}$ and generator A.

By the Feynman-Kac formula, $v^{(1)}$ solves
$$\frac{\partial}{\partial t}v^{(1)}(z,t) = (A - \mathcal{K}(z))v^{(1)}(z,t), \quad t > 0,$$
$$v^{(1)}(z,0) = e^{-f(z)},$$

and $v^{(2)}$ solves
$$\frac{\partial}{\partial t}v^{(2)}(z,s,t) = (A - \mathcal{K}(z))v^{(2)}(z,s,t), \quad t > s,$$
$$v^{(2)}(z,s,s) = \mathcal{K}(z)\mathcal{JF}(z,\cdot,v_f(\cdot,s))(z).$$

Differentiating (1) we obtain
$$\frac{\partial}{\partial t}v_f(z,t) = (A - \mathcal{K}(z))v_f(z,t) + \mathcal{K}(z)\mathcal{JF}(z,\cdot,v_f(\cdot,t))(z)$$
$$= Av_f(z,t) - \mathcal{K}(z)\mathcal{J}[\mathcal{F}(z,\cdot,v_f(\cdot,t)) - v_f(z,t)](z).$$

It follows that $u_f(z,t) = 1 - v_f(z,t)$ solves equation (2.5), whose evolution form is equation (2.4).

Acknowledgement. We thank Ted Cox for a stimulating discussion during the Montreal Workshop, and Don Dawson for illuminating discussions over the past years and for his kind hospitality during our visits to Canada. L. G. G. also thanks CIMAT for hospitality. This research is partially supported by CONACyT grant 2059-E9302.

Bibliography

M. Bramson, J. T. Cox and A. Greven, *Ergodicity of a critical spatial branching process in low dimensions*, 1992 (preprint).

B. Chauvin, A. Rouault, and A. Wakolbinger, *Growing conditioned trees*, Stoch. Proc. Appl. **39** (1991), 117–130.

J. T. Cox, *On the ergodic theory of critical branching Markov chains*, Stoch. Proc. Appl. (1992) (to appear).

D. A. Dawson, *The critical measure diffusion*, Z. Wahr. verw. Geb. **40** (1977), 125–145.

_____, *Measure-Valued Markov Processes*, École d'Été de Probabilités de Saint Flour 1991, Lect. Notes in Math. **1541** (1993), 1–260, Springer-Verlag.

D. A. Dawson and K. Fleischmann, *Critical dimension for a model of branching in a random medium*, Z. Wahr. verw. Geb. **70** (1985), 315–334.

D. A. Dawson, K. Fleischmann, R. D. Foley and L. A. Peletier, *A critical measure-valued branching process with infinite mean*, Stoch. Proc. Appl. **4** (1988), 117–129.

D. A. Dawson, K. Fleischmann and L.G. Gorostiza, *Stable hydrodynamic limit fluctuations of a critical branching particle system in a random medium*, Ann. Prob. **17** (1989), 1083–1117.

D. A. Dawson and E. A. Perkins, *Historical Processes*, Mem. Amer. Math. Soc., vol. 454, 1991.

R. Durrett, *An infinite particle system with additive interactions*, Adv. Appl. Prob. **11** (1979), 255–383.

E. B. Dynkin, *Three classes of infinite dimensional diffusions*, J. Funct. Anal. **86** (1989), 75–110.

_____, *Superprocesses and partial differential equations*, The 1991 Wald Memorial Lectures, Ann. Probab. **21** (1993), 1185–1262.

M. Escobedo and O. Kavian, *Asymptotic behavior of positive solutions of a nonlinear heat equation*, Houston J. Math. **13** (1987), 39–50.

A. M. Etheridge, *Almost sure convergence of measure-valued branching processes: a critical exponent*, (preprint).

_____, *Conditioned superprocesses and a semilinear heat equation*, (preprint).

A. Gmira and L. Veron, *Large time behavior of the solution of a semilinear parabolic equation in $\mathbb{R}^n$*, J. Diff. Equations **53** (1984), 259–276.

L. G. Gorostiza, *A measure valued process arising from a branching particle system with changes of mass*, (this volume).

L. G. Gorostiza, S. Roelly-Coppoletta, and A. Wakolbinger, *Sur la persistence du processus de Dawson-Watanabe stable. L'interversion de la limite en temps et de la renormalisation*, Sém. de Probabilités XXIV, Lect. Notes in Math., vol. 1426, Springer-Verlag, Berlin, 1990, pp. 275–281.

L. G. Gorostiza, S. Roelly, and A. Wakolbinger, *Persistence of critical multitype particle and measure branching processes*, Probab. Theory Relat. Fields **92** (1992), 313–335.

L. G. Gorostiza and A. Wakolbinger, *Persistence criteria for a class of critical branching particle systems in continuous time*, Ann. Probab. **19** (1991), 266–288.

_____, *Convergence to equilibrium of critical branching particle systems and superprocesses, and related nonlinear partial differential equations*, Acta Appl. Math. **27** (1992), 269–291.

_____, *Asymptotic behavior of a reaction-difussion system, a probabilistic approach*, Radom and Comput. Dynamics **1** (1993), 445–463.

I. Iscoe, *A weighted occupation time for a class of measure-valued branching processes*, Prob. Theory Relat. Fields **71** (1986), 85–116.

I. Iscoe and T.-Y. Lee, *Large deviations for occupation times of measure-valued branching Brownian motions*, (preprint).

N. Ikeda, M. Nagasawa, and S. Watanabe, *Branching Markov processes I*, J. Math. Kyoto Univ. **8** (1968), 233–278.

M. Jiřina, *Branching processes with measure valued states*, Transactions of the Third Prague Conference on Information Theory, Statistical Decision Functions, Random Processes, Prague, Czech. Academy of Sciences (1964), 333–357.

O. Kallenberg, *Stability of critical cluster fields*, Math. Nachr. **77** (1977), 7–43.

J. F. Le Gall, *On a path-valued process and its connections with a class of nonlinear partial differential equations*, (preprint).

Z.-H. Li, *A note on the multitype measure branching process*, Adv. Appl. Prob. **24** (1992), 496–498.

A. Liemant, *Kritische Verzweigungsprozesse mit allgemeinen Phasenraum, IV*, Math. Nachr. **102** (1981), 235–354.

A. Liemant, K. Matthes, and A. Wakolbinger, *Equilibrium Distributions of Branching Populations*, Akademie-Verlag, Berlin, and Kluwer Academic Publ., Dordrecht, 1988.

A. Pazy, *Semigroups of Linear Operators with Applications to Partial Differential Equations*, Springer-Verlag, New York, 1983.

Yu. M. Rhyzhov and A. V. Skorokhod, *Homogeneous branching processes with a finite number of types and continuously varying mass*, Teo. Verojanost. i Primenen. **15** (1970), 704–707.

C. Stone, *On a theorem by Dobrushin*, Ann. Math. Stat. **39** (1968), 1391–1401.

A. Stöckl and A. Wakolbinger, *Convergence to equilibrium in independent particle systems with migration and mutation*, Tech. Rep. **446** (1991), Institut für Mathematik, Johannes Kepler Universität Linz.

———, *On clan recurrence and transience in time stationary branching Brownian particle systems*, (this volume)

S. Watanabe, *On two dimensional Markov processes with branching property*, Trans. Amer. Math. Soc. **136** (1969), 447–466.

Y. Wu, *Asymptotic behavior of two level measure branching process*, Ann. Probab. (1993) (to appear).

ABSTRACT. Persistence and convergence towards equilibrium results are presented for a general critical branching particle system in a Polish space. Applications include the asymptotic behavior of a system of nonlinear partial differential equations.

DEPARTAMENTO DE MATEMÁTICAS, CENTRO DE INVESTIGACIÓN Y DE ESTUDIOS AVANZADOS, 07000 MÉXICO, D. F., MÉXICO

E-mail address: gortega@redvax1.dgsca.unam.mx

FACHBEREICH MATHEMATIK, JOHANN WOLFGANG GOETHE-UNIVERSITÄT, D-60054 FRANKFURT/M., GERMANY

E-mail address: wakolbin@math.uni-frankfurt.de

Newtonian Particle Mechanics and Stochastic Partial Differential Equations

Peter Kotelenez and Keming Wang

1. Introduction

The irregular movement of particles immersed into a fluid is often referred to as Brownian motion. The rigorous foundation to the study of this phenomenon has been laid by Einstein, Smoluchowski, Langevin, Ornstein, Uhlenbeck, Wiener, Levy, and Kolmogorov (see [Nelson 1972]). Physically there are mainly two approaches to study Brownian motion. The first one is by introducing a random force and applying Newton's law $f = ma$ to get a stochastic ordinary differential equation (SODE) for the positions and momenta of the particles (see [Gardiner 1990]). The other approach is by using Hamiltonian formalism to derive the distribution of the positions and momenta (see e.g., [Lebowitz and Rubin 1963], where these equations have been derived without a priori introduction of a stochastic term).

The object of this paper is to extend the first approach by introducing more complicated and, in our view, more realistic random forces. Traditionally the random force acting on the position of a particle is assumed to be Gaussian white noise (in t), and the Gaussian white noises associated with different particles are assumed to be independent. Or this random force is the finite sum of independent Gaussian white noises (in t) coupled by some state dependent diffusion coefficient (see e.g., [Vaillancourt 1988] for the latter model). The latter version is one way of introducing correlations into the fluctuation forces. Another way of defining correlated fluctuation forces is by using the convolution of space-time white noise with an appropriate correlation function. This was done by [Kotelenez 1992b] and in a less explicit form by [Borkar 1984][1]. One difference between Kotelenez and [Borkar 1984] is that in the latter reference a spatially correlated Brownian sheet is introduced (leading to the term Brownian medium) and the calculus for SODE's driven by that space-time field has to be developed. Further, in [Borkar 1984] only the position of one particle is considered. On the other hand, in [Kotelenez 1992b], the standard Brownian sheet is used and correlations are introduced through convolution with a correlation functional. Using Walsh's work [Walsh 1986], no particular stochastic analysis for SODE's driven by that space-time field has to be developed. However, the main difference is that in [Kotelenez 1992b] many particles are considered and most of the work is concentrated on deriving mezoscopic (i.e., stochastic) and macroscopic (i.e., deterministic) partial differential equations (PDE's) for the

---

1991 *Mathematics Subject Classification.* Primary 60H15; Secondary 76D05.

This is the final form of the paper.

[1]The first named author wants to thank J. Vaillancourt for bringing Borkar's paper to his attention.

mass, resp. vorticity distribution. We will use here Kotelenez' approach [Kotelenez 1992b]. The main difference of our paper from [Kotelenez 1992b] is that here we define a dynamical theory for the irregular movement of particles with fluctuation forces as in [Kotelenez 1992b], i.e., a generalization of the Ornstein-Uhlenbeck model. In [Kotelenez 1992b] the equations for the positions of the particles are the generalization of the Einstein-Smoluchowski model (Similarly in [Borkar 1984], cf. [Nelson 1972] for the classical model).

Our generalization of the Ornstein-Uhlenbeck model is given in Section 2. The main results of that section is the "adiabatic elimination", or, more precisely, that the positions satisfy approximately an SODE driven by Brownian sheets (the generalization of the Einstein-Smoluchowski model), as the friction tends to ∞. In Section 3 we introduce the stochastic partial differential equation (SPDE) for the mass distribution of our system of N particles and show that the empirical process is a solution of the SPDE. Further we indicate how to extend this particular solution of the SPDE to a solution starting in arbitrary initial measures (with total mass conservation), how to show that this extension is the unique solution of the SPDE, when it has density valued solutions, and how to obtain the macroscopic PDE as the correlation length of the fluctuation forces tends to 0. This is clearly different from the classical approach, where one assumes uncorrelated fluctuation forces, represented by i.i.d. Brownian motion. In that model the fluctuations disappear as N, the number of particles, tends to ∞ (cf. Remark 3.4).

In Section 4 we construct a generalization to "generalized" particles, i.e., "particles" with "positive" and "negative" masses, such as point vortices in a two dimensional fluid (the actual content of [Kotelenez 1992b]) and we discuss the applicability of our approach to other particle models with diffusion, in particular to those, which include a creation-annihilation mechanism.

A final remark about notation, constants will be denoted by c, c_T, etc. and they may have a different meaning in the course of a proof. "$\cdot$" denotes the inner product on $\mathbb{R}^d$. Moreover, let us suppose that there is a stochastic basis $(\Omega, \mathcal{F}, \mathcal{F}_{t\geq 0}, P)$ with right continuous filtration and that all our stochastic processes are defined on Ω and are $\mathcal{F}_t$-adapted (including initial conditions). Moreover, the processes are assumed to be $dP \bigoplus dt$-measurable, where dt is the Lebesgue measure on $[0, \infty)$.

2. The N-particle system and adiabatic elimination

Let us consider N particles in $\mathbb{R}^d$, where $N \in \mathbb{N}$. The position of the i-th particle will be denoted by $\tilde{r}^i \in \mathbb{R}^d$ and its mass m_i. β is the friction constant. We assume that there is no creation or annihilation of particles. The motion of the system is governed by the following SODE:

$$(2.1) \quad m_i \frac{d^2 \tilde{r}^i}{dt^2} = -\beta m_i \frac{d\tilde{r}^i}{dt} + \beta \sum_{j=1}^{N} m_i m_j K(\tilde{r}^i, \tilde{r}^j) + \beta f_i(\tilde{r}^i) + \sqrt{2kT}\beta \frac{M_i(\tilde{r}^{(N)}, dt)}{dt}.$$

Here, k is the Boltzman constant, T the temperature, and $\tilde{r}^{(N)} = (\tilde{r}^1, \ldots, \tilde{r}^N) \in \mathbb{R}^{d \cdot N}$. K describes the pair interaction. Finally f_i and $M_i(\cdot, dt)/dt$ are the forces of the surrounding medium which act on the i-th particle. f_i can be thought of as a mean value of all those forces, whereas $M_i(\cdot, dt)/dt$ represents the rapidly varying random part of those forces. We assume that $M_i(\tilde{r}^{(N)}, t)$ is an $\mathbb{R}^d$ valued square integrable continuous martingale. The expression $M_i(\cdot, dt)/dt$ is to be understood as a derivative of generalized functions, and has to be replaced by an Itô differential, provided we rewrite (2.1) as an Itô stochastic ordinary differential equation (SODE). It is important that M_i can depend both on the position of the i-th particle and on the distribution of all other particles. Therefore the fluctuation forces acting on two or more particles may be correlated (at least if particles are close to one another). Setting

$$v^i = \frac{d\tilde{r}^i}{dt},$$

we obtain from (2.1) an equivalent system of SODE's:

$$(2.2) \quad \begin{cases} d\tilde{r}^i = v^i dt \\ m_i dv^i = \left[-\beta m_i v^i + \beta \sum_{j=1}^{N} m_i m_j K(\tilde{r}^i, \tilde{r}^j) + \beta f_i(\tilde{r}^i)\right] dt \\ \qquad\qquad + \sqrt{2kT}\beta M_i(\tilde{r}^{(N)}, dt). \end{cases}$$

Let us now make the following assumptions on (2.2):

Assumption 2.1.   1.   (a) There is a bounded function $\tilde{\Gamma}_\epsilon : \mathbb{R}^d \times \mathbb{R}^d \to \mathbb{R}^d$ such that

$$\int \tilde{\Gamma}_\epsilon^2(r, p) \, dp = 1$$

$$\int \tilde{\Gamma}_\epsilon(r, p) \tilde{\Gamma}_\epsilon(q, p) \, dp \leq c e^{-|r-q|^2/8\epsilon}.$$

Here ϵ is a fixed positive constant (the correlation length, and the integration is taken over $\mathbb{R}^d$. Also in what follows we will not indicate the integration domain if it is $\mathbb{R}^d$).

(b) There are d i.i.d. Brownian sheets on $\mathbb{R}^d \times [0, \infty)$ $w_l(p,t)(l=1,\ldots,d)$ such that for any Borel sets A and B from $\mathbb{R}^d \times [0, \infty)$

$$E \int_A w_l(dp, dt) \int_B w_l(dp, dt) = |A \cap B|,$$

where $|A \cap B|$ is the $(d+1)$-dimensional Lebesgue measure of $A \cap B$ (cf. [Kotelenez 1992a], [Kotelenez 1992b], and [Walsh 1986].)

(c) Define a $d \times d$ matrix $\hat{\Gamma}_\epsilon = (\hat{\Gamma}_{kl}(r,p))$ by $\hat{\Gamma}_{kl} = 0$ if $k \neq l$ and $\hat{\Gamma}_{ll}(r,p) = \tilde{\Gamma}_\epsilon(r,p)$ and set $w(p,t) := (w_1(p,t), \ldots, w_l(p,t))^T$, where T denotes the transpose. We suppose

$$M_i(r^{(N)}, dt) = m_i \sqrt{2D} \int \hat{\Gamma}_\epsilon(r^i, p) w(dp, dt),$$

where $D = kT > 0$ is a fixed diffusion constant.

2.   (a) $f(r^i) := f_i(r^i)/m_i$ is a bounded function of all arguments independent of m_i.

(b) There is an $L < \infty$ s.t. for all r^i, q^i,

$$|f(r^i) - f(q^i)| \leq L|r^i - q^i|,$$

where $|\cdot|$ is the Euclidian norm on $\mathbb{R}^d$ (and also $\mathbb{R}^{d \cdot N}$).

3. $K(r^i, r^j)$ is a bounded function of both arguments, and there is an $L < \infty$ s.t.
$$|K(r^i, p) - K(q^i, p)| \le L|r^i - q^i|$$
uniformly in r^i, q^i and $p \in \mathbb{R}^d$.
4. f, K and $\widetilde{\Gamma}_\epsilon$ are infinitely often differentiable in all arguments with bounded partial derivatives of all orders.

Example 2.2. Choosing
$$\widetilde{\Gamma}_\epsilon(r, q) := \left((2\pi\epsilon)^{-d/2} e^{-|r-q|^2/2\epsilon}\right)^{1/2}$$
we easily check that $\widetilde{\Gamma}_\epsilon$ has the properties required in 2.1 1.(a) with $c = 1$.

Remark 2.3. 1. Let $\{\phi_n\}_{n \in \mathbb{N}}$ be a complete orthonormal system (CONS) in $L_2(\mathbb{R}^d, dr)$, the space of real valued square integrable functions on $\mathbb{R}^d$, where dr is the d-dimensional Lebesgue measure. For each $n \in \mathbb{N}$ let ϕ_n be the $d \times d$-matrix whose entries on the main diagonal are all equal to ϕ_n and whose other entries are equal to 0. Then
$$\int \widehat{\Gamma}_\epsilon(r, p) w(dp, t) = \sum_{n=1}^\infty \int \widehat{\Gamma}_\epsilon(r, q) \phi_n(q)\, dq \beta^n(t),$$
where $\beta^n(t) := \int \phi_n(q) w(dq, t)$ are $\mathbb{R}^d$-valued i.i.d. Wiener processes whose components are $\sim \mathcal{N}(0, 2Dt)$.

2. Assume that there are two $\mathbb{R}^d$-valued adapted stochastic processes $q^1(t)$ and $q^2(t)$. Then we easily see that the $\int_0^t \int \widehat{\Gamma}_\epsilon(q^i(s), p) w(dp, ds)$ are $\mathbb{R}^d$-valued square integrable continuous martingales ($i = 1, 2$) and their mutual quadratic variation is given by
$$\left\langle\!\!\left\langle \int_0^t \int \widetilde{\Gamma}_\epsilon(q^1(s), p) w_k(dp, ds), \int_0^t \int \widetilde{\Gamma}_\epsilon(q^2(s), p) w_l(dp, ds) \right\rangle\!\!\right\rangle$$
$$= \int_0^t \int \widetilde{\Gamma}_\epsilon(q^1(s), p) \widetilde{\Gamma}_\epsilon(q^2(s), p)\, dp\, ds \cdot \delta_{k,l}$$
$k, l = 1, \ldots, d$, with $\delta_{k,l} = 1$ if $k = l$, and $= 0$ otherwise. Moreover, our assumption 2.1.1 1.(a) implies that correlations are negligible if $|q^1(s) - q^2(s)|^2 \gg \epsilon$ and that they may be observable if $|q^1(s) - q^2(s)|^2 \sim \epsilon$.

THEOREM 2.4. *To each $\mathcal{F}_0$-adapted square integrable initial condition $(\tilde{r}^{(N)}(0), v^{(N)}(0)) \in \mathbb{R}^{2d \cdot N}$ (2.2) has a unique $\mathcal{F}_t$-adapted solution $(\tilde{r}^{(N)}(t), v^{(N)}(t)) \in C([0, \infty); \mathbb{R}^{2d \cdot N})$ a.s. which is an $\mathbb{R}^{2d \cdot N}$-valued Markov process.*

PROOF. (i) Note that
$$1 - e^{-|r-q|^2/8\epsilon} \le c_\epsilon |r - q|^2.$$
Hence if $q_l^{(N)}, l = 1, 2$, are two adapted $\mathbb{R}^{d \cdot N}$-valued processes, Doob's inequality implies for each $i = 1, \ldots, N$:
$$E \sup_{0 \le t \le T} \left| \int_0^t \int [\widehat{\Gamma}_\epsilon(q_1^i(s), p) - \widehat{\Gamma}_\epsilon(q_2^i(s), p)] w(dp, ds) \right|^2$$
$$\le c \cdot c_\epsilon \int_0^T E|q_1^i(s) - q_2^i(s)|^2\, ds.$$

The explicit Lipschitz assumptions on the other coefficients of (2.2) now allows us to derive existence and uniqueness by the contraction mapping principle.

(ii) The existence of a continuous version follows from the fact that (2.2) is driven by a continuous $\mathbb{R}^{2d \cdot N}$-valued square integrable semimartingale.

(iii) The Markov property follows since the Brownian sheets have independent increments. □

Let us now consider the following SODE on $\mathbb{R}^{d \cdot N}$

$$(2.3) \quad dr^i = \left(\sum_{j=1}^{N} m_j K(r^i, r^j) + f(r^i) \right) dt + \sqrt{2D} \int \widehat{\Gamma}_\epsilon(r^i, p) w(dp, dt),$$

$$i = 1, \ldots, N.$$

By exactly the same argument as in Theorem 2.4, we obtain:

THEOREM 2.5. *To each $\mathcal{F}_0$-adapted square integrable initial condition $r^{(N)}(0) \in \mathbb{R}^{d \cdot N}$ (2.3) has a unique $\mathcal{F}_t$-adapted solution $r^{(N)}(\cdot) \in C\big([0,\infty); \mathbb{R}^{d \cdot N}\big)$ a.s. which is an $\mathbb{R}^{d \cdot N}$-valued Markov process.*

Remark 2.6. (i) For each i $M_i(t) := \int_0^t \int \widehat{\Gamma}_\epsilon(r^i(s), p) w(dp, ds)$ is an $\mathbb{R}^d$-valued standard Brownian motion. This follows for the one-dimensional components from P. Levi's theorem (cf. [Liptcer and Shiryayev 1974, Ch. 4, §1] and by independence of the scalar valued Brownian sheets and the diagonal structure of $\widehat{\Gamma}_\epsilon$ it follows then for $M_i(t)$. However, even if $K \equiv 0$ and $f \equiv 0$ the $M_i's$ are correlated and their joint distribution is not Gaussian. This last observation follows from the form of the generator (see the following (2.4)).

(ii) To avoid cumbersome notation we give the generator for (2.3) under the assumption $d = 1$, $N = 2$ and for $\widehat{\Gamma}_\epsilon$ as in Ex. 2.2. Let $\phi \in C^2(\mathbb{R}^2, \mathbb{R})$ (twice continuously differentiable from $\mathbb{R}^2$ to $\mathbb{R}$). Then the generator for (2.3) is given by

$$(2.4) \quad (A\phi)(r^{(N)}) = \sum_{l=1}^{2} \frac{\partial}{\partial r_l} \phi(r^{(N)}) \cdot \left[\sum_{j=1}^{2} m_j K(r^l, r^j) + f(r^l) \right]$$

$$+ D\Delta\phi(r^{(N)}) + 2D \frac{\partial^2}{\partial r_1 \partial r_2} \phi(r^{(N)}) \cdot e^{-|r^1 - r^2|^2/8\epsilon},$$

where Δ is the Laplacian on $C^2(\mathbb{R}^2, \mathbb{R})$ and $r^{(N)} = (r^1, r^2)$.

(iii) As in (ii) for $N = 1$, $d = 2$, $K \equiv 0$, $f \equiv 0$ we obtain from Itô's formula

$$(2.5) \quad E|r^1(t) - r^2(t)|^2 = 2D \int_0^t \left(1 - e^{-|r^1(s) - r^2(s)|^2/8\epsilon}\right) ds < 2Dt.$$

Hence, even if $K \equiv 0$ and $f \equiv 0$ in our model the particles have a tendency to stay closer to one another than in the case of independent Gaussian fluctuation forces (which would give $E|r^1(t) - r^2(t)|^2 = 2Dt$). A possible physical interpretation of this phenomenon is the transport of momentum in a viscous fluid (cf. [Kestin and Dorfman 1971, Ch. 12.5.3] and also our following Theorem 2.7).

Next we derive (2.3) from (2.2) by adiabatic elimination (or Ornstein-Uhlenbeck approximation — cf. [Nelson 1972, Ch. 9 and 10]).

THEOREM 2.7. *Let $\tilde{r}^{(N)}$ be the position of the N-particle system described by* (2.2) *and $r^{(N)}$ the solution of* (2.3). *Suppose*
 (i) $\beta \to \infty$;
 (ii) $E|\tilde{r}^{(N)}(0) - r^{(N)}(0)|^2 \to 0$, *as* $\beta \to \infty$;
 (iii) $\max_{i=1,\ldots,N} \sup_\beta E|v^i(0)|^2 < \infty$.
Then for any $T > 0$

(2.6) $$E \sup_{t \leq T} |\tilde{r}^{(N)}(t) - r^{(N)}(t)|^2 \to 0 \quad as \quad \beta \to \infty.$$

PROOF. (i) Let us denote the macroscopic parts on the right hand sides (r.h.s) of (2.2) and (2.3) for the i-th particle by G_i. Note that by the usual and the stochastic Fubini theorems (see [Walsh 1986, Th. 2.6] for the latter),

$$\int_0^t \int_0^s e^{-\beta(s-u)} \beta G_i(\tilde{r}^{(N)}(u)) \, du \, ds$$
$$+ \int_0^t \int_0^s e^{-\beta(s-u)} \beta \int \widehat{\Gamma}_\epsilon(\tilde{r}^i(u,p)) w(dp, du) \, ds$$
$$= \int_0^t (1 - e^{-\beta(t-u)}) G_i(\tilde{r}^{(N)}(u)) \, du$$
$$+ \int_0^t (1 - e^{-\beta(t-u)}) \int \widehat{\Gamma}_\epsilon(\tilde{r}^i(u)) w(dp, du).$$

(ii) Set
$$y_i(t) = \int_0^t e^{-\beta(t-s)} \, dM_i(s)$$

with
$$M_i(t) = \int_0^t \int \widehat{\Gamma}_\epsilon(\tilde{r}^i(s), p) w(dp, ds).$$

Itô's formula yields
$$|y_i(t)|^2 = 2z_i(t) + \bar{v}_i(t)$$

with
$$z_i(t) := \int_0^t e^{-2\beta(t-s)} \left(\int_0^s e^{-\beta(s-u)} \, dM_i(u) \right) \cdot dM_i(s)$$
$$\bar{v}_i(t) := d \int_0^t e^{-2\beta(t-s)} \, ds.$$

We have

$$E \sup_{0 \leq t \leq T} |z_i(t)| \leq \left\{ E \sup_{0 \leq t \leq T} |z_i(t)|^2 \right\}^{1/2}$$
$$\leq c_T \left\{ E \left| \int_0^T \left(\int_0^s e^{-\beta(s-u)} \, dM_i(u) \right) dM_i(s) \right|^2 \right\}^{1/2}$$

(by [Kotelenez 1984, Th. 2.1] with $c_T < \infty$ independent of M_i and β)

$$\leq c_T \left\{ d \int_0^T \left(\int_0^s e^{-2\beta(s-u)} \, du \right) ds \right\}^{1/2}$$

(since $M_i(t)$ is an $\mathbb{R}^d$-valued square integrable mean zero continuous martingale with mutually uncorrelated components — see [Ikeda and Watanabe 1981, Ch. 2])

$$= c_T \left\{ \int_0^T \frac{1 - e^{-2\beta s}}{2\beta} \, ds \right\} \frac{1}{2} \to 0, \text{ as } \beta \to \infty.$$

Moreover,

$$\bar{v}_i(t) = \frac{d}{2\beta}(1 - e^{-2\beta t})$$

(d is the dimension). Hence,

$$\sup_{0 \le t \le T} \bar{v}_i(t) = \frac{d}{2\beta}(1 - e^{-2\beta T}) \to 0, \text{ as } \beta \to \infty.$$

(iii) The boundedness of G_i implies

$$\sup_{0 \le t \le T} \left| \int_0^t e^{-\beta(t-s)} G_i(\tilde{r}^{(N)}(s)) \, ds \right|^2 \le c \left(\frac{1 - e^{-\beta T}}{\beta} \right) \to 0, \text{ as } \beta \to \infty.$$

(iv) Therefore,

$$E \sup_{t \le T} |\tilde{r}^i(t) - r^i(t)|^2$$

$$\le cE|\tilde{r}^i(0) - r^i(0)|^2 + cE \left| \int_0^t \left(G_i(\tilde{r}^{(N)}(s)) - G_i(r^{(N)}(s)) \right) ds \right|^2$$

$$+ cE \left| \int_0^t \int [\widehat{\Gamma}_\epsilon(\tilde{r}^i(s), p) - \widehat{\Gamma}_\epsilon(r^i(s), p)] w(dp, ds) \right|^2 + o(1) \quad (\text{as } \beta \to \infty)$$

(where we incorporated $E \left| \int_0^t e^{-\beta s} v^i(0) \, ds \right|^2$ in the $o(1)$).

(v) The Lipschitz assumption on the macroscopic coefficients and on $\widehat{\Gamma}_\epsilon$ (cf. step (i) in the proof of Theorem 2.4 now imply

$$E \sup_{t \le T} |\tilde{r}^N(t) - r^N(t)|^2 \le c \int_0^t E \sup_{s \le t} |\tilde{r}^N(s) - r^N(s)|^2 ds + o(1).$$

Hence, application of the Gronwall lemma finishes the proof. □

Remark 2.8. Following [Nelson 1972, §10], we certainly could weaken the assumptions on the macroscopic coefficients in Theorem 2.7. We will not pursue this now since at this stage we want to be conceptual. A more general treatment of the adiabatic elimination for (2.2) and (2.3) will be given in a subsequent paper [Kotelenez and Wang 1993].

3. A stochastic partial differential equation (SPDE) for the mass distribution of finitely and infinitely many particles and the macroscopic limit

Set $m_+ = \sum_{i=1}^N m_i$ (the total mass of the particle system (2.2)/(2.3)). Further, define

$$\mathbb{M}_+ := \{\mu : \mu \text{ is a Borel measure on } \mathbb{R}^d \text{ s.t. } \mu(\mathbb{R}^d) = m_+\}.$$

The empirical process associated with (2.3) is given by

$$\mathcal{X}_N(t) := \sum_{i=1}^N m_i \delta_{r^i(t)}, \tag{3.1}$$

where $\delta_r(A) = 1$ if $r \in A$ and $= 0$ otherwise for $A \subset \mathbb{R}^d$. Obviously, $\mathcal{X}_N(t) \in \mathbb{M}_+$ for all $t \geq 0$ and all ω. Set

$$\Gamma_\epsilon(q,p) := \epsilon^{-d/4}\widehat{\Gamma}_\epsilon(q,p),$$

and consider the following SPDE on $\mathbb{M}_+$:

$$d\mathcal{X}_\epsilon = \left[D\Delta\mathcal{X}_\epsilon - \nabla \cdot \left[\mathcal{X}_\epsilon \int K(\cdot,p)\mathcal{X}_\epsilon(dp)\right] - \nabla \cdot [\mathcal{X}_\epsilon f(\cdot)]\right] dt \tag{3.2}$$
$$- \sqrt{2D}\epsilon^{d/2}\nabla \cdot \left[\mathcal{X}_\epsilon \int \Gamma_\epsilon(\cdot,p)w(dp,dt)\right],$$

$$\mathcal{X}_\epsilon(\mathbb{R}^d,t) \stackrel{\text{a.s.}}{=} \mathcal{X}(\mathbb{R}^d,0) = m_+. \tag{3.3}$$

Here the coefficients from (2.3) are to be interpreted as densities of $\mathcal{X}_\epsilon$. This last relation reduces to pointwise multiplication if $\mathcal{X}_\epsilon$ has a density with respect to the Lebesgue measure and (3.2) becomes an SPDE for this density. The following lemma is the bridge between (2.3) and (3.2)/(3.3).

LEMMA 3.1. *The empirical process $\mathcal{X}_N(t)$ given by (3.1) is a weak solution of (3.2)/(3.3).*

PROOF. (i) Itô's formula implies that $\mathcal{X}_N(t)$ satisfies (3.2) weakly.

(ii) The conservation of total mass required by (3.3) follows from the construction. □

For the formulation of the following theorem we need more notation.

First of all we introduce a bounded metric on $\mathbb{R}^d$:

$$\rho(r,q) := (c_\epsilon |r-q|) \wedge 1, \tag{3.4}$$

where c_ϵ is the constant appearing in step (i) of the proof of Theorem 2.4.

If $\mu, \tilde{\mu} \in \mathbb{M}_+$ the set of all joint representations of μ and $\tilde{\mu}$ will be denoted by $C(\mu, \tilde{\mu})$. Then

$$\gamma_2(\mu,\tilde{\mu}) := \left\{\inf_{Q \in C(\mu,\tilde{\mu})} \int Q(dr,dq)\rho^2(r,q)\right\}^{1/2}$$

is a metric on $\mathbb{M}_+$ and $(\mathbb{M}_+, \gamma_2)$ is a complete metric space and the set of finite (positive) linear combinations of point measures (denoted $\mathbb{M}^d$) in $\mathbb{M}_+$ is dense in $\mathbb{M}_+$ (cf. [Dudley 1989, 11.5.5 and 11.8.2], [De Acosta 1982, Appendix, Lemma 4], and [Kotelenez 1992b]). Set

$$\widetilde{\mathcal{M}}_0 := L_2(\Omega; \mathbb{M}^d),$$
$$\mathcal{M}_0 := L_2(\Omega; \mathbb{M}_+),$$
$$\mathcal{M}_{[0,T]} := L_2(\Omega; C([0,T]; \mathbb{M}_+)).$$

Note that $\mathcal{M}_0$ and $\mathcal{M}_{[0,T]}$ are complete metric spaces, since $\mathbb{M}_+$ is complete, where the metric on $\mathcal{M}_{[0,T]}$ is given by $(E\sup_{0\leq t\leq T}\gamma_2^2(\mu_t,\tilde{\mu}_t))^{1/2}$ for $\mu_t, \tilde{\mu}_t \in \mathcal{M}_{[0,T]}$. Further let $C(\mathbb{R}^d \times [0,\infty); \mathbb{R})$ be continuous real valued functions on $\mathbb{R}^d \times [0,\infty)$.

THEOREM 3.2.   (i) *Suppose $\mathcal{X}_0 \in \mathcal{M}_0$. Then there is a mild solution $\mathcal{X}_\epsilon(\cdot, \mathcal{X}_0)$ of (3.2)/(3.3) such that*

$$\mathcal{X}_\epsilon(\cdot, \mathcal{X}_0) \in \mathcal{M}_{[0,T]}.$$

(ii) *Suppose $\mathcal{X}_0 \in \mathcal{M}_0$ and $\mathcal{X}_0$ has a density X_0 with respect to the Lebesgue measure such that*

$$\operatorname*{ess\,sup}_{(r,\omega)} X_0(r,\omega) < \infty$$

and $X_0(r)$ continuous in r a.s. Then $\mathcal{X}_\epsilon(\cdot, \mathcal{X}_0) \in C(\mathbb{R}^d \times [0, \infty); \mathbb{R})$ a.s.

PROOF [IDEA]. For $d = 2$ this result was derived for the (otherwise more general) class of signed measure valued processes (with application to the distribution of vortices in a two dimensional fluid) in [Kotelenez 1992b]. The proof for general d is exactly the same and is a special case of the corresponding theorems in [Kotelenez 1993]. Existence is shown by showing that $\mathcal{X}_\epsilon(t, \mathcal{X}_N(0)) := \mathcal{X}_N(t)$ (the empirical process) has a unique extension from $\widetilde{\mathcal{M}}_0$ to $\mathcal{M}_0$ with values in $\mathcal{M}_{[0,T]}$. Smoothness are shown by semigroup techniques. □

Consider the semilinear partial integro-differential equation on $\mathbb{H}_0 := L_2(\mathbb{R}^d, dr)$

(3.5)
$$\frac{\partial}{\partial t} X = D\Delta X - \nabla \cdot \left(X \int K(\cdot, p) X(p) dp\right) - \nabla \cdot [X f(\cdot)]$$
$$X(0) = X_0 \geq 0, X_0 \in \mathbb{H}_0 \text{ and } \int X_0(r)\, dr = m_+.$$

Let us assume that (3.5) has a unique mild solution $X(\cdot, X_0) \in C([0, T]; \mathbb{H}_0)$ for and $T > 0$, where $C([0, T]; \mathbb{H}_0)$ is the space of continuous functions from $[0, T]$ into $\mathbb{H}_0$. Denote the duality between distributions and smooth functions which extends the $\mathbb{H}_0$ inner product by $\langle \cdot, \cdot \rangle$.

THEOREM 3.3 (MACROSCOPIC LIMIT).   (i) $X_0 \in \mathbb{H}_0$ and $\mathcal{X}_0 \in \mathcal{M}_0 \cap \mathbb{H}_0$ *and $\mathcal{X}_0 \to X_0$ in $\mathbb{H}_0$ in mean square as $\epsilon \to 0$.*

(ii) $\widehat{\Gamma}_\epsilon$ *is as in Example 2.2.*

Let ϕ be infinitely differentiable with compact support in $\mathbb{R}^d$. Then for any $T > 0$

$$E \sup_{0 \leq t \leq T} \langle \mathcal{X}_\epsilon(t, \mathcal{X}_0) - X(t, X_0), \phi \rangle^2 \to 0, \text{ as } \epsilon \to 0$$

where $\mathcal{X}_\epsilon(t, \mathcal{X}_0)$ is the solution of (3.3) starting at $\mathcal{X}_0$ and $X(t, X_0)$ the solution of (3.5) starting at X_0.

Again, for $d = 2$ and the (more general) signed measure valued case a slightly stronger version was proved in [Kotelenez 1992b]. For general d we refer to [Kotelenez 1993].

Remark 3.4. Note that our approach does not give the macroscopic limit as $N \to \infty$ but only as $N \to \infty$ and $\epsilon \to 0$ unlike in the traditional approach, with uncorrelated fluctuation forces represented by i.i.d. Brownian motions.

Let us for the sake of simplicity assume $d = 1$ and $K \equiv 0 \equiv f$. In the classical case we have

$$dr^i(t) = d\beta^i(t),$$

and hence by Itô's formula the quadratic variation of

$$\langle \mathcal{X}_N(t), \phi \rangle = \frac{1}{N} \sum_{i=1}^{N} \phi(r^i(t))$$

becomes

$$(3.6) \qquad \frac{1}{N^2}\sum_{i=1}^{N}[\dot{\phi}]^2(r^i(t)) \sim O\left(\frac{1}{N}\right),$$

where $\dot{\phi}$ is the derivative of ϕ. Clearly, the noise must disappear as $N \to \infty$, and $\mathcal{X}_N(t)$ must tend to macroscopic (i.e. deterministic) space-time field which is in this case the solution of the heat equation.

On the other hand under the same assumptions as above for our model we have

$$dr^i(t) = \int \widetilde{\Gamma}_\epsilon(r^i,p)w(dp,dt),$$

hence for this model the quadratic variation of $\mathcal{X}_N(t)$ becomes

$$(3.7) \qquad \approx \frac{1}{N^2}\sum_{i,j=1}^{N}\dot{\phi}(r^i(t))\cdot\dot{\phi}(r^j(t))e^{-|r^i(t)-r^j(t)|^2/8\epsilon} \sim O(\epsilon^{d/2}).$$

(3.7) cannot be $O\left(\frac{1}{N}\right)$ since we still have N^2 terms. So as $N \to \infty$ the fluctuation do not have to disappear and $\mathcal{X}_N(t)$ does not become a macroscopic field. That the correct order is $O(\epsilon^{d/2})$ has been shown for $d = 2$ in [Kotelenez 1992b] and for general d in [Kotelenez 1993].

4. Extension to signed measure valued problems and other space-time models described by SPDE's

First of all let us mention that our approach is not restricted to particle system in the classical sense. Instead with the same effort we can introduce a "generalized" particle system, where we allow both "positive" and "negative" masses. The corresponding SPDE for the distribution of the "generalized" mass would have to conserve both the total "positive" and the total "negative" masses. A physical example is the vorticity distribution of a two dimensional fluid. Then the "positive" and "negative" masses are the positive and negative intensities, and the approach of Section 2 and 3 (except the adiabatic elimination) had been developed first for this model (cf. [Kotelenez 1992b]). Another model, where our approach might be applicable but has not yet been tested for is the Hodgkin-Huxley equation with stochastic input for the voltage potential of a neuron. A more ad-hoc type approach to this equation leading to a linear SPDE had been developed first by [Walsh 1981] and then extended by [Kallianpur and Wolpert 1984].

It is not clear at this stage whether our approach could lead to a derivation of the Dawson-Watanabe process (an SPDE obtained by a diffusion approximation to a system of diffusing and branching particles — cf. [Dawson 1975] and [Konno and Shiga 1988]). The main difficulty is the creation and annihilation of particles or mass which means that either the number of particles becomes dependent on chance and time or equations (2.2) and (2.3) have to be supplemented by a stochastic equation for the varying mass. Another difference is that our fluctuation term is driven by spatially correlated noise, whereas in branching the usual assumption is that the driving noise is space-time white noise (i.e. both in time and space uncorrelated).

Finally let us briefly comment on the results of [Vaillancourt 1988] and [Dawson and Vaillancourt 1993], who consider an N-particle system with correlated fluctuation forces (no adiabatic elimination) and the SPDE for the associated mass distribution. In these papers, for each N the correlated fluctuation forces are represented by *finitely* many independent Brownian motions coupled by diffusion coefficients which depend on the position and distribution of all particles. The fact that at each step there are only finitely many Brownian motions involved makes it impossible to derive an SPDE via a direct extension of the empirical process as done in [Kotelenez 1992b]. Thus the SPDE in Dawson and Vaillancourt is derived by a martingale problem approach.

Bibliography

V. S. Borkar, *Evolution of interacting particles in a Brownian medium*, Stochastics **14** (1984), 33-79.

D. A. Dawson, *Stochastic evolution equations and related measure processes*, J. Multivariate Anal. **5** (1975), 1-52.

D. A. Dawson and J. Vaillancourt, *Stochastic McKean-Vlasov limits*, (manuscript).

A. De Acosta, *Invariance principles in probability for triangular arrays of B-valued random vectors and some applications*, Ann. Probab. **2** (1982), 346-373.

R. M. Dudley, *Real analysis and probability*, Wadsworth and Brooks, Belmont, California, 1989.

Gardiner, *Handbook of stochastic methods*, Springer-Verlag, 1990.

N. Ikeda and S. Watanabe, *Stochastic differential equations and diffusion processes*, Amsterdam Oxford New York, North Holland, 1981.

G. Kallianpur and R. Wolpert, *Infinite dimensional differential equation models for spatially distributed neurons*, Appl. Math. Optim. **12** (1984), 125-172.

J. Kestin and R. J. Dorfman, *A course in statistical thermodynamics*, Academic Press, New York, 1971.

N. Konno and T. Shiga, *Stochastic partial differential equations for some measure-valued diffusions*, Probab. Theor. Relat. Fields **79** (1988), 201-225.

P. Kotelenez, *A stopped Doob inequality for stochastic convolution integrals and stochastic evolution equations*, Stochastic Anal. Appl. **2** (1984), 245-265.

_____, *Existence, uniqueness and smoothness for a class of function valued stochastic partial differential equations*, Stochastics and Stochastic Rep. **41** (1992), 177-199.

_____, *A stochastic Navier-Stokes equation for the vorticity of a two-dimensional fluid*, Case Western Reserve University, Dept. of Mathematics, Preprint #92-115, 1992.

_____, *Stochastic partial differential equations for particles and vortices*, (in preparation).

P. Kotelenez and K. Wang, (manuscript in preparation).

J. C. Lebowitz and E. Rubin, *Dynamical study of Brownian motion*, Phys. Rev. **131** (1963), 2381-2396.

P. Liptcer and A. N. Shiryayev, *Statistics of random processes*, "Nauka", Moscow, 1974 (Russian); English translation in Springer.

E. Nelson, *Dynamical theories of Brownian motion*, Princeton University Press, Princeton, N. J., 1972.

J. Vaillancourt, *On the existence of randon McKean-Vlasov limits for triangular arrays of exchangeable diffusions*, Stochastic Anal. Appl. (1988).

J. B. Walsh, *A stochastic model of neural response*, Adv. in Appl. Probab. **13** (1981), 231-281.

_____, *An introduction to stochastic partial differential equations*, École d'Été de Probabilités de Saint-Flour XIV-1984 (P. L. Hennequin ed.), Lecture Notes in Math., vol. 1180, Springer Berlin-Heidelberg-New York, 1986.

DEPARTMENT OF MATHEMATICS, CASE WESTERN RESERVE UNIVERSITY, CLEVELAND, OH 44106

E-mail address: pxk@po.cwru.edu

DEPARTMENT OF MATHEMATICS, CASE WESTERN RESERVE UNIVERSITY, CLEVELAND, OH 44106

E-mail address: kxw16@po.cwru.edu

Occupation Time Limit Theorems for Independent Random Walks

Tzong-Yow Lee and Bruno Remillard

1. Introduction

In this paper we review many results on the asymptotic behavior of functionals of the occupation time for systems of independent random walks. Some extensions to general Markov processes are also mentioned and given references. We do not intend to detail the corresponding literature for the branching models, neither for the interacting models; we refer the reader to [Cox and Griffeath 1985a, Lee 1993a, Iscoe and Lee 1993] for the branching models, and to [Cox and Griffeath 1983, Bramson, Cox, and Griffeath 1988, Cox 1988, Landim 1992], for the interacting models. A quick look at the few results obtained so far for the branching and the interacting models reveals that qualitatively, the same results as in the case of independent random walks turn out to be true. For results obtained before 1985, we refer the reader to [Cox and Griffeath 1985b] for a nice review. Of course, results for the branching and interacting models are more difficult to prove, but the results presented in this paper will however tell us what results to anticipate in these models.

By posing corresponding problems for branching and interacting models, one has many *conjectures* to verify. We believe that this direction of research will be challenging and fruitful, and hope that our summary could arouse some interests for these topics.

Let $\{\omega_j(\cdot)\}_{j \geq 1}$ be independent simple random walks on $X = \mathbb{Z}^d$, and define

$$\xi_n(x) = \sum_{j=1}^{\infty} 1_x(\omega_j(n)),$$

which represents the number of particles at time n at site x. We will suppose that the initial configuration $\{\xi_0(x)\}_{x \in X}$ satisfies one of the following conditions:
  (1) $\{\xi_0(x)\}_{x \in \mathbb{Z}^d}$ are independent and identically distributed (i.i.d. for short) Poisson random variables with mean 1.
  (2) $\xi_0(x) = 1$ for all $x \in \mathbb{Z}^d$.

1991 *Mathematics Subject Classification*. Primary: 60B12; Secondary: 60F05, 60F10, 60J15.

Key words and phrases. Large deviations, convergence in law, infinite particle systems, random walks.

The first author is supported in part by NSF Grant No. DMS 9207928.

The second author is supported in part by the Fonds Institutionnel de Recherche, Université du Québec à Trois-Rivières and by the Natural Sciences and Engineering Research Council of Canada, Grant No. OGP0042137.

This is the final form of the paper.

We will denote by P_i the law of the process $\{\xi_N(\cdot)\}_{N\geq 0}$ under initial configuration i, $i = 1, 2$. Integration with respect to P_i will be denoted by E_i, $i = 1, 2$.

Remark 1.1. Initial distribution (1) is the one that is commonly used in the literature, while initial configuration (2) is a particular case of the following initial distribution which was studied in [Cox and Durrett 1990] and [Remillard 1990]: $\{\xi_0(x)\}_{x \in X}$ is non-random and $\lim_{M \to \infty} \frac{1}{(2M)^d} \sum_{\substack{x; |x_j| \leq M \\ 1 \leq j \leq d}} \xi_0(x) = 1$.

All results stated in this paper can be expressed in terms of the so-called occupation time density process $D_N(\cdot)$ defined by

$$D_N(x) = N^{-1} \sum_{s=0}^{N-1} \xi_s(x)$$

First, the strong law of large numbers

$$P_i \left\{ \lim_{N \to \infty} D_N(x) = 1 \text{ for all } x \in \mathbb{Z}^d \right\} = 1, \quad i = 1, 2$$

is not difficult to prove. Then it is natural to attempt at large deviations from this strong law. The first surprise along this line is the following dimensional dependence property [Cox and Griffeath 1984]. For any $x \in \mathbb{Z}^d$

$$\log P_1 \{D_N(x) \geq b > 1\} \sim \begin{cases} -h_1(b)N^{1/2}, & \text{for } d = 1, \\ -h_2(b)N/\log N, & \text{for } d = 2, \\ -h_d(b)N, & \text{for } d \geq 3, \end{cases}$$

as $N \to \infty$, where the h_d's are positive functions on $(1, \infty)$ and where "$\sim$" means asymptotic to, that is RHS/LHS tends to 1. So the large deviation tails are fat in dimensions 1 and 2. The functions h_1 and h_2 can be expressed respectively in terms of $f_{1/2}$ and f_0, where f_β is the Laplace transform of the Mittag-Leffler distribution of index β, $\beta \in [0, 1)$, It is well-known, cf. [Darling and Kac 1957], that

$$f_\beta(s) = \begin{cases} \sum_{k=0}^{\infty} \frac{s^k}{\Gamma(k\beta+1)}, & 0 < \beta < 1, \\ \frac{1}{1-s}, & s < 1 \text{ and } \beta = 0, \\ +\infty, & s \geq 1 \text{ and } \beta = 0. \end{cases}$$

In Section 2, we will study the behavior of D_N when $X = \mathbb{Z}$. Then various generalizations will be discussed in Section 3. In Section 4 we will study the case $X = \mathbb{Z}^2$, while Section 5 will be devoted to higher dimensions.

2. Limit theorems for D_N when $X = \mathbb{Z}$

Before stating the first theorem, we need to introduce functions $\psi_i(\cdot)$ defined by

$$\psi_i(s) = \begin{cases} \int_{-\infty}^{\infty} E^{W_x} \left\{ 1_{\{\tau \leq 1\}} \left(f_{1/2}\left(s\sqrt{\frac{1-\tau}{2}}\right) - 1 \right) \right\} dx, & i = 1, \\ \int_{-\infty}^{\infty} \log \left(1 + E^{W_x} \left\{ 1_{\{\tau \leq 1\}} \left(f_{1/2}\left(s\sqrt{\frac{1-\tau}{2}}\right) - 1 \right) \right\} \right) dx, & i = 2, \end{cases}$$

where E^{W_x} denotes integration with respect to Brownian motion starting at x, and where τ is the first hitting time of the origin. Thus the density of τ is given by

$$P_x(\tau \in dt) = \frac{|x|e^{-x^2/2t}}{(2\pi t^3)^{1/2}} dt$$

Remark 2.1. Note that $\psi_1 \geq \psi_2$, they are equal only at 0,

$$\psi_1'(0) = 1 = \psi_2'(0) \text{ and } \sigma_1^2 = \psi_1''(0) > \sigma_2^2 = \psi_2''(0)$$

Moreover, ψ_1 can be simplified as $\psi_1(s) = s \int_0^1 f_{1/2}\left(s\sqrt{\frac{t}{2}}\right) dt$

We now state a large deviation result for $D_N(x)$, $x \in \mathbb{Z}$; the result follows esily from [Cox and Griffeath 1984] for P_1 and from [Remillard 1990] for P_2.

THEOREM 2.1 (LARGE DEVIATIONS). *If* $x \in \mathbb{Z}$ *then*

$$\log P_i \{D_N(x) \geq b\} \sim -\psi_i^*(b) N^{1/2}, \quad i = 1, 2, \quad b > 1$$

and

$$\log P_i \{D_N(x) \leq b\} \sim -\psi_i^*(b) N^{1/2}, \quad i = 1, 2, \quad b < 1$$

as $N \to \infty$, *where* $\psi_i^*(b) = \sup_s sb - \psi_i(s)$, *i.e.* ψ_i^* *is the Legendre transform of* ψ_i.

Remark 2.2. The functions ψ_i^*, $i = 1, 2$, are continuous for $b \geq 0$ and are $+\infty$ for $b < 0$. Moreover $\psi_1^*(b) < \psi_2^*(b)$ for all $b \geq 0$, except at $b = 1$, where $\psi_1^*(1) = 0 = \psi_2^*(1)$.

The whole proof in this theorem is to show that

$$\lim_{N \to \infty} \frac{1}{\sqrt{N}} \log E_i \left\{ e^{sN^{1/2} D_N(x)} \right\} = \psi_i(s), \quad i = 1, 2$$

The following Gärtner-Ellis theorem, cf. [Ellis 1985], then takes care of the rest.

THEOREM. *Let* X_n *be random variables on* $\mathbb{R}^d$ *such that for some sequence* a_n *increasing to* ∞,

$$\lim_{n \to \infty} \frac{1}{a_n} \log E \left\{ e^{a_n \langle \lambda, X_n \rangle} \right\} = c(\lambda)$$

exists for all $\lambda \in \mathbb{R}^d$. *If* $c(\cdot)$ *is differentiable everywhere, then for any closed set* C *and any open set* $O \subset \mathbb{R}^d$, *we have*

$$\limsup_{n \to \infty} \frac{1}{a_n} \log P\{X_n \in C\} \leq -\inf_{x \in C} c^*(x)$$

and

$$\liminf_{n \to \infty} \frac{1}{a_n} \log P\{X_n \in O\} \geq -\inf_{x \in O} c^*(x),$$

where c^* *is the Legendre transform of* c, *i.e.*

$$c^*(x) = \sup_{\lambda \in \mathbb{R}^d} \langle \lambda, x \rangle - c(\lambda)$$

Next, we point out that Theorem 2.1 also holds true when the number b is replaced by a sequence $\{b_N\}$, where either $b_N \to 1$ or $b_N \to \infty$ as $N \to \infty$ at appropriate rates. More precisely, the following two theorems, cf. [Remillard 1990, Lee 1993b, Cox and Griffeath 1984], hold true.

THEOREM 2.2. *The large deviation rate functions $\psi_i^*(b)$ satisfy*

$$\psi_i^*(b) \sim \frac{1}{2\sigma_i^2}(b-1)^2 \quad \text{as } b \to 1 \text{ and } i = 1, 2$$

$$\psi_i^*(b) \sim \begin{cases} b(2\log b)^{1/2}, & i = 1 \\ 4(b/3)^{3/2}, & i = 2, \end{cases} \quad \text{as } b \to \infty$$

Moreover, if $b > 0$,
 (i) *Enormous deviations*

$$\log P_i\{D_N(x) \geq bN^\theta\} \sim \begin{cases} -b(2\theta \log N)^{1/2} N^{(1+2\theta)/2}, & i = 1,\ 0 < \theta < \infty \\ -4\left(\frac{b}{3}\right)^{3/2} N^{(1+3\theta)/2}, & i = 2,\ 0 < \theta < 1 \end{cases}$$

as $N \to \infty$.
 (ii) *Moderate deviations*

$$\log P_i\{D_N(x) \geq 1 + bN^{-\theta}\} \sim -N^{1/2}\psi_i^*(1 + bN^{-\theta}) \sim -\frac{b^2}{2\sigma_i^2} N^{(1-4\theta)/2}$$

as $N \to \infty$, for $i = 1, 2$ and $0 < \theta < \frac{1}{4}$.

Remark 2.3. (1) As in Theorem 2.1, one uses the Gärtner-Ellis theorem for proving (ii) and $i = 2$ of (i). The proof of appropriate limits of cumulant generating functions comprises most of the work. The enormous deviation result for P_1 requires some different techniques. The term enormous deviations has its origin in [Cox and Durrett 1990].

(2) The processes P_i, $i = 1, 2$, have very different large deviation behavior although their moderate deviation behavior is quite similar. The enormous deviations are much more likely when the initial distribution is random (P_1).

THEOREM 2.3 (WEAK CONVERGENCE). *For $x \in \mathbb{Z}$ and $a \in \mathbb{R}$*

$$\lim_{N \to \infty} P_i\{D_N(x) \geq 1 + aN^{-1/4}\} = \int_a^\infty \frac{1}{\sqrt{2\pi}\,\sigma_i} e^{-\frac{u^2}{2\sigma_i^2}} du$$

That is, under P_i, $i = 1, 2$, $N^{1/4}(D_N(x) - 1) \xrightarrow[N \to \infty]{\text{Law}} \sigma_i Z$, Z being a standard normal random variable.

With a little more calculations, one can find that under P_i, $i = 1, 2$,

$$N^{1/4}(D_N(0) - 1, D_N(1) - 1) \xrightarrow[N \to \infty]{\text{Law}} (\sigma_i Z, \sigma_i Z)$$

This tells us that $L_N = D_N(0) - D_N(1)$ is small compared with $D_N(0)$ as $N \to \infty$. It requires a different scale to study L_N. A hierarchy of limiting results for L_N is the subject of the next theorem; see [Cox and Durrett 1990, Remillard 1990, Lee and Remillard 1993, Lee 1993b] for more details.

THEOREM 2.4. *Suppose $b > 0$ and $a \in \mathbb{R}$. Then*
 (i) *Large deviations*

$$\log P_i\{L_N \geq bN^{-1/4}\} \sim -J_i(b)N^{1/2} \quad \text{as } N \to \infty,$$

where J_i is the Legendre transform of $\psi_i(s^2)$, that is

$$J_i(b) = \sup_s sb - \psi_i(s^2)$$

(ii) *Enormous deviations*

$$\log P_i \left\{ L_N \geq bN^{(4\theta-1)/4} \right\} \sim -J_i(bN^\theta)N^{1/2}$$
$$\sim \begin{cases} -b(2\theta)^{1/4}N^{(2\theta+1)/2}(\log N)^{1/4}, & i=1, \ 0 < \theta < \infty \\ -\frac{5}{4}\left(\frac{2b}{3}\right)^{6/5} N^{(6\theta/5)+(1/2)}, & i=2, \ 0 < \theta < \frac{5}{4} \end{cases}$$

as $N \to \infty$.

(iii) *Moderate deviations*

$$\log P_i \left\{ L_N \geq bN^{-(4\theta+1)/4} \right\} \sim -J_i(bN^{-\theta})N^{1/2} \sim -\frac{b^2}{4}N^{(1-4\theta)/2},$$

as $N \to \infty$, where $i = 1, 2$ and $0 < \theta < \frac{1}{4}$.

(iv) *Weak convergence*

$$\lim_{N \to \infty} P_i \left\{ L_N \geq aN^{-1/2} \right\} = \int_a^\infty \frac{1}{2\sqrt{\pi}} e^{-\frac{u^2}{4}} du$$

That is, under P_i, $i = 1, 2$, $N^{1/2}L_N \xrightarrow[N \to \infty]{\text{Law}} 2^{1/2}Z$, where Z is a standard normal random variable.

Remark 2.4. We see that this time, P_1 and P_2 agree in the regime of moderate deviations and weak convergence, but disagree in the regime of large deviations and enormous deviations.

3. Generalizations

So far, we have results on the univariate behavior of D_N. So one can ask what is the joint behavior of $(D_N(x_1), D_N(x_2), \ldots, D_N(x_k))$. To answer this question, let $\mathcal{V} = \{V : X \to \mathbb{R}$ and V has finite support$\}$ and let $\overline{V} = \sum_x V(x)$. We also need to define $\mathcal{V}_0 = \{V \in \mathcal{V} : \overline{V} = 0\}$.

This section is devoted to generalizations of the previous results to $\langle D_N, V \rangle$ and also to function spaces.

Remark 3.1. (1) If $V(\cdot) = 1_{\{x\}}(\cdot)$, then $V \in \mathcal{V} \setminus \mathcal{V}_0$ and $\langle D_N, V \rangle = D_N(x)$. If $V = 1_{\{0\}} - 1_{\{1\}}$, then $V \in \mathcal{V}_0$ and $\langle D_N, V \rangle = L_N$.

(2) Note that $D_N(\cdot)$ belongs to $\Lambda = \{\lambda : X \to \mathbb{R}\}$, and if Λ is endowed with the product topology, then $\mathcal{V}$ can be identified with the dual of Λ through the relation

$$\langle \lambda, V \rangle = \sum_x \lambda(x) V(x), \quad \lambda \in \Lambda, V \in \mathcal{V}$$

In the next theorem, we restrict ourself to $\langle D_N, V \rangle$ with $V \in \mathcal{V} \setminus \mathcal{V}_0$. As one can guess from the previous results, the behavior of $\langle D_N, V \rangle$ is quite different depending on whether $V \in \mathcal{V} \setminus \mathcal{V}_0$ or $V \in \mathcal{V}_0$.

THEOREM 3.1. *Let $V \in \mathcal{V} \setminus \mathcal{V}_0$ be fixed. Then we have*

(i) *Large deviations*

$$N^{-1/2} \log P_i\{\langle D_N, V\rangle \in S\} \approx - \inf_{s \in S} \psi_i^*\left(\frac{s}{\overline{V}}\right),$$

where $\approx$ *means* $\begin{cases} \limsup LHS \leq RHS, & S \text{ closed,} \\ \liminf LHS \geq RHS, & S \text{ open.} \end{cases}$

(ii) *Enormous deviations*
If $b_N \to \infty$ then

$$\left[N^{1/2} b_N (\log b_N)^{1/2}\right]^{-1} \log P_1\{b_N^{-1}\langle D_N, V\rangle \in S\} \approx - \inf_{s \in S} I_1(s)$$

where $I_1(s) = \begin{cases} 2^{1/2} \frac{s}{\overline{V}} & s\overline{V} \geq 0 \\ +\infty & s\overline{V} < 0 \end{cases}$.

If $b_N \to \infty$ in such a way that $b_N/N \to 0$ then

$$\left(N^{1/2} b_N^{3/2}\right)^{-1} \log P_2\{b_N^{-1}\langle D_N, V\rangle \in S\} \approx - \inf_{s \in S} I_2(s),$$

where $I_2(s) = \begin{cases} 4\left(\frac{s}{3\overline{V}}\right)^{3/2} & s\overline{V} \geq 0 \\ +\infty & s\overline{V} < 0 \end{cases}$.

(iii) *Moderate deviations*
If $b_N \to 0$ in such a way that $b_N N^{1/4} \to \infty$, then

$$\left(N^{1/2} b_N^2\right)^{-1} \log P_i\{b_N^{-1}\left(\langle D_N, V\rangle - \overline{V}\right) \in S\} \approx - \inf_{s \in S} l_i(s/\overline{V})$$

where $l_i(s) = \frac{s^2}{2\sigma_i^2}$, $i = 1, 2$

(iv) *Weak convergence*

$$N^{1/4}\left(\langle D_N, V\rangle - \overline{V}\right) \xrightarrow[N \to \infty]{\text{Law}} |\overline{V}|\sigma_i Z, \quad i = 1, 2$$

Remark 3.2. (1) The proof for the enormous deviations in the case of P_1 is similar to the proof of Theorem 1 in [Lee 1993b]. The proof in the case of P_2 can be found in [Remillard 1990]. The proofs for the other limiting results are similar to the proofs of the results found in [Lee and Remillard 1993].

(2) Replacing $\overline{V}$ by $E_2(\langle D_N, V\rangle)$ in (iii) and (iv), one can prove that the last theorem also holds for the initial non-random distribution studied in [Cox and Durrett 1990] and in [Remillard 1990].

Let $\mathbf{1} \in \Lambda$ be such that $\mathbf{1}(\cdot) \equiv 1$. Using the fact that Λ is the projective limit of the spaces $\{\mathbb{R}^A, A \text{ finite }\}$, we can apply the Dawson-Gärtner Theorem ([Dawson and Gärtner 1987]) to prove the following function space extension of Theorem 3.1.

COROLLARY 3.1. *Let S be a subset of Λ. Then we have*
(i) *Large deviations*

$$N^{-1/2} \log P_i\{D_N \in S\} \approx - \inf_{\lambda \in S} \tilde{\psi}_i^*(\lambda),$$

where $\tilde{\psi}_i^*(\lambda) = \begin{cases} \psi_i^*(s), & \text{if } \lambda = s\mathbf{1} \\ +\infty, & \text{otherwise.} \end{cases}$

(ii) *Enormous deviations*
If $b_N \to \infty$ then
$$\left[N^{1/2}b_N(\log b_N)^{1/2}\right]^{-1} \log P_1\{b_N^{-1}D_N \in S\} \approx - \inf_{\lambda \in S} \tilde{I}_1(\lambda),$$
where $\tilde{I}_1(\lambda) = \begin{cases} 2^{1/2}s, & \text{if } \lambda = s\mathbf{1},\ s \geq 0 \\ +\infty, & \text{otherwise} \end{cases}$.

If $b_N \to \infty$ in such a way that $b_N/N \to 0$ then
$$\left(N^{1/2}b_N^{3/2}\right)^{-1} \log P_2\{b_N^{-1}D_N \in S\} \approx - \inf_{\lambda \in S} \tilde{I}_2(\lambda),$$
where $\tilde{I}_2(\lambda) = \begin{cases} 4\left(\frac{s}{3}\right)^{3/2} & \text{if } \lambda = s\mathbf{1},\ s \geq 0 \\ +\infty, & \text{otherwise}. \end{cases}$

(iii) *Moderate deviations*
If $b_N \to \infty$ in such a way that $b_N N^{1/4} \to \infty$, then
$$\left(N^{1/2}b_N^2\right)^{-1} \log P_i\{b_N^{-1}(D_N - \mathbf{1}) \in S\} \approx - \inf_{\lambda \in S} \tilde{l}_i(\lambda), \quad i = 1, 2$$
where $\tilde{l}_i(\lambda) = \begin{cases} \frac{s^2}{2\sigma_i^2}, & \text{if } \lambda = s\mathbf{1} \\ +\infty, & \text{otherwise}. \end{cases}$

(iv) *Weak convergence*
$$N^{1/4}(D_N - \mathbf{1}) \xrightarrow[N \to \infty]{\text{Law}} \sigma_i Z \mathbf{1}, \quad i = 1, 2$$
where Z is a standard normal random variable.

REMARK 3.3. $\tilde{\psi}_i^*(\lambda) = 0$ iff $\lambda = \mathbf{1}$.

Before stating the next theorem, we need to introduce some notations. For $V \in \mathcal{V}_0$, let $B_V = \frac{1}{2}\sum_x (G(x) + G(x-1))^2$, where $G(x) = \sum_{y \leq x} V(y)$. Then $B_V = 0$ iff $V \equiv 0$, and $B_{\lambda V} = \lambda^2 B_V$. For example, if $V = 1_{\{0\}} - 1_{\{1\}}$, then $V \in \mathcal{V}_0$ and $B_V = 1$.

The following theorem is the companion of Theorem 3.1. This time, we are interested in the limiting behavior of $\langle D_N, V \rangle$ for $V \in \mathcal{V}_0$.

THEOREM 3.2. *Let $V \in \mathcal{V}_0, V \not\equiv 0$ be fixed.*
(i) *Large deviations*
$$N^{-1/2} \log P_i\{N^{1/4}\langle D_N, V \rangle \in S\} \approx - \inf_{s \in S} J_i\left(\frac{s}{B_V^{1/2}}\right), \quad i = 1, 2$$

(ii) *Enormous deviations*
If $b_N \to \infty$ then
$$\left[N^{1/2}b_N(\log b_N)^{1/4}\right]^{-1} \log P_1\{b_N^{-1}N^{1/4}\langle D_N, V \rangle \in S\} \approx - \inf_{s \in S} 2^{1/4}\frac{s}{B_V^{1/2}}$$

If $b_N \to \infty$ in such a way that $b_N N^{-5/4} \to 0$ then
$$\left(N^{1/2}b_N^{6/5}\right)^{-1} \log P_2\{b_N^{-1}N^{1/4}\langle D_N, V \rangle \in S\} \approx - \inf_{s \in S} \frac{5}{4}\left(\frac{2s}{3\sqrt{B_V}}\right)^{6/5}$$

(iii) *Moderate deviations*

If $b_N \to 0$ in such a way that $b_N N^{1/4} \to \infty$, then

$$\left(N^{1/2}b_N^2\right)^{-1} \log P_i\{b_N^{-1}N^{1/4}\langle D_N, V\rangle \in S\} \approx -\inf_{s \in S} \frac{s^2}{4B_V}, \quad i = 1, 2$$

(iv) *Weak convergence*

$$N^{1/2}\langle D_N, V\rangle \xrightarrow[N\to\infty]{\text{Law}} \sqrt{2B_V}\, Z, \quad i = 1, 2,$$

where Z is a standard normal random variable.

Remark 3.4. (1) The processes P_1 and P_2 have the same moderate deviation behavior. However their behavior differ for large and enormous deviations.

(2) The proof for the enormous deviations in the case of P_1 is similar to the proof of Theorem 1 in [Lee 1993b]. The proof in the case of P_2 can be found in [Remillard 1990].

To state the function space results, let $\widehat{D}_N(x) = D_N(x) - D_N(0)$ and $\hat{\Lambda} = \{\lambda : \mathbb{Z}\setminus\{0\} \to \mathbb{R}\}$. Define $\|V\|^2 = 2B_V$ for $V \in \mathcal{V}_0$ and the corresponding inner product

$$(V_1, V_2) = \frac{1}{4}\left(\|V_1 + V_2\|^2 + \|V_1 - V_2\|^2\right)$$

COROLLARY 3.2. *Let S be a subset of $\hat{\Lambda}$. Then we have*

(i) *Large deviations*

$$N^{-1/2}\log P\{N^{1/4}\widehat{D}_N \in S\} \approx -\inf_{\lambda \in S} J_i\left(\sqrt{2}\,\|\lambda\|_*\right),$$

where $\|\lambda\|_* = \sup_{\|V\|=1} \sum_{x \neq 0} \lambda(x)V(x)$

(ii) *Enormous deviations*

If $b_N \to \infty$ then

$$\left[N^{1/2}b_N(\log b_N)^{1/4}\right]^{-1}\log P_1\{b_N^{-1}N^{1/4}\widehat{D}_N \in S\} \approx -\inf_{\lambda \in S} 2^{3/4}\|\lambda\|_*$$

If $b_N \to \infty$ in such a way that $b_N N^{-5/4} \to 0$ then

$$\left(N^{1/2}b_N^{6/5}\right)^{-1}\log P_2\{b_N^{-1}N^{1/4}\widehat{D}_N \in S\} \approx -\inf_{\lambda \in S} 10\left(\frac{\|\lambda\|_*}{6}\right)^{6/5}$$

(iii) *Moderate deviations*

If $b_N \to 0$ in such a way that $b_N N^{1/4} \to \infty$, then

$$\left(N^{1/2}b_N^2\right)^{-1}\log P_i\{b_N^{-1}N^{1/4}\widehat{D}_N \in S\} \approx -\inf_{\lambda \in S} \frac{\|\lambda\|_*^2}{2}, \quad i = 1, 2$$

(iv) *Weak convergence*

$$N^{1/2}\widehat{D}_N \xrightarrow[N\to\infty]{\text{Law}} \mathcal{D}, \quad i = 1, 2,$$

where $\mathcal{D}$ is a mean zero Gaussian field with covariance function given by

$$\text{Cov}\bigl(\mathcal{D}(x), \mathcal{D}(y)\bigr) = (1_y - 1_0, 1_y - 1_0), \quad x, y \in \mathbb{Z}\setminus\{0\}$$

Remark 3.5. In [Lee and Remillard 1992], the authors also studied large deviations, moderate deviations and weak convergence for the difference density $\{\widehat{D}_N\}_{N\geq 1}$ of systems of independent Markovian particles starting from Poisson initial distribution. These results complement the results obtained in [Lee 1989] for the density $\{D_N\}_{N\geq 1}$ of such particle systems.

4. Limits theorems for D_N when $X = \mathbb{Z}^2$

The counterparts of Theorems 2.1, 2.2 and 2.3 are the contents of the following theorems. The results are stated only for the Poisson initial distribution (P_1), but we conjecture that they are also valid for P_2. The proofs for Theorems 4.1 and 4.2 can be found in [Cox and Griffeath 1984].

Let us define ϕ and its Legendre transform ϕ^* in the following way:

$$\phi(s) = \begin{cases} \frac{\pi s}{\pi - s} & s < \pi \\ +\infty & s \geq \pi \end{cases}$$

$$\phi^*(b) = \sup_s sb - \phi_i(s) = \begin{cases} \pi(\sqrt{b}-1)^2 & b \geq 0 \\ +\infty & b < 0 \end{cases}$$

THEOREM 4.1 (LARGE DEVIATIONS). *If* $x \in \mathbb{Z}^2$ *we have*

$$\log P_1\{D_N(x) \geq b\} \sim -\phi^*(b)N/\log N, \quad b > 1$$

and

$$\log P_i\{D_N(x) \leq b\} \sim -\phi^*(b)N/\log N, \quad b < 1$$

as $N \to \infty$.

THEOREM 4.2 (MODERATE DEVIATIONS AND WEAK CONVERGENCE). *Suppose* $x \in \mathbb{Z}^2$, $b > 0$ *and* $a \in \mathbb{R}$. *Then*
  (i) *Moderate deviations*

$$\log P_1\{D_N(x) \geq 1 + bN^{-\theta}\} \sim -\phi^*(1+bN^{-\theta})N/\log N$$

$$\sim -\frac{\pi b^2}{4} N^{(1-2\theta)}/\log N$$

  as $N \to \infty$, *for* $0 < \theta < 1/2$.
  (ii) *Weak convergence*

$$\lim_{N\to\infty} P_1\left\{D_N(x) \geq 1 + a(N/\log N)^{-1/2}\right\} = \int_a^\infty \frac{1}{2} e^{-\frac{\pi}{4}u^2} du$$

That is $\sqrt{N/\log N}\,(D_N(x) - 1) \xrightarrow[N\to\infty]{\text{Law}} \sqrt{2/\pi}\, Z$ *for the process* P_1, *where* Z *is a standard normal random variable.*

Remark 4.1. It is not known whether the enormous deviations have the following property:

$$\log P_1\{D_N(x) \geq bN^\theta\} \sim -\phi^*(bN^\theta)N/\log N \sim -bN^{1+\theta}/\log N$$

as $N \to \infty$, for $b > 0$ and for some $\theta > 0$.

Let $e_1 = (1,0) \in \mathbb{Z}^2, o = (0,0)$, and let $L_N = D_N(o) - D_N(e_1)$. As in section 2, we have

$$\left(\sqrt{N/\log N}(D_N(o) - 1), \sqrt{N/\log N}(D_N(e_1) - 1)\right) \xrightarrow[N \to \infty]{\text{Law}} \left(\sqrt{2/\pi}\, Z, \sqrt{2/\pi}\, Z\right),$$

where Z is a standard normal random variable. So it requires a different scaling to study L_N. We will summarize the known results for L_N in the next theorem. The proofs of these results can be found in [Lee and Remillard 1993].

First, we define $I(b) = \sup_{s \in \mathbb{R}} bs - \phi(s^2)$. It is easy to check that $I(b) \sim b^2/4$ as $b \to 0$.

THEOREM 4.3. *There exists $b_0 > 0$ such that*
(i) *Large deviations*

$$\log P_1\left\{L_N \geq b(\log N)^{-1/2}\right\} \sim -I(b)N/\log N, \quad \text{as } N \to \infty,$$

for all $0 < b < b_0$.
(ii) *Moderate deviations*

$$\log P_1\left\{L_N \geq bN^{-\theta}\right\} \sim -\frac{b^2}{4}N^{1-2\theta}, \quad b > 0,\ 0 < \theta < \frac{1}{2}$$

(iii) *Weak convergence*

$$\lim_{N \to \infty} P_1\left\{L_N \geq aN^{-1/2}\right\} = \int_a^\infty \frac{1}{\sqrt{4\pi}} e^{-\frac{u^2}{4}}\, du \quad a \in \mathbb{R},$$

i.e. $\sqrt{N} L_N \xrightarrow[N \to \infty]{\text{Law}} \sqrt{2}\, Z$, *where Z is a standard normal random variable.*

CONJECTURE 4.1. We think that (i) is true for all $b > 0$ but we don't have a proof. It is unknown whether

$$\log P_1\left\{L_N \geq bN^\theta\right\} \sim -I\left(bN^\theta\sqrt{\log N}\right) N/\log N, \quad \text{as } N \to \infty$$

for $b > 0$ and for some $\theta > 0$. We also conjecture that Theorems 4.1, 4.2 and 4.3 remain true for P_2.

Finally, we state two theorems on the asymptotic behavior of $D_N \in \Lambda$ and of $\widehat{D}_N \in \hat{\Lambda}$, where $\widehat{D}_N(x) = D_N(x) - D_N(o)$, $x \in \mathbb{Z}^2 \setminus \{o\}$.

THEOREM 4.4. *Let S be a subset of Λ. Then we have*
(i) *Large deviations*

$$(\log N/N) \log P_1\{D_N \in S\} \approx -\inf_{\lambda \in S} \tilde{\phi}^*(\lambda),$$

where $\tilde{\phi}^*(\lambda) = \begin{cases} \pi(\sqrt{s} - 1)^2 & \text{if } \lambda = s\mathbf{1} \\ +\infty & \text{otherwise} \end{cases}$

(ii) *Moderate deviations*
If $b_N \to 0$ in such a way that $b_N(N/\log N)^{1/2} \to \infty$, then

$$(b_N^2 N/\log N)^{-1} \log P_1\{b_N^{-1}(\log N)^{1/2}(D_N - 1) \in S\} \approx -\inf_{\lambda \in S} \tilde{L}(\lambda),$$

where $\tilde{L}(\lambda) = \begin{cases} \dfrac{\pi s^2}{4} & \text{if } \lambda = s\mathbf{1} \\ +\infty & \text{otherwise} \end{cases}$

(iii) *Weak convergence*
Under P_1
$$\sqrt{N/\log N}(D_N - 1) \xrightarrow[N\to\infty]{\text{Law}} \sqrt{2/\pi}\, Z\mathbf{1},$$
where Z is a standard normal random variable.

Before stating the last result of this section, let
$$a(x) = \sum_{n=0}^{\infty} \left(P(\omega(n) = o \mid \omega(0) = o) - P(\omega(n) = x \mid \omega(0) = o)\right), \quad x \in \mathbb{Z}^2,$$

where $\omega(n)$, $n \geq 0$, is a symmetric simple random walk on $\mathbb{Z}^2$. Then it is easy to see that $a(e_1) = 1$. Define
$$\|V\|^2 = -2 \sum_x \sum_y V(x)V(y)a(x-y) - \sum_x V^2(x), \quad V \in \mathcal{V}_0$$

For example,
$$\|1_x - 1_o\|^2 = -2\sum_z \sum_y \left(1_x(z) - 1_o(z)\right)\left(1_x(y) - 1_o(y)\right)a(z-y)$$
$$- \sum_y \left(1_x(y) - 1_o(y)\right)^2 = 2.$$

We also define corresponding inner product:
$$(V_1, V_2) = \frac{1}{4}\left(\|V_1 + V_2\|^2 + \|V_1 - V_2\|^2\right)$$

Further let $\|\lambda\|_* = \sup_{\|V\|=1} \sum_{x \neq 0} \lambda(x)V(x)$.

THEOREM 4.5. *Let S be a subset of $\widehat{\Lambda}$. Then we have*
(i) *Large deviations: There exists $s_0 > 0$ such that if $S \subset \{\lambda; \|\lambda\|_* < s_0\}$, then*
$$(\log N/N) \log P_1\{\sqrt{\log N}\, \widehat{D}_N \in S\} \approx -\inf_{\lambda \in S} I(\|\lambda\|_*)$$

(ii) *Moderate deviations*
If $b_N \to 0$ in such a way that $b_N(N/\log N)^{1/2} \to \infty$, then
$$(b_N^2 N/\log N)^{-1} \log P_1\{b_N^{-1}(\log N)^{1/2} \widehat{D}_N \in S\} \approx -\inf_{\lambda \in S} \frac{\|\lambda\|_*^2}{2}$$

(iii) *Weak convergence*
Under P_1
$$\sqrt{N}\widehat{D}_N \xrightarrow[N\to\infty]{\text{Law}} \mathcal{D},$$
where $\mathcal{D}$ is a nondegenerate Gaussian random field with covariance given by
$$\operatorname{Cov}(\mathcal{D}(x), \mathcal{D}(y)) = (1_x - 1_o, 1_y - 1_o), \quad x, y \in \mathbb{Z}^2 \setminus \{o\}$$

Remark 4.2. The last theorem follows mainly from the following result:
$$\lim_{N\to\infty} \log E_1\left\{e^{N/\sqrt{\log N} \sum_{x \in \mathbb{Z}^2 \setminus \{o\}} V(x)\widehat{D}_N(x)}\right\} = \phi(\|V\|^2/2),$$

for all $V \in \mathcal{V}_0$ such that $\|V\| < \sqrt{\pi/2}$.

5. Limit theorems in high dimensions: $X = \mathbb{Z}^d$, $d \geq 3$

When $X = \mathbb{Z}^d$, $d \geq 3$, the large deviation probabilities $P_1\{D_N(x) \geq b > 1\}$, $x \in X$, have the usual exponential decay as $N \to \infty$. Define

$$q = q_d = P\{\omega(n) = o \text{ for some } n \geq 1 \mid \omega(0) = o\},$$

$$g(s) = \begin{cases} (1-q)(e^s - 1)/(1 - qe^s), & s < -\log q \\ +\infty, & s \geq -\log q \end{cases},$$

$$g^*(b) = \sup_{s \in \mathbb{R}}(bs - g(s)),$$

where $\omega(n)$, $n \geq 0$, is a simple symmetric random walk on $\mathbb{Z}^d$, and $o \in \mathbb{Z}^d$ is the origin. In the next theorem, we summarize some results that can be found or that are easy consequences of results in [Cox and Griffeath 1984, Donsker and Varadhan 1987].

THEOREM 5.1. *For $x \in \mathbb{Z}^d$, $d \geq 3$, we have*
(i) *Large deviations*

$$\log P_1\{D_N(x) \geq b\} \sim -g^*(b)N \text{ as } N \to \infty, \text{ for } b > 1$$

(ii) *Moderate deviations*

$$\log P_1\{D_N(x) \geq 1 + bN^{-\theta}\} \sim -g^*(1 + bN^{-\theta})N \sim -\frac{(1-q)b^2}{2(1+q)}N^{1-2\theta},$$

as $N \to \infty$, where $0 < \theta < 1/2$ and $b > 0$.
(iii) *Weak convergence*

$$\lim_{N \to \infty} P_1\{D_N(x) \geq 1 + aN^{-1/2}\} = \int_a^\infty \frac{1}{\sqrt{2\pi\frac{1+q}{1-q}}} e^{-\frac{(1-q)u^2}{2(1+q)}} du, \quad a \in \mathbb{R},$$

i.e. $\sqrt{N}(D_N(x) - 1) \xrightarrow[N \to \infty]{\text{Law}} \sqrt{\frac{1+q}{1-q}} Z$, *where Z is a standard normal random variable.*

Remark 5.1. Enormous deviations have not been studied yet. We conjecture that

$$\log P_1\{D_N(x) \geq bN^\theta\} \sim -g^*(bN^\theta)N, \text{ as } N \to \infty$$

for $b > 0$, and for some $\theta > 0$. It is an interesting problem how the processes P_1 and P_2 differ in the behavior of enormous deviations in the cases of $X = \mathbb{Z}^d$, $d \geq 2$. For this we know much less, compared with the case of $X = \mathbb{Z}$, cf. Theorems 2.2 and 2.4.

Finally, if $e_1 = (1, 0, \ldots, 0)$, then one can prove that $\sqrt{N}(D_N(o) - D_N(e_1))$ converges in law to a nondegenerate Gaussian random variable. This means that $L_N = D_N(o) - D_N(e_1)$ behaves qualitatively as $D_N(x)$, as $N \to \infty$. That is, there is no need to consider $\mathcal{V} \setminus \mathcal{V}_0$ and $\mathcal{V}_0$ separately.

Bibliography

M. Bramson, J. T. Cox, and D. Griffeath, *Occupation time large deviation of the voter model*, Probab. Theory Related Fields **77** (1988), 401–413.

J. T. Cox, *Some limit theorems for voter model occupation times*, Ann. Probab. **16** (1988), 1559–1569.

J. T. Cox and R. Durrett, *Large deviations for independent random walks*, Probab. Theory Related Fields **84** (1990), 67–82.

J. T. Cox and D. Griffeath, *Occupation time limit theorems for the voter model*, Ann. Probab. **11** (1983), 876–893.

_____, *Large deviations for Poisson systems of independent random walks*, Z. Wahrsch. Verw. Gebiete **69** (1984), 543–558.

_____, *Occupation times for critical branching Brownian motions*, Ann. Probab. **13** (1985), 1108–1132.

_____, *Large deviations for some infinite particle systems occupation times*, Particle Systems, Random Media and Large Deviations (R. Durrett, ed.), Contemp. Math., vol. 41, Amer. Math. Soc. Providence, 1985, pp. 43–54.

D. A. Dawson and J. Gärtner, *Large deviations for McKean-Vlasov limit of weakly interacting diffusions*, Stochastics **20** (1987), 247–308.

D. A. Darling and M. Kac, *On occupation times for Markov processes*, Trans. Amer. Math. Soc. **84** (1957), 444–458.

M. D. Donsker and S. R. S. Varadhan, *Large deviation for noninteracting infinite particle systems*, J. Statist. Phys. **46** (1987), 1195–1232.

R. S. Ellis, *Entropy, large deviations and statistical mechanics*, Grundlehren Math. Wiss., vol. 271, Springer-Verlag, 1985.

I. Iscoe and T.-Y. Lee, *Large deviations for occupation times of measure-valued branching Brownian motions*, Stochastics Stochastics Rep., 1993 (to appear).

C. Landim, *Occupation time large deviations for the symmetric simple exclusion process*, Ann. Probab. **20** (1992), 206–231.

T.-Y. Lee, *Large deviations for systems of noninteracting recurrent particles*, Ann. Probab. **17** (1989), 46–57.

_____, *Some limit theorems for super-Brownian motion and semilinear differential equations*, Ann. Probab. **21** (1993), 979–995

_____, *Large deviations for independent random walks on the line*, 1993 (submitted for publication).

T.-Y. Lee and B. Remillard, *Occupation times in systems of null recurrent Markov processes*, C. R. Math. Rep. Acad. Sci. Canada **14** (1992), 2–6.

_____, *Occupation times in systems of null recurrent Markov processes*, 1993, to appear in Probab. Theor. Relat. Fields.

B. Remillard, *Asymptotic behaviour of the Laplace transform of weighted occupation times of random walks and applications*, Diffusion Processes and Related Problems in Analysis, I (Mark A. Pinsky, ed.), Progr. Probab., vol. 22, Birkhäuser, Boston, 1990, pp. 497–519.

ABSTRACT. We summarize many limit theorems for systems of independent simple random walks in $\mathbb{Z}^d$. These theorems are classified into four classes: weak convergence, moderate deviations, large deviations and enormous deviations. A hierarchy of relations is pointed out and some open problems are posed. Extensions to function spaces are also mentioned.

DEPT. OF MATHEMATICS, UNIVERSITY OF MARYLAND, COLLEGE PARK, MD 20742,, USA
E-mail address: tyl@math.umd.edu

DÉP. DE MATHÉMATIQUES ET D'INFORMATIQUE, UNIVERSITÉ DU QUÉBEC À TROIS-RIVIÈRES, TROIS-RIVIÈRES (QUÉBEC), CANADA G1K 7P4
E-mail address: bruno_remillard@uqtr.uquebec.ca

Large Deviations and Boltzmann Equation

Christian Léonard

1. Introduction

The aim of this paper is to study large deviations for a large particle system associated to the spatially homogeneous Boltzmann equation. The detailed proofs of the results stated below will appear elsewhere [Léonard 1993]; in the present paper, are only given sketches of the proofs.

About Boltzmann equation. Let $u_t(x,z)$ stand for the density of molecules of a gas at time t, with location $x \in \mathbb{R}^3$ and velocity $z \in \mathbb{R}^3$. We clearly have

$$u_t(x,z) \geq 0 \quad \text{and} \quad \int_{\mathbb{R}^3 \times \mathbb{R}^3} u_t(x,z)\,dx\,dz = 1, \quad \forall 0 \leq t \leq T.$$

The evolution of a dilute gas is well described by the following Boltzmann equation

$$(1.1) \quad \frac{\partial u_t}{\partial t} + z \cdot \frac{\partial}{\partial x} u_t = \int_{\mathbb{R}^3 \times S_2} \{u_t(x,z^*)u_t(x,z'^*) - u_t(x,z)u_t(x,z')\}$$
$$\times q(|z-z'|, (z-z') \cdot n)\,dz'\,dn,$$

where

$$z^* = z - (z-z') \cdot n\, n,$$
$$z'^* = z' - (z'-z) \cdot n\, n,$$

(z^*, z'^*) and (z,z') respectively standing for the incident and resulting velocities of a biparticle performing a collision which is described by means of a parameter $n \in S_2$: the unit sphere of $\mathbb{R}^3$. This leads us to

$$(1.2) \quad \begin{cases} z^* + z'^* &= z + z' \\ |z-z'| &= |z^* - z'^*| \\ (z-z') \cdot n &= -(z^* - z'^*) \cdot n \end{cases}$$

and to the conservation of kinetic energy equation

$$(1.3) \quad |z^*|^2 + |z'^*|^2 = |z|^2 + |z'|^2.$$

In the special case, where

$$q(z,z') := \int_{S_2} q(|z-z'|, (z-z') \cdot n)\,dn < \infty,$$

1991 *Mathematics Subject Classification.* Primary: 60F10, 60G57; Secondary: 60K35.
The final form of this paper will appear elsewhere.

$q(z, z')$ is the mean intensity of the collisions with incident velocities (z, z') while

$$\frac{q(|z - z'|, (z - z') \cdot n)}{q(z, z')} dn$$

describes the distribution of these random-like collisions. It is precisely the integration with respect to dn which yields the growth of the entropy (Boltzmann H-theorem).

If the initial distribution is spatially homogeneous, that is $\frac{\partial u_0}{\partial x} \equiv 0$, this homogeneity is preserved as time runs : $\frac{\partial u_t}{\partial x} \equiv 0$, $\forall t > 0$ and (1.1) becomes

$$(1.4) \quad \frac{\partial}{\partial t} u_t(z) = \int_{\mathbb{R}^3 \times S_2} \{u_t(z^*) u_t(z'^*) - u_t(z) u_t(z')\} q(|z - z'|, (z - z') \cdot n) \, dz' \, dn$$

which is the *spatially homogeneous Boltzmann equation*. Taking (1.2) into account, one can interprete equation (1.4) as a *flow equation*.

L. Arkeryd and A.S. Sznitmann have proved strong [Arkeryd 1972] and weak [Sznitman 1984] existence and uniqueness of the solutions of (1.4) in the *hard spheres* case which corresponds to $q(z, z', n) = |(z - z') \cdot n|$.

H. Tanaka [Tanaka 1978] also proved the weak existence and uniqueness of the solutions of (1.4) in the *Maxwellian molecules* case which corresponds to $q(z, z', n) = \psi(|\cos(z - z', n)|)$ with $\psi(u) = O_{u \to 0}(u^{-3/2})$.

In both cases, difficulties of the probabilistic approach mostly come from

$$(1.5) \quad \sup_{z, z'} q(z, z') = +\infty,$$

while choosing an analytic approach, one has to deal with the non-linearity (almost quadratic) of equation (1.1).

Carrying out the change of variables

$$\begin{cases} (z, z', n) \to (z, z', \Delta, \Delta') \\ q(|z - z'|, (z - z') \cdot n) \, dn \to \mathcal{L}(z, z', d\Delta d\Delta') \end{cases} \text{ with } \begin{cases} \Delta(z, z', n) = (z - z') \cdot nn, \\ \Delta'(z, z', n) = (z' - z) \cdot nn, \end{cases}$$

equation (1.4) becomes

$$(1.6) \quad \frac{\partial}{\partial t} u_t = A(u_t)^* u_t,$$

where

$$A(u_t) f(z) = \int_{(\Delta)} \{f(z + \Delta) - f(z)\} \left(\int_{(z', \Delta')} \mathcal{L}(z, z', \cdot \times d\Delta') u_t(dz') \right) (d\Delta).$$

The particle system. Let us take a collision (Lévy) kernel $\mathcal{L}$, we want to build a Markov particle system whose empirical measure, as the number N of particles tends to infinity, approaches a weak solution of (1.6). Let us denote

$$\mathcal{Z} = \mathbb{R}^d$$

the space of the z's and $z^N = (z_1, \ldots, z_N) \in \mathcal{Z}^N$ any configuration, the set of all the jumps of the biparticules is $E = \mathcal{Z}^2 \setminus \{(0, 0)\}$, $M_1(\mathcal{Z})$ stands for the set of all probability measures built on the Borel σ-field of $\mathcal{Z}$.

Let $X^N = (X_i^N)_{1 \leq i \leq N}$ be the Markov process on $\mathcal{Z}^N$ with generator

$$(1.7) \quad A_N \Phi(z^N) = \frac{1}{N-1} \sum_{1 \leq i < j \leq N} \int_E \left\{ \Phi(z^N + \Delta_{(i)} + \Delta'_{(j)}) - \Phi(z^N) \right\}$$
$$\times \mathcal{L}(z_i, z_j, d\Delta d\Delta'), \quad \Phi \in C_0^1(\mathcal{Z}^N),$$

where for all $\Delta \in \mathcal{Z}$, $1 \leq i \leq N$, $\Delta_{(i)} = (0, \ldots, 0, \underset{\underset{i^{\text{th}}}{\uparrow}}{\Delta}, 0, \ldots, 0) \in \mathcal{Z}^N$. Let $\bar{z}^N \in M_1(\mathcal{Z})$ stand for the empirical measure of z^N and

$$\overline{X}^N : t \mapsto \overline{X}^N(t) = \frac{1}{N} \sum_{1 \leq i \leq N} \delta_{X_i^N(t)} \in M_1(\mathcal{Z})$$

be the empirical process of the particle system. If $\overline{X}^N(0) \xrightarrow[N \to \infty]{\mathcal{L}} u_0$, then

$$(1.8) \quad \overline{X}^N(\cdot) \xrightarrow[N \to \infty]{\mathcal{L}} u(\cdot),$$

where $u(\cdot)$ is a weak solution of (1.6), with initial condition $u(0) = u_0$. The proof relies upon

$$A_N f(\langle \varphi, \cdot \rangle)(\bar{z}^N) = f'(\langle \varphi, \bar{z}^N \rangle) \langle A(\bar{z}^N) \varphi, \bar{z}^N \rangle + O_{f,\varphi}\left(\frac{1}{N}\right).$$

A.S. Sznitmann [Sznitman 1984] has proved this law of large numbers in the hard spheres case.

A consequence of (1.8) is the *propagation of chaos* which in its weaker form may approximately be stated as follows. If for any $N \geq 1$, the law of $(X_i^N(0))_{1 \leq i \leq N}$ (at time $t = 0$) is $u_0^{\otimes N}$, then for any $k \geq 1$ and any $t > 0$, $u(t)^{\otimes k}$ is the limiting law of $(X_i^N(t))_{1 \leq i \leq k}$.

Large deviations. We would like to get a large deviation principle for the laws of the $\overline{X}^N$'s, that is an estimation of the form

$$\frac{1}{N} \log \mathbb{P}\left(\overline{X}^N \in A\right) \underset{N \to \infty}{\asymp} - \inf_{\mu \in A} I(\mu)$$

for some subsets A of $M_1(\mathcal{Z})^{[0,T]}$. Similar results have already been obtained by F. Comets [Comets 1987] in the case of the long range Ising model, by D.A. Dawson and J. Gärtner [Dawson and Gärtner 1987] for weakly interacting diffusion systems or by C. Kipnis and S. Olla [Kipnis and Olla 1990] for the hydrodynamical limit of independent Brownian motions.

The main troubles one has to deal with in the case of (1.7) are the following ones.

(1.9.i) Because a system lead by equation (1.7) performs simultaneous jumps, there is no noninteracting (i.i.d.) reference (dominating in the sense of absolute continuity) particle system. Indeed, two independent Poisson processes never jump at the same time.

(1.9.ii) Because of (1.2), the cone generated by the support of the measure $\mathcal{L}(z, z', \cdot)$ may not be the whole space. In other words, the diffusion with jumps X^N may be degenerated.

(1.9.iii) The intensity of the jumps is unbounded (see (1.5)).

2. Statement of the results

Let us assume that the initial data and the Lévy kernel $\mathcal{L}$ satisfy the following hypotheses.

Without restriction, we choose $\mathcal{L}$ such that $\mathcal{L}(z, z', d\Delta d\Delta') = \mathcal{L}(z', z, d\Delta' d\Delta)$, $\forall z, z', d\Delta, d\Delta'$.

ASSUMPTIONS. In order to make things simple, let us assume that the initial configuration is deterministic: for any $N \geq 1$ and $1 \leq i \leq N$, $X_i^N(0) = x_i \in \mathcal{Z}$ with $\frac{1}{N} \sum_{i=1}^{N} \delta_{x_i} \xrightarrow[N \to \infty]{} u_0$ in $M_1(\mathcal{Z})$.

We also assume that there exist a C^1 function $\varphi : \mathcal{Z} \to [1, +\infty[$ satisfying $\lim_{|z| \to \infty} \varphi(z) = +\infty$, and a real number $\lambda \geq 0$ such that for all $z, z' \in \mathcal{Z}$:

(A.0) $\int_{\mathcal{Z}} \exp(\alpha \varphi(z)) u_0(dz) < \infty$, for some $\alpha > 0$,

(A.1) $\int_E \{\exp[\varphi \oplus \varphi(z + \Delta, z' + \Delta') - \varphi \oplus \varphi(z, z')] - 1\} \mathcal{L}(z, z', d\Delta d\Delta') \leq \lambda\, \varphi \oplus \varphi(z, z')$,

(A.2) $\int_E \inf(|\Delta| + |\Delta'|, 1)\, \mathcal{L}(z, z', d\Delta d\Delta') \leq \lambda(1 + \varphi(z)\varphi(z'))$,

(A.3) for any continuous bounded function g on E, $(z, z') \in \mathcal{Z}^2 \mapsto \int_E \inf(|\Delta| + |\Delta'|, 1) g(\Delta, \Delta') \mathcal{L}(z, z', d\Delta d\Delta') \in \mathbb{R}$ is continuous, and one assumes that

(A.4) $\sup_{(z,z') \in \mathcal{Z}^2} \int_E \mathbf{1}_{\{(z+\Delta, z'+\Delta') \in K\}} \inf(|\Delta| + |\Delta'|, 1) \mathcal{L}(z, z', d\Delta d\Delta') < +\infty$.

REMARKS. In the case, where $\mathcal{L}$ comes from (1.4), in view of (1.3), one will choose

$$\varphi(z) = 1 + |z|^2.$$

We have introduced the control function φ to overcome the difficulty (1.9iii).

These assumptions are satisfied for the hard spheres and the Maxwellian molecules.

Topology. It is usual to state large deviation principles in terms of upper bounds for closed subsets and lower bounds for open subsets of some topologies. Let us now define the topological space of $M_1(\mathcal{Z})$-valued paths which is relevant for our purpose.

For any function $g : \mathbb{R}^d \to \mathbb{R}$, let

$$\|g\|_\varphi = \sup_{z \in \mathcal{Z}} \frac{|g(z)|}{\varphi(z)}$$

which defines a norm on the function space

$$C_\varphi(\mathcal{Z}) = \{g; g : \mathcal{Z} \to \mathbb{R},\ g \text{ is continuous and } \|g\|_\varphi < +\infty\}.$$

Let $M_\varphi(\mathcal{Z})$ be the set of all probability measures which integrate φ, that is

$$M_\varphi(\mathcal{Z}) = \left\{\nu \in M_1(\mathcal{Z}); \int_{\mathcal{Z}} \varphi(z) \nu(dz) < \infty\right\}.$$

Let us denote

$$C_\varphi = \Big\{h : [0,T] \times \mathcal{Z} \mapsto \mathbb{R}; h \text{ is continuous on } [0,T] \times \mathcal{Z} \text{ and}$$
$$\sup_{0 \leq t \leq T} \|h(t, \cdot)\|_\varphi < \infty\Big\}$$

The path space we are interested in is $D_{M_\varphi} = \{\mu(\cdot) : t \in [0,T] \mapsto \mu(t) \in M_\varphi(\mathcal{Z})$ such that $\mu(0) = u_0$ and $h \in C_\varphi([0,T] \times \mathcal{Z}), t \in [0,T] \mapsto \int_{\mathcal{Z}} h(t, z) \mu(t; dz)$ is càdlàg$\}$.

It is equipped with the weak-$*$ topology $\sigma(D_{M_\varphi}, C_\varphi)$. In particular, $\mu_n(\cdot) \xrightarrow[n\to\infty]{} \mu(\cdot)$ in D_{M_φ} if and only if

$$\int_{[0,T]\times \mathcal{Z}} h(t,z)\mu_n(t;dz)\,dt \xrightarrow[n\to\infty]{} \int_{[0,T]\times \mathcal{Z}} h(t,z)\mu(t;dz)\,dt), \text{ for any } h \in C_\varphi.$$

REMARK. One can prove that the Borel σ-field of D_{M_φ} is equal to the σ-field which is generated by the projections: $\pi_{t,f}(\mu(\cdot)) = \int_\mathcal{Z} f\,d\mu_t$, $0 \le t \le T$, $f \in C_\varphi(\mathcal{Z})$.

Let us denote

$$Df(t,z,\Delta) = f(t,z+\Delta) - f(t,z),$$
$$Df(t,z,z',\Delta,\Delta') = f(t,z+\Delta) - f(t,z) + f(t,z'+\Delta') - f(t,z'),$$
$$\int_0^T \langle f(t,\cdot), \dot\mu_t - A(\mu_t)^*\mu_t\rangle\,dt = \langle f(T,\cdot),\mu_T\rangle - \langle f(0,\cdot),\mu_0\rangle$$
$$- \int_0^T \langle (\frac{\partial}{\partial t} + A(\mu_t))f(t,\cdot), \mu_t\rangle\,dt,$$

and

$$\tau(u) = e^u - u - 1, \quad u \in \mathbb{R}.$$

THEOREM 1 (UPPER BOUND). *Under the above assumptions, for any closed subset C of D_{M_φ}, we have*

$$\limsup_{N\to\infty} \frac{1}{N}\log \mathbb{P}(\overline{X}^N \in C) \le -\inf_{\mu \in C} I(\mu),$$

where

$$I(\mu) = \sup_{f \in C_0^{1,1}(]0,T[\times \mathcal{Z})} \Big\{ \int_0^T \langle f(t,\cdot), \dot\mu_t - A(\mu_t)^*\mu_t\rangle\,dt$$
$$- \int_0^T dt \int_{\mathcal{Z}^2 \times E} \tau\big(Df(t,z,z',\Delta,\Delta')\big)\frac{1}{2}\mathcal{L}(z,z',d\Delta d\Delta')\mu_t^{\otimes 2}(dz\,dz')\Big\}.$$

In order to state our next result at Theorem 2, let us introduce the following definition.

DEFINITION. One says that $\mu(\cdot) \in D_{M_\varphi}$ is a *Boltzmann path* if one can find measurable functions $k: [0,T]\times \mathbb{R}^d \to \mathbb{R}$ and $\theta: [0,T]\times \mathbb{R}^d \to \mathcal{B}(\mathbb{R}^d_*)$, where $\mathcal{B}(\mathbb{R}^d_*)$ is the Borel σ-field of $\mathbb{R}^d_*$ such that $\mu(\cdot)$ is a solution to the Boltzmann weak equation

(2.1) $$\langle f, \frac{\partial}{\partial t}u_t - A_{k,\theta,t}(u_t)^*u_t\rangle = 0, \quad \forall f \in C_0^\infty([0,T]\times \mathcal{Z})$$

with

$$A_{k,\theta,t}(u_t)f(z) = \int_{(\Delta)} \{f(z+\Delta) - f(z)\}\left(\int_{(z',\Delta')} \mathcal{L}_{k,\theta}(t,z,z',\cdot \times d\Delta')u_t(dz')\right)(d\Delta)$$

and

$$\mathcal{L}_{k,\theta}(t,z,z',d\Delta d\Delta') = 1_{\theta(t,z)}(\Delta)\exp\big(Dk(t,z,\Delta)\big)1_{\theta(t,z')}(\Delta')\exp\big(Dk(t,z',\Delta')\big)$$
$$\times \mathcal{L}(z,z',d\Delta d\Delta').$$

Let us denote $\mathcal{K}(\mu)$ the set of all (k,θ) such that equation (2.1) is satisfied.

As a convention, if $\mu \in D_{M_\varphi}$ is not a Boltzmann path, we set $\mathcal{K}(\mu) = \emptyset$. We also set: $\inf_\emptyset F(\cdot) = +\infty$.

Let us introduce the Legendre transform τ^* of the function τ. It is given by

$$\tau^*(u) = \begin{cases} (u+1)\log(u+1) - u & \text{if } u > -1 \\ +1 & \text{if } u = -1 \\ +\infty & \text{if } u < -1. \end{cases}$$

REMARK. The functions τ and τ^* are respectively the log-Laplace and the Cramér transforms of the centred Poisson law with parameter 1: $\mathcal{P}(1) - 1$.

Our next result is a non-variational formulation of the rate function I of Theorem 1.

THEOREM 2. *For any $\mu \in D_{M_\varphi}$, the $[0,+\infty]$-valued rate function I is given by*

$$I(\mu) = \inf_{(k,\theta) \in \mathcal{K}(\mu)} \Big\{ \int_0^T dt \int_{\mathcal{Z}^2 \times E} \tau^*\Big(\frac{d\mathcal{L}_{k,\theta}(t,z,z',\cdot)}{d\mathcal{L}(z,z',\cdot)}(\Delta,\Delta') - 1\Big)$$
$$\times \frac{1}{2} \mathcal{L}(z,z',d\Delta d\Delta')\mu_t^{\otimes 2}(dz\,dz') \Big\}.$$

If $\mu \in D_{M_\varphi}$ is such that $I(\mu)$ is finite, then it is a Boltzmann path and there is a unique (k,θ) in $\mathcal{K}(\mu)$ up to the equivalence relation given by

$$(k_1,\theta_1) \sim (k_2,\theta_2) \stackrel{\text{def}}{\iff} Dk_1(t,z,\Delta) = Dk_2(t,z,\Delta) \text{ and } 1_{\theta_1(t,z)}(\Delta) = 1_{\theta_2(t,z)}(\Delta),$$

where the above equalities hold $\big(\int_{\mathcal{Z}} \mathcal{L}(\cdot,z',\cdot \times \mathcal{Z})\mu_t(dz')\big)(z,d\Delta)\mu_t(dz)\,dt$-almost everywhere.

REMARKS. In particular, if $\mu \in D_{M_\varphi}$ is such that $I(\mu)$ is finite, since $\mathcal{L}_{k,\theta}$ is absolutely continuous with respect to $\mathcal{L}$, if $\mathcal{L}$ is the Lévy kernel of equation (1.4), the conservation equations (1.2) and (1.3) still hold for μ.

It can also be proved that such a μ is absolutely continuous in the sense of [Dawson and Gärtner 1987], Definition 4.1.

3. Outline of the proof of Theorem 1

STEP 0. Under (A.0), (A.1) and (A.2), one proves that the particle system exists up to time T.

STEP 1. Let $(C_\varphi)^\sharp$ denote the algebraic dual space of C_φ. With the duality bracket

$$\langle h, \nu \rangle = \int_{[0,T] \times \mathcal{Z}} h(t,z)\nu(t;dz)\,dt, h \in C_\varphi, \nu \in D_{M_\varphi},$$

one identifies D_{M_φ} as a topological subspace of $(C_\varphi)^\sharp$. This identification is injective, since any $\nu \in D_{M_\varphi}$ is completely determined by its action on C_φ. Indeed, consider functions h in C_φ of the form $h(t,z) = \psi_{t_0,\varepsilon}(t)g(z)$, where $(\psi_{t_0,\varepsilon})_{\varepsilon>0}$ is a regular approximation of δ_{t_0} with $t_0 \in [0,T]$ and g belongs to $C_\varphi(\mathcal{Z})$. Therefore, the topology of D_{M_φ} is the relative topology on D_{M_φ} of C_φ: the space $(C_\varphi)^\sharp$ endowed with the weak-$*$ topology $\sigma\big((C_\varphi)^\sharp, C_\varphi\big) : D_{M_\varphi} \underset{\text{continuous}}{\subset} C_\varphi$.

STEP 2. One obtains the large deviation upper bound in C_φ, for any closed subset A of C_φ:

$$\limsup_{N\to\infty} \frac{1}{N} \log \mathbb{P}(\overline{X}^N \in A) \leq -\inf_{\mu \in A} L(\mu), \tag{3.1}$$

where

$$L(\mu) = \sup_{F \in C_b(C_\varphi)} \{F(\mu) - H(F)\}, \quad \mu \in (C_\varphi)^\sharp$$

with

$$H(F) = \limsup_{N\to\infty} \frac{1}{N} \log \mathbb{E} \exp\left(NF(\overline{X}^N)\right), \quad F \in C_b(C_\varphi).$$

This equality holds thanks to the exponential tightness

$$\forall a > 0, \quad \exists b > 0, \quad \limsup_{N\to\infty} \frac{1}{N} \log \mathbb{P}\left(\sup_{0 \leq t \leq T} \sum_{i=1}^N \varphi\left(X_i^N(t)\right) > Nb\right) \leq -a$$

which is a standard consequence of (A.0) and (A.1).

STEP 3. To show that (3.1) still holds in D_{M_φ} rather than in $(C_\varphi)^\sharp$, it is enough to check that the domain of the rate function L is included in D_{M_φ}. Let μ be such that $L(\mu) < \infty$. One first proves that μ is almost everywhere $M_\varphi(\mathcal{Z})$-valued, by means of the estimate

$$\lim_{|z|\to\infty} \varphi(z) = +\infty$$

and

$$\exists \alpha > 0, \quad \limsup_{N\to\infty} \frac{1}{N} \log \mathbb{E} \exp\left[\alpha \sum_{i=1}^N \int_0^T \varphi(X_i^N(t))\, dt\right] < +\infty$$

which insures that $\mu(t) \in (C_\varphi(\mathcal{Z}))^\sharp$ is σ-additive for almost every $t \in [0,T]$. The regularity of the path is a corollary of Theorem 2 which implies that $t \mapsto \mu(t)$ is continuous.

STEP 4. For any $f \in C_0^{1,1}(]0,T[\times \mathcal{Z})$, (notice that $f(0,\cdot) = f(T,\cdot) = 0$), let us define

$$F_f(\mu) = -\int_{]0,T[\times \mathcal{Z}} f'_t(t,z)\, d\mu(t;dz)\, dt$$

$$- \int_{]0,T[\times \mathcal{Z}^2 \times E} \left(\exp[Df(t,z,z',\Delta,\Delta')] - 1\right) \frac{1}{2} \mathcal{L}(z,z',d\Delta d\Delta') \mu^{\otimes 2}(t;dz\, dz')\, dt$$

$$= \int_{]0,T[} \langle f(t,\cdot), \dot{\mu}_t - A(\mu_t)^* \mu_t \rangle\, dt$$

$$- \int_{]0,T[\times \mathcal{Z}^2 \times E} \tau\bigl(Df(t,z,z',\Delta,\Delta')\bigr) \frac{1}{2} \mathcal{L}(z,z',d\Delta d\Delta') \mu^{\otimes 2}(t;dz\, dz')\, dt.$$

By (A.2), (A.3) and (A.4), F_f is continuous and bounded on D_{M_φ}. Thanks to (3.1) and step 3, one gets for any closed subset C in D_{M_φ}

$$\limsup_{N\to\infty} \frac{1}{N} \log \mathbb{P}(\overline{X}^N \in C) \leq -\inf_{\mu \in C} \sup_{f \in C_0^{1,1}(]0,T[\times \mathcal{Z})} \{F_f(\mu) - H(F_f)\}.$$

But, stating

$$\widetilde{X}_s^N(dz\,dz') = \frac{N}{N-1}\overline{X}_s^{N\otimes 2}(dz\,dz') - \frac{1}{N-1}\overline{X}_s^N(dz)\left(\overline{X}_s^N - \delta_z\right)(dz'),$$

$$Z_t^{N,f} = \exp\left[N\left(\int_{\mathcal{Z}} f(t,z)d\overline{X}_t^N - \int_{\mathcal{Z}} f(0,z)d\overline{X}_0^N - \int_{[0,t]\times\mathcal{Z}} f'_t(s,z)d\overline{X}_s^N(dz)ds\right.\right.$$
$$\left.\left. - \int_0^t ds \int_{\mathcal{Z}^2\times E} \left(\exp[Df(t,z,z',\Delta,\Delta')] - 1\right)\frac{1}{2}\mathcal{L}(z,z',d\Delta d\Delta')\widetilde{X}_s^N(dz\,dz')\right)\right],$$

is a martingale with expectation 1, so that $H(F_f) = 0$. $\square$

4. Outline of the proof of Theorem 2

STEP 1. Let $\mathcal{C}$ stand for $C_0^{1,1}(]0,T[\times\mathcal{Z})$ and $\mathcal{C}^\sharp$ be its algebraic dual. Let μ be such that $I(\mu) < +\infty$, where $I(\mu)$ is given in its variational form at Theorem 1. With μ is associated the linear form q_μ, defined by

$$q_\mu : f \in \mathcal{C} \mapsto \int_{]0,T[} \langle f(t,\cdot), \dot\mu_t - A(\mu_t)^*\mu_t\rangle\, dt \in \mathbb{R}.$$

We define the convex function

$$G_\mu : f \in \mathcal{C} \mapsto \int_{]0,T[\times\mathcal{Z}^2\times E} \tau\big(Df(t,z,z',\Delta,\Delta')\big)$$
$$\times \frac{1}{2}\mathcal{L}(z,z',d\Delta d\Delta')\mu^{\otimes 2}(t;dz\,dz')\,dt \in \mathbb{R}$$

so that one can express $I(\mu)$ as a Legendre transform:

$$I(\mu) = \sup_{f\in\mathcal{C}}\{q_\mu(f) - G_\mu(f)\} = G_\mu^*(q_\mu).$$

To simplify, let us suppose that $\Lambda_\mu(dt\,dz\,dz'd\Delta d\Delta') = \mathcal{L}(z,z',d\Delta d\Delta')\mu^{\otimes 2}(t;dz\,dz')$ is a bounded measure. Now, let us study the function

$$q \in \mathcal{C}^\sharp \mapsto G_\mu^*(q) \in [0,+\infty].$$

Theorem 2 is then a consequence of general results on the Legendre transform, which are established in the Section 4 of [Léonard 1993].

STEP 2 (THE DOMAIN OF G_μ^*). Let $q \in \mathcal{C}^\sharp$ be such that $G_\mu^*(q) < +\infty$. Then, for all $f \in \mathcal{C}$ and all $\lambda > 0$, one gets

$$\lambda q(f) \leq \int \tau\left(\frac{|Df|}{\lambda}\right) d\Lambda_\mu + G_\mu^*(q).$$

Choosing $\lambda = \|Df\|_{\tau,\Lambda_\mu}$, where

$$\|f\|_{\tau,\Lambda_\mu} = \inf\left\{a > 0; \int \tau\left(\frac{|Df|}{a}\right) d\Lambda_\mu \leq 1\right\}$$

is the norm on the Orlicz space $L^\tau(]0,T[\times\mathcal{Z}^2\times E,\Lambda_\mu)$, one obtains

$$|q(f)| \leq \big(1 + G_\mu^*(q)\big)\|Df\|_{\tau,\Lambda_\mu}, \quad \forall f \in \mathcal{C}.$$

Therefore, q is a continuous linear form on the completion of the space $\{Df; f \in \mathcal{C}\}$ with respect to the norm $\|\cdot\|_{\tau,\Lambda_\mu}$. As a consequence, there exists at least one l_q in the dual Orlicz space $L^{\tau^*}(]0,T[\times \mathcal{Z}^2 \times E, \Lambda_\mu)$ such that

$$q(f) = \int l_q Df d\Lambda_\mu, \quad \forall f \in \mathcal{C}.$$

For Orlicz spaces, see for instance [Krasnosel'skii and Rutickii 1961] or the appendix of [Neveu 1972].

STEP 3. One shows that G_μ^* is $\|\cdot\|_{\tau^*,\Lambda_\mu}$-continuous on the intrinsic interior of its domain. Thanks to this continuity, one proves that for any "interior" q, we get the usual Legendre equality

$$G_\mu^*(q) + G_\mu(k_q) = q(k_q),$$

where k_q is a solution of

$$\langle G_\mu'(k_q), f \rangle = \int (e^{Dk_q} - 1) Df d\Lambda_\mu = q(f), \quad \forall f \in \mathcal{C}.$$

The uniqueness of Dk_q (in the sense of Theorem 2) is a consequence of the strict convexity of G_μ.

If q is a "boundary" point, it can be approached by a sequence (q_n) of interior points such that $G_\mu^*(q) = \lim_{n\to\infty} G_\mu^*(q_n)$ and $e^{Dk_{q_n}} \xrightarrow[n\to\infty]{} 1_{\theta_q} e^{Dk_q}$. (Formally, $1_{\theta_q} = 0$ corresponds to $Dk_q = -\infty$).

STEP 4. In the case, where Λ_μ is unbounded, one has to consider the bounded measure

$$\Lambda_\mu'(dt\, dz\, dz'\, d\Delta\, d\Delta') = \inf(|\Delta|^2 + |\Delta'|^2, 1)\Lambda_\mu(dt\, dz\, dz' d\Delta d\Delta')$$

instead of Λ_μ and to consider

$$\frac{Df(t,z,z',\Delta,\Delta')}{\inf(|\Delta|+|\Delta'|,1)}$$

instead of $Df(t,z,z',\Delta,\Delta')$. The same proof as before still works.

References

L. Arkeryd, *On the Boltzmann equation* I, Arch. Rational Mech. Anal. **45** (1972), 1–16; II, 17–34.

F. Comets, *Nucleation for a long range magnetic model*, Ann. Inst. Henri Poincaré. Probab. Statist. **23** (1987), no. 2, 137–178.

D.A. Dawson and J. Gärtner, *Large deviations from the McKean-Vlasov limit for weakly interacting diffusions*, Stochastics **20** (1987), 247–308.

C. Kipnis and S. Olla, *Large deviations from the hydrodynamical limit for a system of independent Brownian particles*, Stochastics **33** (1990), 17–25.

M. A. Krasnosel'skii and Ya. B. Rutickii, *Convex functions and Orlicz spaces*, P. Noordhoff Ltd, 1961.

C. Léonard, *On large deviations for particle systems associated with spatially homogeneous Boltzmann type equations* (to appear).

J. Neveu, *Martingales à temps discret*, Masson, 1972.

A. S. Sznitman, *Équations de type Boltzmann, spatialement homogènes*, Z. Wahrsch. Verw. Gebiete **66** (1984), 559–592.

H. Tanaka, *Probabilistic treatment of the Boltzmann equation of Maxwellian molecules*, Z. Wahrsch. Verw. Gebiete **46** (1978), 67–105.

ÉQUIPE DE MODÉLISATION STOCHASTIQUE ET STATISTIQUE. U.R.A. CNRS D 0743 UNIVERSITÉ DE PARIS-SUD, DÉPARTEMENT DE MATHÉMATIQUES, BÂTIMENT 425, 91405 ORSAY CEDEX, FRANCE

E-mail address: leonard@stats.matups.fr

A Stochastic PDE Arising as the Limit of a Long-Range Contact Process, and its Phase Transition

Carl Mueller and Roger Tribe

1. Introduction

The study of hydrodynamic limits is a popular topic. Usually, one starts with a particle system, and after rescaling, obtains the solution of a partial differential equation. The particle system may represent the laws of nature of the atomic scale, or some simplification of them, and the partial differential equation represents the large scale laws of nature. Usually, the differential equations are deterministic when random limits are obtained, they result from rescaling the difference between the particle system and the limiting partial differential equation, in the style of the central limit theorem. By contrast, we consider a particle system whose hydrodynamic limit is a stochastic partial differential equation (SPDE). Our particle system is balanced at the point between survival and extinction, and this is the source of randomness in our limit.

Our second goal is to study the phase transitions of our system. While phase transitions have long been studied for particle systems, they are less well known for SPDE. It turns out that our SPDE has the same type of phase transition as our particle system. The 2 phases involve survival and extinction. Our proof involves construction embedded particle system in the SPDE, and showing that these systems survive or go extinct, depending on the value of our parameter. Our work was inspired by some questions of R. Durrett. We are also indebted to E. Perkins, whose ideas helped us prove the hydrodynamic limit.

Consider the equation

(1.1) $$u_t = \frac{1}{6}u_{xx} + \theta u - u^2 + u^{1/2}\dot{W}, \quad t > 0, x \in \mathbb{R}, \theta > 0$$
$$u(0, x) = u_0(x) \geq 0$$

where $\dot{W} = \dot{W}(t, x)$ is spacetime white noise. R. Durrett suggested that (1.1) should arise as a limit of the long-range contact process studied in [Bramson, Durrett, and Swindle 1989]. In addition, Durrett guessed that (1.1) exhibits a phase transition as θ varies. For small values of θ, $u(t, x)$ should die out to 0 in finite time. For large values of θ, $u(t, x)$ survives with non zero probability.

The purpose of our work is to prove both of these properties. In proving the first property, we are heavily indebted to E. Perkins, who showed in an unpublished

---

1991 *Mathematics Subject Classification.* Primary: 60H15; Secondary: 35R60.

The first author is supported by an NSA grant, and by the Army, through Cornell's Mathematical Sciences Institute.

The final form of this paper will appear elsewhere.

paper [Perkins 1988] that (1.1) arises from the limit of a discrete-time particle system. We have adapted his proof to the case of continuous time.

Here is the model considered by [Bramson, Durrett, and Swindle 1989]. We give here an informal description, leaving an exact construction to a future paper. Note that the model in [Bramson, Durrett, and Swindle 1989] is defined on $\mathbb{Z}^d$, but we only consider $\mathbb{Z}$. In fact, as shown in [Walsh 1986], (1.1) should have distribution solutions in higher dimensions, but then it is difficult to give meaning to nonlinear terms such as u^2.

Fix a nonnegative integer μ. Sites in $\mu^{-2}\mathbb{Z}$ are either vacant or occupied by a particle. Particles die at Poisson rate μ leaving their sites vacant. Particles give birth at rate $\mu + \theta$. If a particle gives birth, it chooses one of the $2[\mu^{3/2}]$ nearest neighbor sites, each with equal probability. If that site is already occupied, the birth is aborted. If that site is vacant, it becomes occupied at the time of the birth.

Let M_F be the space of finite measures on $\mathbb{R}$ with the topology of weak convergence. Define

$$(1.2) \quad F = \left(m \in M_F : \iint \log_+(1/|x-y|) m(dx)\, m(dy) < \infty \right).$$

Define a M_F valued process X_t^μ by

$$(1.3) \quad X_t^\mu(A) = \mu^{-1} \sum_{x \in \mu^{-2}\mathbb{Z}} \mathbb{I}(x \in A, x \text{ is occupied}) \quad \text{for } t \geq 0, \text{ Borel } A.$$

THEOREM 1. *If $X_0^\mu \to m \in F$ then the processes X^μ converge in distribution to a continuous M_F valued process X. If also $m(dx) = u_0(x)dx$ then X_t has a density $u(t,x)$ for all time $t \geq 0$ which is a solution to (1.1).*

For our second result, we assume that the initial function $u_0(x)$ in (1.1) is continuous with compact support, nonnegative, and not identically 0. We write $u_0 \in C_c^t$. We say that $u(t,x)$ survives if for all $t > 0$, $u(t,0)$ is not identically 0.

THEOREM 2. *There exists a constant $\theta_c > 0$, not depending on u_0, such that*
  (i) *If $\theta < \theta_c$, then $P(u(t,x) \text{ survives}) = 0$.*
  (ii) *If $\theta > \theta_c$, then $P(u(t,x) \text{ survives}) > 0$.*

For future use, we let T be the first time such that $u(t,x)$ is identically 0. Let $T = \infty$ if there is no such time. In the rest of this survey, we give a sketch of the proofs.

The proof of theorem 1 has two parts; a proof of the tightness of $(X^\mu)_{\mu=1,2,\ldots}$ and then a characterisation of all limit points. The limit points are characterised by a martingale problem, in the sense of Stroock and Varadhan. We also set up a martingale problem satisfied by X^μ. We regard the martingale problem for X^μ as an approximation to the martingale problem for the limit; the former will resemble the latter, but will have several error terms. In the first part of the proof, we analyze these error terms. Our analysis uses a comparison. Indeed, if we allow births on occupied sites in the long-range contact process, we may define a random measure $\overline{X}_t^\mu$ just as in (1.3). Furthermore, we can define both X_t^μ and $\overline{X}_t^\mu$ on a common probability space such that $X_t^\mu \leq \overline{X}_t^\mu$ for all $t \geq 0$. Since $\overline{X}_t^\mu$ is based on a process in which particles evolve independently, it is much easier to analyze and exact moment calculations are possible. This particle system is essentially independent branching random walks and the sequence $\overline{X}^\mu$ converges to the widely studied super-Brownian motion, with mass creation at rate θ and with a different time scale. The analysis of the error terms is enough to prove tightness and the remainder of the proof is spent passing to the limit in the approximate martingale problems. This shows that all limit points satisfy a martingale problem for which uniqueness holds and hence convergence is proved.

We summarise the differences between our proof and that for the discrete time model considered in [Perkins 1988]. Our proof of tightness follows that of [Perkins 1988]. The majority of the rest of proof is spent in characterising the overcrowding term $\int X_s((dX_s/dx)\phi)ds$ in the limiting martingale problem. For this an approximate density $V_t(x)$ is defined for the contact processes. The key step is to establish some regularity of the path $x \to V_t(x)$. We use a new method for this which seems easier. In our continuous time setting we also need regularity of $t \to V_t(x)$. Once these regularity properties are obtained the method of [Perkins 1988] goes through.

Next we discuss the proof of theorem 2. First consider case (i), in which $\theta < \theta_c$, and we must show $P\{u(t,x) \text{ survives}\} = 0$. We compare $u(t,x)$ to a continuous time branching process with expected offspring size $\mu < 1$. Such a branching process dies out with probability 1. To implement this comparison, we write $u_0(x) = \sum_{i=1}^N u_0^{(i)}(x)$, where $u_0^{(i)}$ is supported on interval $[z_i, z_i + 1]$, $\int_{z_i}^{z_i+1} u_0^{(i)}(x)dx \leq 1$, and $u_0^{(i)}(x) \geq 0$. We show that one can think of the $u^{(i)}(t,x)$ as almost evolving independently, starting from $u_0^{(i)}(x)$, and such that $u(t,x) \geq \sum_{i=1}^N u^{(i)}(t,x)$. The $u^{(i)}$ satisfy an equation similar to (1.1). We call the $u^{(i)}$ "bricks" that make up u. We choose a stopping time τ, and again majorize $u_i(\tau, x)$ by a sum of $N(i, \tau)$ bricks. These are the offspring of the original brick. If θ is small, we show that the expected offspring size $EN(i, \tau) < 1$, completing the proof. This final step involves scaling (1.1), thereby transforming it into

$$(1.4) \qquad v_t = \frac{\theta^3}{6}v_{xx} + v - v^2 + v^{1/2}\dot{W}.$$

Here, θ is small, so the solution does not spread quickly. Furthermore, the term $-v^2$ keeps the solution from becoming too large. Assume that we start with a brick: $v(0,x) = 1(n \leq x \leq n+1)$. If θ is small, then with high probability, the noise will drive $v(t,x)$ to 0 before it spreads out too much.

Secondly, consider theorem 2, case (ii). Here $\theta > \theta_c$, and we must prove that $P\{u(t,x) \text{ survives}\} > 0$. We compare $u(t,x)$ to oriented percolation, using an idea of [Durrett 1991]. We begin with another scaling of $u(t,x)$, obtaining

$$(1.5) \qquad v_t = \frac{1}{6}v_{xx} + v - v^2 + \theta^{-3/4}v^{1/2}\dot{W}.$$

If θ is large, the noise in (1.4) is small, and we almost have

$$(1.6) \qquad w_t = \frac{1}{6}w_{xx} + w - w^2.$$

But (1.6) is the classical Kolmogorov-Petrovskii-Piscuinov equation for which many solutions converge to travelling waves. In particular, if $w(0,x) > \frac{1}{2}1(n \leq x \leq n+1)$ then for some future time t_0, $w(t_0, x) \geq \frac{1}{2}1(n-1 \leq x \leq n+2)$. The same holds true for $v(t,x)$ satisfying (1.4), with high probability. Again, we can show approximate independence of the bricks. By this method, we compare $v(t,x)$ to an oriented site percolation process. This process is similar to ordinary percolation, except that only paths having certain directions (the forward time direction) are allowed. There is some mild dependence in our percolation process, but this causes no trouble. We show that if $\theta > \theta_c$, then there is a positive probability that, in the percolation model, the origin is part of an infinite connected cluster. This in turn implies that $u(t, x)$ survives with positive probability.

Finally, we note that there is a further connection between (1.1) and the Kolmogorov-Petrovskii-Piscuinov equation. In a future paper, R. Tribe will show that (1.1) possesses random travelling wave solutions. The proof will use techniques of [Durrett 1984]. In particular, $u(t, x)$ will be compared to the rightmost occupied site in oriented percolation.

Bibliography

S. Albevario, R. Høegh-Krohn, J. E.Fenstad, and T. Lindstrom, *Nonstandard methods in stochastic analysis and mathematical physics*, Academic Press, 1986.

K. Athreya and P. Ney, *Branching Processes*, Springer-Verlag, 1972.

R. M. Anderson and S. Rashid, *A nonstandard characterization of weak convergence*, Proc. Amer. Math. Soc. **69** (1978), 327–332.

R. N. Bhattacharya and R. R. Rao, *Normal approximation and asymptotic expansions*, John Wiley & Sons, 1976.

M. Bramson, R. Durrett and G. Swindle, *Statistical mechanics of crabgrass*, Ann. Probab. **17** (1989), 444–481.

R. Durrett, *Oriented percolation in two dimensions*, Ann. Probab. **12** (1984), 999–1040.

_____, *A new method for proving phase transitions*, Spatial Stochastic Processes (K. Alexander and J. Watkins eds.), Birkhäuser, Boston, 1991, pp. 141–170.

R. Durrett and C. Neuhauser, *Particle systems and reaction-diffusion equations*, 1992 (preprint).

S. N. Evans and E. A. Perkins, 1993 (preprint).

D. N. Hoover and E. A. Perkins, *Nonstandard construction of the stochastic integral and applications to stochastic differential equations* I, Trans. Amer. Math. Soc. **275** (1983), 1–36.

I. Iscoe, *On the supports of measure-valued critical branching Brownian motion*, Ann. Probab. **16** (1988), 200–221.

P. Kotelenez, *Comparison methods for a class of function valued stochastic partial differential equations*, Probab. Theor. Relat. Fields (1992) **93** (1), 1–29.

C. Mueller, *On the support of solutions to the heat equation with noise*, Stochastics **37** (1991), 225–246.

C. Mueller and R. Sowers, *Blowup for the heat equation with a noise term*, Probab. Theor. Relat. Fields, (1992), (to appear).

E. A. Perkins, *On a problem of Durrett*, 1988, (handwritten manuscript).

T. Shiga, *Two contrastive properties of solutions for one-dimensional stochastic partial differential equations*, 1990 (preprint).

R. Sowers, *Large deviations for a reaction-diffusion equation with non-Gaussian perturbations*, Ann. Probab. **20** (1992), 504–537.

J. B. Walsh, *An introduction to stochastic partial differential equations*, Lecture Notes in Math., vol. 1180, Springer, Berlin-Heidelberg-New York, 1986, pp. 265–439.

DEPARTMENT OF MATHEMATICS, UNIVERSITY OF ROCHESTER, ROCHESTER, NY 14627, USA
E-mail address: cmlr@troi.cc.rochester.edu

MATHEMATICS INSTITUTE, UNIVERSITY OF WARWICK, COVENTRY CV4 7AL, ENGLAND

Some Aspects of the Martin Boundary of Measure-Valued Diffusions

Ludger Overbeck

1. Introduction

In this paper we want to discuss some aspects of the Martin boundary of measure-valued diffusions. The *Martin boundary* of a Markov process (X, P) on D_E is usually define extreme points in the set of positive *space-time harmonic* functions H of (X, P). A positive function H on $[0, \infty) \times D_E$ is space-time harmonic iff $\bigl(H(t, X_t)\bigr)_{t \geq 0}$ is a martingale under P. Every space-time harmonic function H yields a new process, the H-transform P^H. It is the measure on D_E given by

$$dP^H\bigr|_{\mathcal{F}_t} = H(t, X_t) dP\bigr|_{\mathcal{F}_t},$$

where $(\mathcal{F}_t)$ is the canonical filtration on D_E. The main result in the paper is the characterization of a measure-valued diffusion as the unique solution of a martingale problem. The martingale generator of P^H has the additional interaction term $H'_x(s, \mu)/H(s, \mu)$ of a particle at position x, time s, which depends on the state μ of the process, cf. Theorem 2.3. In [Overbeck 1992b] this was proved for the super-Brownian motion. Now we consider more general measure-valued diffusions, including the Fleming-Viot process (with selection) and the Historical process, Sections 3.1 and 3.2. The class of processes which we investigate is introduced in the lecture notes of [Dawson 1992], to which we also refer for further background and notation.

In Section 3.3, I recall the definition of *conditioned processes* which appear in the probabilistic notion of "Martin boundary" cf. [Föllmer 1975, Föllmer 1990, Dynkin 1971, Dynkin 1978]. I give an example of an *additive* conditioned process of the Historical process which is not an H-transform of the Historical process.

2. The martingale problem of an H-transform

Since we want to apply Dawson's Girsanov transformation we briefly recall his setting.

2.1. Stochastic calculus of measure-valued diffusions. Let us consider a probability measure P on the canonical filtered space, $\bigl(D_M, \mathcal{F}, (\mathcal{F}_t)\bigr)$, where M is either the space of probability measures M_1 or the space of finite measures M_F over a polish space E, both equipped with the weak topology.

Assume that $P[X_0 = \mu_0] = 1$ with $\mu_0 \in M$.

1991 *Mathematics Subject Classification.* Primary: 60G57, 60J50; Secondary: 60K35, 60J80.
This is the final form of the paper.

Let $A : D(A) \to bB(E)(=$ bounded measurable function) be a linear operator with domain $D(A) \subset bB(E)$.

The *quadratic functional* $Q(\mu; dx, dy)$ determines for every $\mu \in M$ a signed symmetric measure on $E \times E$ which is positive and finite on the diagonal. Let r be a measurable function on $[0, \infty) \times E \times M$, such that the *interaction term* $R(s, \mu)$ defined by

$$R(s, \mu)(dx) = \int_E r(s, y, \mu) Q(\mu; dx, dy),$$

is a σ-finite measure on E for every $s \geq 0$, $\mu \in M$.

The directional derivative of a function F on M in direction δ_x is denoted by $F'_x(\mu)$.

A function F on $[0, \infty) \times M \to \mathbb{R}$ belongs to the class $\mathcal{FB}$ of finitely-based function if

$$F(t, \mu) = \phi\big(t, \langle \mu, f_1(t) \rangle, \ldots, \langle \mu, f_k(t) \rangle\big)$$

with $\phi \in C^{1,2}$, $f_i \in D(A)$, $k \in \mathbb{N}$, and $\langle \mu, f \rangle := \mu(f) := \int f(x) \mu(dx)$

DEFINITION. The measure P is called (the distribution of) the measure-valued diffusion associated with (A, Q, R) iff the process $M[F]$ defined by

(2.1) $$M_t[F] := F(t, X_t) - F(0, X_0) - \int_0^t \mathcal{A}F(s, X_s) ds$$

is a local martingale for all $F \in \mathcal{FB}$, where

(2.2) $\mathcal{A}F(s, \mu) :=$

$$\left\langle \mu, \left(A + \frac{\partial}{\partial s}\right) F'(s, \mu) \right\rangle + \langle R(s, \mu), F' \rangle + \int_E \int_E (F'_x)'_y Q(\mu; dx, dy).$$

EXAMPLES.   1. Let A be the generator of a Feller diffusion on $E = \mathbb{R}^d$.
(a) The quadratic functional of the Dawson-Watanabe process equals $Q^{DW}(\mu; dx, dy) := \delta_x(dy) \mu(dx)$, $\mu \in M_F$. Processes with this quadratic functional may have, for example, the following interactions:
   (i) If we set $r = 1/\mu(1)$, we obtained the superprocess conditioned on non-extinction.
   (ii) If $(A + \partial/\partial s) h(s, x) = 0, h \geq 0$, we obtain an additive H-transform by setting $r(s, x, \mu) = h(s, x)/\mu(h(s))$, cf. [Overbeck 1992b].
   (iii) If r is uniformly bounded and does not depend on time, and if $r(\cdot, \mu)$ is a continuous function in x for every μ, then we obtain "branching with interaction" as in [Dawson 1992, 10.1.2].
(b) The quadratic functional of Fleming-Viot process is defined by

$$Q^{FV}(\mu; dx, dy) := \delta_x(dy)\mu(dx) - \mu(dx)\mu(dy), \quad \mu \in M_1.$$

For the Fleming-Viot process without selection we have $r = 0$. The Fleming-Viot process with selection is given by

$$r(x, \mu) = \int_E v(x, y) \mu(dy),$$

where v is the fitness function, cf. [Dawson 1992, 10.1.1].

2. The Historical Brownian motion is based on "martingale generator"
$$(A, D(A))$$
of the path process of the Brownian motion on $\mathbb{R}^d$, cf. [Dawson 1992, 12.3.3]. Furthermore it has the quadratic functional Q^{DW} and $r = 0$, as in 1.(a).

In all cases P is uniquely defined by (2.1) and (2.2).

The interaction of an H-transform P^H of P is described in Theorem 2.3 in terms of the gradient square operator
$$\Gamma(F, G) := \mathcal{A}FG - F\mathcal{A}G - G\mathcal{A}F$$
associated with $\mathcal{A}$. An easy calculation yields.

LEMMA 2.1.
$$\Gamma(F, G) = \int_E \int_E G'_x F'_y Q(\,\cdot\,; dx, dy)$$
for all $F, G \in \mathcal{FB}$.

2.2. Representation of space-time harmonic functions by martingale measures.
Let $M(ds, dx)$ be the martingale measure associated with the measure-valued process P (cf. [Dawson 1992, 7.1]). Then the local martingale $M[F]$ in (2.1) takes the form

$$(2.3) \qquad M_t[F] = \int_0^t \int_E F'_x(s, X_s) M(ds, dx)$$

for all $F \in \mathcal{FB}$.

Now, we want to extend the formulas (2.1)–(2.3) to slightly more general functions F which are only locally in $\mathcal{FB}$.

Let $F : [0, \infty) \times M \to \mathbb{R}$ and T be a stopping time. Assume there is a sequence $\{F_k\}_{k \in \mathbb{N}} \subset \mathcal{FB}$ such that up to T,

$$(2.4) \qquad F_k(s, X_s) \xrightarrow[k \to \infty]{} F(s, X_s) \quad \text{a.s. bounded or monoton}$$
$$(2.5) \qquad \mathcal{A}F_k(s, X_s) \xrightarrow[k \to \infty]{} \mathcal{A}F(s, X_s) \quad \text{a.s. bounded or monoton}$$
$$(2.6) \qquad (F_k)'_x(s, X_s) \xrightarrow[k \to \infty]{} F'_x(s, X_s) \quad \text{in } \mathcal{P}_M,$$

where $\mathcal{P}_M$ is the set of M-integrable functions (cf. [Dawson 1992, 7.1]). Assumption (2.6) is e.g. satisfied if $(F_k)'_x(s, X_s) \to F'_x(s, X_s)$ pointwise and
$$\sup_{\omega, x} \sup_{s \leq T(\omega)} (F_k)'_x\bigl(s, X_s(\omega)\bigr) < \infty.$$

DEFINITION. We write $F \in \mathcal{FB}_{\mathrm{loc}}$ if there is a sequence of stopping times $\{T_n\}_n$ such that $T_n \to \infty$ and F admits an approximation as in (2.4)–(2.6) for every T_n.

LEMMA 2.2. Let $H \in \mathcal{FB}_{\mathrm{loc}}$ such that $\bigl(H(t, X_t)\bigr)_{t \geq 0}$ is a local martingale. Then

$$(2.7) \qquad H(t, X_t) = \int_0^t \int_E H'_x(s, X_s) M(ds, dx),$$

and the quadratic variation is given by

$$(2.8) \quad \langle\langle H(t, X_t)\rangle\rangle = \int_0^t \int_E \int_E H'_x(s, X_s) H'_y(s, X_s) Q(X_s; dx, dy) \, ds.$$

PROOF. Fix $n \in \mathbb{N}$. If $\{F_k\}_{k \in \mathbb{N}} \in \mathcal{FB}$ is the approximating sequence of H up to T_n then by (2.3)

$$M_t[F_k] = \int_0^t \int_E (F_k)'_x(s, X_s) M(ds, dx).$$

By (2.6) and space-time harmonicity of H it follows that $M_t[F_k]$ converges to $M_t[H] = H(t, X_t)$. This implies by (2.6) that

$$H(t \wedge T_n, X_{t \wedge T_n}) = \int_0^{t \wedge T_n} \int_E H'_x(s, X_s) M(ds, dx).$$

Letting $n \to \infty$ we can prove the assertion. $\square$

2.3. The martingale problem of an H-transform.

THEOREM 2.3. *Assume that the martingale problem in (2.1) and (2.2) has a unique solution P. Let $H \in \mathcal{FB}_{\text{loc}}$ be space-time harmonic. Define the stopping time*

$$T_n := \inf \{t \geq 0 \mid H(t, X_t) \notin [n^{-1}, n]\}.$$

The H-transform P^H is the unique measure on D_M which solves the local martingale problem associated with the operator

$$\mathcal{A}^H F(s, \mu) := \mathcal{A} F(s, \mu) + \frac{1}{H(s, \mu)} \Gamma(H, F)(s, \mu)$$

$$= \mathcal{A} F(s, \mu) + \int_E \int_E \frac{H'_x(s, \mu)}{H(s, \mu)} F'_y(s, \mu) Q(\mu; dx, dy)$$

on $\mathcal{FB}$ up to T_n for every $n \in \mathbb{N}$.

PROOF. In order to apply Dawson's Girsanov transformation we shall write the martingale $\bigl(H(t, X_t)\bigr)_{\geq 0}$ as an exponential martingale. By Itô's lemma and Lemma 2.2 we have

$$\log H(t, X_t) = \int_0^t \frac{1}{H(s, X_s)} dM[H]_s - \frac{1}{2} \int_0^t \frac{1}{H^2(s, X_s)} d\langle M[H]\rangle_s$$

$$= \int_0^t \int_E \frac{H'_x(s, \mu)}{H(s, \mu)} M(ds, dx)$$

$$- \frac{1}{2} \int_0^t \int_E \int_E \frac{H'_x(s, \mu) H'_y(s, \mu)}{H^2(s, \mu)} Q(X_s; dx, dy) ds.$$

Therefore we get by [Dawson 1992, 7.2.2] that $P^H\big|_{\mathcal{F}_{T_n}}$ is the unique measure such that $(M^H_{t \wedge T_n}[F])_{t \geq 0}$ is a local martingale up to the stopping time

$$T_n = \inf\{t \geq 0 \mid H(t, X_t) \notin [n^{-1}, n]\}.$$

Since $\sup_n T_n = \infty$ P^H a.s. the assertion follows. $\square$

3. Examples

3.1. H-transforms of the Fleming-Viot process.
Let P be the Fleming-Viot process with $r = 0$ as in Example 1.(b). By Theorem 2.3 the H-transform P^H exhibits the state-dependent *selection* term $H'_x(s,\mu)/H(s,\mu)$.

Let us consider the additive case $H(s,\mu) = \mu(h(s))$. The additive function H is space-time harmonic iff $Ah + \partial/\partial_s h = 0$. The selection term is $H'(s,\mu)/H(s,\mu) = h(s,x)/\mu(h(s))$. A fitness function $v(x,y)$ defined in [Dawson 1992, 10.1.1], cf. Example 1.(b), compares the fitness of a particle at place x with the fitness of a particle at place y. The fitness function $v^h(x,\mu) := h(s,x)/\mu(h(s))$ of an additive H-transform is independent of y but depends on μ. So the fitness of a particle at place x is described relative to the total fitness of the present population. Gene types in the region where h has greater values have a better chance to survive. Thus, a transformation with an additive space-time harmonic function can be seen as a selection in the space $\mathbb{R}^d$ weighted with the harmonic function h. Under this selection the process looks like a Fleming-Viot process, where the gene type of at least one family mutates as an h-transform of the one particle motion. This interpretation is plausible if we use the well known connection between super- and Fleming-Viot processes [Dawson 1992, 8.1], and the interpretation of an additive H-transform of a superprocesses as a superprocess with an immortal particle which moves as the h-transform of the one particle motion, cf. [Overbeck 1992b].

3.2. H-transforms of superprocesses which depend only on the total mass.
Let P be a superprocess as in 2.1 Example 1.(a) or 2. We consider the class

$$\{P^H \mid H(s,\mu) = \eta(t,\mu(1)), \eta \text{ space-time harmonic for the squared Bessel }(0)\}$$

which is isomorph to the Martin boundary of the squared Bessel (0) process; this class may be viewed as a *face* (cf. [Dynkin 1971]) of the convex set $\mathcal{Q}_P$ introduced below. The extremal space-time harmonic functions $\eta^c, c \in [0,\infty) \cup \{\varnothing\}$ of this process are calculated in [Overbeck 1992a]. The corresponding $\eta(s,\mu(1))$-transforms P^c arise as conditioned processes for $c \geq 0$

$$P[\,\cdot\, \mid X_{t_n}(1) = c_{t_n}] \to P^{\eta^c} \text{ as } n \to \infty,$$

if $0 < c_{t_n}/t_n^2 \to c$, $t_n \to \infty$. This result also implies the following limit behaviour of the total mass.

For $c \in [0,\infty)$, we have $X_t(1) > 0 \,\forall t$ and $X_t(1)/t^2 \to c$ as $t \to \infty$ P^{η^c} a.s.

The Martin point "$\varnothing$" corresponds to the squared Bessel (0) which has extinction as its limit behaviour.

3.3. Additive conditioned processes.

3.3.1. *Conditioned processes.*
For a Markov process (X,P) on D_E we consider the class of all processes which have the same "bridges" as P:

$$\mathcal{Q}_P := \left\{Q \in \mathcal{P}(D_E) \mid Q[\,\cdot\, \mid \widehat{\mathcal{F}}_t](\omega) = P[\,\cdot\, \mid \widehat{\mathcal{F}}_t](\omega), Q \text{ a.s } \forall t \geq 0\right\},$$

where the σ-field $\widehat{\mathcal{F}}_t$ denotes the future after time t, cf. [Föllmer 1975, Dynkin 1978]. The Markov property and the definition of P^H imply that $\{P^H \mid H \text{ space-time harmonic for } P\}$ is contained in $\mathcal{Q}_P$. Every $Q \in \mathcal{Q}_P$ has a unique integral

representation
$$Q = \int_{\mathcal{Q}_P^e} Q^e \, m^Q(dQ^e),$$
where $\mathcal{Q}_P^e$ is the set of extreme points of $\mathcal{Q}_P$. The latter are those elements of
$$\{Q \in \mathcal{P}(D_E) \mid Q = \lim_{t_n \to \infty} P[\,\cdot\,\mid \widehat{\mathcal{F}}_{t_n}](\omega_n)\{t_n, \omega_n\}_{n \in \mathbb{N}} \text{ s.t. the weak limit exists}\}$$
which are trivial on the tail-σ-field. An element $Q \in \mathcal{Q}_P$ can therefore be viewed as the process P conditioned to have the exit distribution m^Q. This justifies to call $\mathcal{Q}_P$ the set of *conditioned processes*. This definition of $\mathcal{Q}_P^e$ was motivated by the notion of Gibbs states in statistical mechanics. It coincides with the definition of Martin boundary iff every measure $Q \in \mathcal{Q}$ is locally absolutely continuous with respect to P. In this case the density $dP/dQ|_{\mathcal{F}_t}$ equals $H(t, X_t)$ with a space-time harmonic function H.

[Föllmer 1990] and [Röckner 1992] give examples of Markov processes on infinite-dimensional state space, where elements of $\mathcal{Q}_P^e$ are not H-transforms. The reason for that is the absence of a reference measure for the transition functions of these infinite dimensional processes, cf. [Kuznecov 1974]. However, for superprocesses, a result of [Evans and Perkins 1991] implies that every conditioned process of a super-process is an H-transform of the superprocess if the same holds for the one-particle motion, cf. [Overbeck 1992b].

3.3.2. *Relation between the conditioned processes of the superprocess and of its one particle motion.* The intensity measure of the superprocess is

(3.1) $$P[X_t(f) \mid X_s = \mu] = \mu(\mathcal{P}_{s,\cdot}[f(\xi_t)]),$$

where $\mathcal{P}$ is the distribution of the one-particle-motion on D_E. On the one hand this shows that $\mu(h(t))$ is space-time harmonic if h is space-time harmonic for the one particle motion. On the other hand it implies that a typical particle moves as the one particle motion. Picking a typical particle and describing its position is usually done by the Campbell measure. A Campbell-type process (Y, ζ) with values in $M \times E$ is defined by its transition function

(3.2) $$\mathbb{P}\left[e_f(Y_t)g(\zeta_t) \mid (Y_r, \zeta_r) = (\nu, x)\right]$$
$$= e^{-\nu(V_t^r f)} \mathcal{P}_{r,x}\left[\exp\left(-\int_r^t V_t^s f(\xi_s) ds\right) g(\xi_t)\right]$$

where $e_f(\mu) = e^{-\mu(f)}$ (cf. [Evans 1992]).

If we condition the second component to exit at a point a in its Martin boundary

(3.3) $$\lim_{(\zeta_{t_n}, t_n) \stackrel{\mathcal{M}}{\to} a} \mathbb{P}[\,\cdot\,\mid \zeta_{t_n}](\omega),$$

we get the measure $\mathbb{P}^a$ with transition function ($0 \leq r < t$)

(3.4) $$\mathbb{P}^a\left[e_f(Y_t)g(\zeta_t) \mid (Y_r, \zeta_r) = (\nu, x)\right]$$
$$= e^{-\nu(V_t^r f)} \mathcal{P}_{r,x}^a\left[\exp\left(-\int_r^t V_t^s f(\xi_s) ds\right) g(\xi_t)\right]$$

as a limit. Here V_t^s denotes the non-linear two parameter semigroup associated with the superprocess (cf. [Dawson 1992, 12.3.4] if we consider the Historical process). The measure $\mathcal{P}^a$ arises as an extremal conditioned process in $\mathcal{Q}_\mathcal{P}^e$. The distribution

P^a of Y under $\mathbb{P}^a$ can be seen as the superprocess with an immortal particle whose motion is governed by $\mathcal{P}^a$, cf. [Overbeck 1992b].

Obviously, the face

$$(3.5) \qquad \mathcal{Q}_P^{\mathrm{add}} := \left\{ Q \mid \exists m \text{ s.t. } Q = \int_{\mathcal{Q}_P^e} P^a\, m(da) \right\}$$

of $\mathcal{Q}_P$ is isomorph (as a simplex, cf. [Dynkin 1971, Dynkin 1978]) to the set $\mathcal{Q}_\mathcal{P}$. We call $\mathcal{Q}_P^{\mathrm{add}}$ the set of *additive conditioned processes*.

If every $Q \in \mathcal{Q}_\mathcal{P}$ is an h-transform of $\mathcal{P}$, then P^a is the H-transform P^{H^a} with $H(s,\mu) = \mu\bigl(h^a(s)\bigr)$ where h^a is the extremal space-time harmonic function which corresponds to a in the Martin boundary of $\mathcal{P}$.

But in the case of the Historical process the one-particle-motion $\mathcal{P}$ is the path process. The measure $\mathcal{P}$ is a measure on $D\bigl([0,\infty), D_E\bigr)$. Because its conditional distribution with respect to the future $\widehat{\mathcal{F}}_t$ satisfies

$$(3.6) \qquad \mathcal{P}[\,\cdot\,\mid \widehat{\mathcal{F}}_t](\omega) = \delta_\omega,$$

we have

$$(3.7) \qquad \mathcal{Q}_\mathcal{P}^e = \{\delta_\omega \mid \omega \in D_E\}.$$

If we construct now $\mathbb{P}^\omega$ we see that it is no more an H-transform of $\mathbb{P}$. Let P^ω be the distribution of Y under $\mathbb{P}^\omega$. We have

$$(3.8) \qquad P^\omega[e_f(X_t) \mid X_r] = \exp\left(-\int_r^t V_t^s f\bigl(\omega(s)\bigr) ds\right) e^{-X_r(V_t^r f)}.$$

Clearly,

$$(3.9) \qquad P^\omega[\,\cdot\,\mid \widehat{\mathcal{F}}_t] = P[\,\cdot\,\mid \widehat{\mathcal{F}}_t],$$

because $\mathbb{P}^\omega$ satisfies the analogous equation. As in [Evans 1992, Theorem 1.4] formula (3.8) implies that the tail-σ field is P^ω-trivial. Hence P^ω is an extreme point of $\mathcal{Q}_P$. Consider for $r < t$ the function f on D_E defined by

$$f = 1_{\{y \mid \exists \rho > r \text{ s.t. } y(\sigma) = \omega(\sigma)\, \forall\, r < \sigma \leq \rho\}}.$$

Then

$$P_{r,\mu}^\omega[X_t(f)] = \int_r^t T_t^s f\bigl(\omega(s)\bigr) ds + \mu(T_t^r f) > 0,$$

but

$$P_{r,\mu}[X_t(f)] = \mu(T_t^r f) = 0$$

(where T_t^r is the transition semigroup of the path process). Hence

$$P_{r,\mu}^\omega\big|_{\mathcal{F}_t} \not\ll P_{r,\mu}\big|_{\mathcal{F}_t}.$$

Note also that the support of the representing measure of any additive H-transform is contained in the set

$$\{P^\omega \mid \omega \in D_E\},$$

which does not contain any H-transform.

Bibliography

D.A. Dawson, *Measure-valued Markov Processes*, Lecture Notes, École d'Été de Probabilités de Saint Flour 1991, (preprint) 1992.

E.B. Dynkin, *Integral representation of exessive measures and exessive functions*, Russian Math. Surveys **26** (1971), 165–185.

_____, *Sufficient Statistics and Extreme Points*, Annals of Probability **6** (1978), 705–730.

S.N. Evans, *Two representations of conditioned superprocesses*, 1992 (preprint).

S.N. Evans and E. Perkins, *Absolute continuity results for superprocesses with some applications*, Trans. Amer. Math. Soc. **325** (1991), 661–681.

H. Föllmer, *Phase Transition and Martin Boundary*, Séminaire de Probabilités IX, Lecture Notes in Math. **465**, Springer, Berlin, 1975, pp. 305–318.

_____, *Martin Boundaries on Wiener Spaces*, Diffusion Processes and Related Problems in Analysis, vol. I, Hrsg. Pinsky, M.A., 1990, pp. 3–16.

S.E. Kuznecov, *On decomposition of excessive functions*, Dokl. Akad. Nauk. SSSR **214** (1974), 276–278.

L. Overbeck, *Martin boundaries of some branching processes*, Ann. Inst. H. Poincaré Probab. Statist. (1993) (to appear).

_____, *Conditioned Super-Brownian Motion*, Probab. Theor. Relat. Fields **96** (1993), 545–570.

M. Röckner, *On the parabolic Martin boundary of the Ornstein-Uhlenbeck operator*, Ann. of. Prob. **20** (1992), 1063–1085.

Institut für Angewandte Mathematik, Rheinische Friedrich-Wilhelms-Universität Bonn, Wegelerstrasse 6, 53115 Bonn, Federal Republic of Germany

E-mail address: unm30a@ibm.rhrz.uni-bonn.de

A Dirichlet Form Primer

Byron Schmuland

A Dirichlet form $(\mathcal{E}, D(\mathcal{E}))$, like a Feller semigroup, is an analytic object that can be used to construct and study a certain Markov process $\{X_t\}_{t\geq 0}$. Unlike the Feller semigroup approach, which uses a pointwise analysis, the Dirichlet form approach uses a quasi-sure analysis, meaning that we are permitted to ignore certain exceptional sets which are not visited by the process. This slight ambiguity, where some of the definitions and results only hold quasi-everywhere, has certain advantages and certain disadvantages. On the one hand, even when studying as familiar a process as Brownian motion in $\mathbb{R}^n$, we can do some things more generally. For instance, M. Fukushima [Fukushima 1980, Example 5.1.1.] shows that when $n \geq 2$ and for any $\alpha \in \mathbb{R}$, the additive functional $A_t(\omega) = \int_0^t |X_s(\omega)|^\alpha \, ds$ makes perfect sense in the context of Dirichlet forms, even though when $\alpha \leq -2$ we have

$$(1) \qquad P_0\left(\int_0^t |X_s(\omega)|^\alpha \, ds = \infty\right) = 1.$$

This is because the state $\{0\}$, which is causing all the headaches, is polar and so can be completely removed from consideration. On the other hand, this slight ambiguity also means that when we use Dirichlet forms to solve stochastic differential equations for example, as in [Albeverio and Röckner 1991], the solutions are defined only from quasi-every starting point in the state space, rather than from every starting point. The absence of pointwise results is the price we pay for the increased generality of cases that can be covered using Dirichlet form theory.

Dirichlet forms have emerged as a powerful tool of stochastic analysis especially in the case, where the state space is infinite dimensional, where it is useful to throw away (topologically) large exceptional sets, even for very standard examples. In this paper we shall look at some recent developments in the general theory of Dirichlet forms, especially at those aspects that allow us to consider infinite dimensional cases. In particular, we will try to explain the condition of *quasi-regularity*, how it is defined and what it implies for a Dirichlet form. Our main reference is the recent book by Z. M. Ma and M. Röckner [Ma and Röckner 1992] whose treatment and notation we will try to follow throughout. The reader interested in other recent work in the area is encouraged to look, as well, at [Albeverio and Ma 1991], [Albeverio and Ma 1992, 1993], [Albeverio and Röckner 1989, 1990, 1991], [Bouleau and Hirsch 1991], [Röckner and Zhang 1992], and the references therein.

Let E be a Hausdorff topological space equipped with its Borel σ-field $\mathcal{B}(E)$, and m a σ-finite Borel measure on E.

1991 *Mathematics Subject Classification*. Primary: 60J45; Secondary: 60H15, 60G17.
This is the final form of the paper.

DEFINITION 1. A pair $(\mathcal{E}, D(\mathcal{E}))$ is called a *closed form* on (real) $L^2(E; m)$ if $D(\mathcal{E})$ is a dense linear subspace of $L^2(E; m)$ and if $\mathcal{E} : D(\mathcal{E}) \times D(\mathcal{E}) \to \mathbb{R}$ is a bilinear form such that the following conditions hold:
(i) $\mathcal{E}(u, u) \geq 0$ for all $u \in D(\mathcal{E})$.
(ii) $D(\mathcal{E})$ is a Hilbert space when equipped with the inner product $\tilde{\mathcal{E}}_1(u, v) := (1/2)\{\mathcal{E}(u, v) + \mathcal{E}(v, u)\} + (u, v)_{L^2}$.

DEFINITION 2. Given a closed form $(\mathcal{E}, D(\mathcal{E}))$ on $L^2(E; m)$ we define we following closed forms all of whose domains coincide with $D(\mathcal{E})$:
(i) $\tilde{\mathcal{E}}(u, v) = (1/2)\{\mathcal{E}(u, v) + \mathcal{E}(v, u)\}$.
(ii) $\widehat{\mathcal{E}}(u, v) = \mathcal{E}(v, u)$.
(iii) $\mathcal{E}_1(u, v) = \mathcal{E}(u, v) + (u, v)_{L^2}$.

DEFINITION 3. A closed form $(\mathcal{E}, D(\mathcal{E}))$ is called *coercive* if it satisfies the sector condition (see 9 below).

DEFINITION 4. A coercive, closed form $(\mathcal{E}, D(\mathcal{E}))$ on $L^2(E; m)$ is called a *Dirichlet form* if it has the following (unit) contraction property: for all $u \in D(\mathcal{E})$, we have $u^+ \wedge 1 \in D(\mathcal{E})$ and

$$(2) \quad \mathcal{E}(u + u^+ \wedge 1, u - u^+ \wedge 1) \geq 0 \quad \text{and} \quad \mathcal{E}(u - u^+ \wedge 1, u + u^+ \wedge 1) \geq 0.$$

Our hope would be to try to establish a relation of the following type between this form $\mathcal{E}$ and some Markov process $\{X_t\}$.

DEFINITION 5. A right process M cf. [Sharpe 1988] with state space E and transition semigroup $(p_t)_{t>0}$ is said to be *associated with* $(\mathcal{E}, D(\mathcal{E}))$ if $p_t f$ is an m-version of $T_t f$ for all $t > 0$ and every bounded, Borel measurable f in $L^2(E; m)$.

The construction of the process associated with a given Dirichlet form, when it is possible, follows a path that looks something like this:

$$(3) \qquad (\mathcal{E}, D(\mathcal{E})) \to (T_t)_{t \geq 0} \to (p_t)_{t \geq 0} \to \{X_t\}_{t \geq 0}.$$

The first step, to get the L^2 generator $(T_t)_{t \geq 0}$ is usually not difficult, and follows from the theory of semigroups of linear operators applied to the the Hilbert space $L^2(E; m)$. Condition (2) guarantees that T_t behaves like a sub-Markov transition kernel in the sense that $0 \leq f \leq 1$ implies $0 \leq T_t f \leq 1$ m-a.e. Also, once you are able to get a reasonable kernel $(p_t)_{t \geq 0}$, the last step follows from standard constructions of a Markov process from $(p_t)_{t \geq 0}$ (or from the resolvent kernel $(R_\alpha)_{\alpha > 0}$) on general state spaces, as done for instance in Sharpe's book [Sharpe 1988]. The middle step relates the L^2 semigroup $(T_t)_{t \geq 0}$ with the transition semigroup $(p_t)_{t \geq 0}$ on E, and this is the most difficult, especially if you have a particular process $\{X_t\}$ in mind. It is in this step, where your decision on an appropriate state space E and measure m plays the biggest role.

Once such a relationship between $(\mathcal{E}, D(\mathcal{E}))$ and M has been established we can use analysis of $(\mathcal{E}, D(\mathcal{E}))$ to get information about the behaviour of the process $\{X_t\}$. For instance, if A is a Borel subset of E with $m(A) = 0$, then $p_t 1_A(z) = 0$ m-a.e. $z \in E$. That means from m-almost every starting point, at a fixed time t, we won't catch X_t visiting the set A, i.e., $P_m(X_t \in A) = 0$. A stronger notion of smallness of sets is needed if we want to conclude that X_t absolutely never visits A, i.e., $P_m(\exists t \geq 0 \text{ so that } X_t \in A) = 0$. We will discuss this in Proposition 15.

Until recently, the general theory of Dirichlet forms had been restricted to the case omit where the underlying space E is locally compact. In M. Fukushima's book [Fukushima 1980], which is the standard reference in the area, the local compactness is used throughout and is crucial in the construction of the associated Markov processes. Fukushima assumes that E is a locally compact, separable metric space and that m is a positive Radon measure on $\mathcal{B}(E)$ with full support. He then constructs a Markov process, indeed a Hunt process, associated with any symmetric, regular Dirichlet form.

DEFINITION 6. A Dirichlet form is called *symmetric* if $\mathcal{E}(u,v) = \mathcal{E}(v,u)$ for all $u,v \in D(\mathcal{E})$.

DEFINITION 7. A Dirichlet form is called *regular* if $D(\mathcal{E}) \cap C_0(E)$ is $\tilde{\mathcal{E}}_1^{1/2}$-dense in $D(\mathcal{E})$, and is uniformly dense in $C_0(E)$. Here $C_0(E)$ is the space of continuous real-valued functions with compact support.

Now the local compactness assumption, of course, eliminates the possibility of using Fukushima's theory in the study of infinite-dimensional processes. Also the symmetry assumption confines us to treating only symmetrizable processes. Nevertheless, in the years since the publication of [Fukushima 1980] several authors have been able to modify Fukushima's construction in special cases and obtain processes in infinite-dimensional state spaces. cf. [Albeverio and Køegh-Krohn 1975, 1977], [Kusuoka 1982], [Albeverio and Röckner 1989, 1991], [Schmuland 1990], [Bouleau and Hirsch 1991], and see also the reference list in [Ma and Röckner 1992]. The theory of non-symmetric forms and the construction of the associated process has also been considered by a number of authors cf. [Carrillo Menendez 1975], [Feyel and de La Pradelle 1978], [LeJan 1982], [Oshima 1988]. It turns out that while neither assumption can be dropped, both are too strong.

EXAMPLE 8. Why we can't just drop symmetry.

Let's look at the least symmetric process possible. Consider the space $E = \mathbb{R}$ equipped with the usual topology and with m equal to Lebesgue measure. Let $\{X_t\}$ be the process that just moves to the right uniformly. That is, $p_t(z,dy) = \epsilon_{z+t}(dy)$. This yields the strongly continuous semigroup $(T_t)_{t \geq 0}$ on $L^2(E;m)$ which acts on f by the formula $(T_t f)(z) = f(z+t)$. The generator L is then $Lf = f'$ and $D(L) = \{f \in L^2 : f \text{ is absolutely continuous and } f' \in L^2\}$. Therefore the usual pre-Dirichlet form defined by $D(\mathcal{E}) = D(L)$ and $\mathcal{E}(f,g) = \int (-Lf) g \, dz$, is not closable. In fact, $\mathcal{E}$ is anti-symmetric so the Hilbert space norm in Definition 1 (ii) is just the L^2-norm. An unclosable form fails to give us the kind of probabilistic potential theory needed to connect the form $\mathcal{E}$ and the process $\{X_t\}$. So, in this example, a reasonable Markov process leads to an unreasonable form. This kind of situation can be avoided by assuming the sector condition, which allows us to control the form $\mathcal{E}$ by its symmetric part $\tilde{\mathcal{E}}$.

CONDITION 9. The substitute for symmetry: The sector condition.

The form $(\mathcal{E}, D(\mathcal{E}))$ satisfies the *sector condition* if there exists a constant $K > 0$ such that

(4) $$|\mathcal{E}_1(u,v)| \leq K \, \mathcal{E}_1(u,u)^{1/2} \mathcal{E}_1(v,v)^{1/2},$$

for all $u,v \in D(\mathcal{E})$.

EXAMPLE 10. Why we can't just drop regularity.

Take $E = [0,1)$ equipped with $m=$ Lebesgue measure, and let $(\mathcal{E}, D(\mathcal{E}))$ be the Dirichlet form associated with reflecting Brownian motion on $[0,1]$. That is,

$$D(\mathcal{E}) = \{u : u \text{ is absolutely continuous on } (0,1) \text{ and } u' \in L^2((0,1);dz)\},$$

(5)
$$\mathcal{E}(u,v) = (1/2)\int u'(z)\,v'(z)\,dz.$$

This form is not regular, in fact, the closure of $C_0([0,1))$ in $(D(\mathcal{E}), \tilde{\mathcal{E}}_1^{1/2})$ consists only of those functions f in $D(\mathcal{E})$ with $f(z) \to 0$ as $z \to 1$. And of course it is impossible to construct a reasonable process on $[0,1)$ associated with the form $(\mathcal{E}, D(\mathcal{E}))$ because the appropriate process is reflecting Brownian motion on $[0,1]$. The space $E = [0,1)$ is adequate to define the form but not as a state space for the process. So this time we had a reasonable Dirichlet form lead to an unreasonable process.

Now in this example the problem is obviously that the point $\{1\}$ is missing from the space, and if we put it back, we recover the usual regular form on $E = [0,1]$ corresponding to reflecting Brownian motion. However, the problem of "missing boundary points" can occur even when the space E is topologically complete. This can happen when the measure m lives on one Hilbert space, but the process $\{X_t\}$ needs an even bigger Hilbert space as a state space. Such an example is discussed in [Röckner and Schmuland 1993]. We'd like to find a condition that is not as restrictive as regularity, but which would let us know that our space is large enough to accommodate the associated process. But before we can get to our substitute for regularity we need to learn a bit more about exceptional sets for Dirichlet forms.

DEFINITION 11.
(i) For a closed subset $F \subseteq E$ we define

(6)
$$D(\mathcal{E})_F := \{u \in D(\mathcal{E}) | \ u = 0 \ m - \text{ a.e. on } E \backslash F\}.$$

Note that $D(\mathcal{E})_F$ is a closed subspace of $D(\mathcal{E})$.

(ii) An increasing sequence $(F_k)_{k \in \mathbb{N}}$ of closed subsets of E is called an $\mathcal{E}$-nest if $\cup_{k \geq 1} D(\mathcal{E})_{F_k}$ is $\tilde{\mathcal{E}}_1^{1/2}$-dense in $D(\mathcal{E})$.

(iii) A subset $N \subset E$ is called $\mathcal{E}$-exceptional if $N \subseteq \cap_k F_k^c$ for some $\mathcal{E}$-nest $(F_k)_{k \in \mathbb{N}}$. A property of points in E holds $\mathcal{E}$-quasi-everywhere (abbreviated $\mathcal{E}$-q.e.), if the property holds outside some $\mathcal{E}$-exceptional set. It can be seen that every $\mathcal{E}$-exceptional set has m-measure zero.

(iv) An $\mathcal{E}$-q.e. defined function $f : E \to \mathbb{R}$ is called $\mathcal{E}$-quasi-continuous if there exists an $\mathcal{E}$-nest $(F_k)_{k \in \mathbb{N}}$ so that $f|_{F_k}$ is continuous for each $k \in \mathbb{N}$.

(v) Let $(f_n)_{n \in \mathbb{N}}$ and f be ($\mathcal{E}$-q.e. defined) functions on E. We say that $(f_n)_{n \in \mathbb{N}}$ converges $\mathcal{E}$-quasi-uniformly to f if there exists an $\mathcal{E}$-nest $(F_k)_{k \in \mathbb{N}}$ such that $f_n \to f$ uniformly on each F_k.

The following result is very important in Dirichlet form analysis cf. [Ma and Röckner 1992, Chapter III, Proposition 3.5.].

LEMMA 12. *Let $(\mathcal{E}, D(\mathcal{E}))$ be a Dirichlet form on $L^2(E; m)$ and suppose that $(u_n)_{n \in \mathbb{N}} \in D(\mathcal{E})$, having $\mathcal{E}$-quasi-continuous m-versions $(\tilde{u}_n)_{n \in \mathbb{N}}$, converges to $u \in D(\mathcal{E})$ in $\tilde{\mathcal{E}}_1^{1/2}$-norm. Then there exists a subsequence $(u_{n_k})_{k \in \mathbb{N}}$ and an $\mathcal{E}$-quasi-continuous m-version $\tilde{u}$ of u so that $(\tilde{u}_{n_k})_{k \in \mathbb{N}}$ converges $\mathcal{E}$-quasi-uniformly to $\tilde{u}$.*

CONDITION 13. The substitute for regularity: Quasi-regularity.

A Dirichlet form $(\mathcal{E}, D(\mathcal{E}))$ on $L^2(E; m)$ is called *quasi-regular* if:

(QR1) There exists an $\mathcal{E}$-nest $(F_k)_{k \in \mathbb{N}}$ consisting of compact sets.

(QR2) There exists an $\tilde{\mathcal{E}}_1^{1/2}$-dense subset of $D(\mathcal{E})$ whose elements have $\mathcal{E}$-quasi-continuous m-versions.

(QR3) There exists a countable collection $(u_n)_{n \in \mathbb{N}} \in D(\mathcal{E})$, having $\mathcal{E}$-quasi-continuous m-versions $(\tilde{u}_n)_{n \in \mathbb{N}}$, and an $\mathcal{E}$-exceptional set $N \subset E$ such that $(\tilde{u}_n)_{n \in \mathbb{N}}$ separates the points of $E \setminus N$.

The fundamental existence result in the framework of quasi-regular forms is found in [Ma and Röckner 1992, Chapter IV, Theorem 6.7.] and it says the following:

THEOREM 14. *Let E be a metrizable Lusin space. Then a Dirichlet form $(\mathcal{E}, D(\mathcal{E}))$ on $L^2(E; m)$ is quasi-regular if and only if there exists a pair $(\boldsymbol{M}, \widehat{\boldsymbol{M}})$ of normal, right continuous, strong Markov processes so that $\boldsymbol{M}$ is associated with $\mathcal{E}$, and $\widehat{\boldsymbol{M}}$ is associated with $\widehat{\mathcal{E}}$.*

This says that the class of quasi-regular Dirichlet forms is the correct setting for the study of those forms associated with nice Markov processes. L. Overbeck and M. Röckner [Overbeck and Röckner 1993] have recently proved a one-sided version of the existence result for quasi-regular *semi*-Dirichlet forms; i.e., for those closed, coercive forms $(\mathcal{E}, D(\mathcal{E}))$ such that only the first relation in (2) holds. In this case, of course, we don't get a pair of processes but only the process $\boldsymbol{M}$.

The next result is the heart of the matter, as it establishes the connection between the probabilistic potential theory of the process $\boldsymbol{M}$ and the analytic potential theory of the form $(\mathcal{E}, D(\mathcal{E}))$. cf. [Ma and Röckner 1992, Chapter IV, Theorem 5.29.].

PROPOSITION 15. *Suppose the right process $\boldsymbol{M}$ is associated with the quasi-regular Dirichlet form $(\mathcal{E}, D(\mathcal{E}))$. Then a subset N of E is $\mathcal{E}$-exceptional if and only if $N \subset \tilde{N}$, where $\tilde{N}$ is a Borel set such that*

$$(7) \qquad P_m(\tau_{\tilde{N}} < \zeta) = 0,$$

where ζ is the lifetime of the process and τ means the touching time of the set, i.e., for any Borel set B we define $\tau_B(\omega) := \inf\{0 \le t < \zeta(\omega) : X_t(\omega) \in B$ or $X_{t-}(\omega) \in B\} \wedge \zeta(\omega)$.

It is easy to see that this new concept of a quasi-regular Dirichlet form includes the classical concept of a regular Dirichlet form. This follows from the next proposition cf. [Ma and Röckner 1992, Chapter IV, Section 4a].

PROPOSITION 16. *Assume E is a locally compact, separable, metric space and m is a positive Radon measure on $\mathcal{B}(E)$. If $(\mathcal{E}, D(\mathcal{E}))$ is a regular Dirichlet form on $L^2(E; m)$, then $(\mathcal{E}, D(\mathcal{E}))$ is quasi-regular.*

PROOF. We only show (QR1), as (QR2) and (QR3) are easy exercises. By the topological assumptions on E, we may write $E = \cup_{k=1}^{\infty} F_k$, where $(F_k)_{k \in \mathbb{N}}$ is an increasing sequence of compact sets in E so that F_k is contained in the interior of F_{k+1} for all $k \ge 1$. It is then easy to see that

$$(8) \qquad C_0(E) \cap D(\mathcal{E}) \subseteq \cup_{k=1}^{\infty} D(\mathcal{E})_{F_k},$$

which concludes the proof. □

By Theorem 14, we know that if we have a Dirichlet form $(\mathcal{E}, D(\mathcal{E}))$, even on an infinite dimensional Banach space E, and we want to show the existence of an associated process, all we have to do is prove that $(\mathcal{E}, D(\mathcal{E}))$ is quasi-regular. Typically the form $(\mathcal{E}, D(\mathcal{E}))$ is defined as the closure of some pre-Dirichlet form over a core of continuous functions that separates points in E. So usually the conditions (QR2) and (QR3) of Condition 13 are easily checked. It is (QR1) that is usually most troublesome, and that is because it is not always easy to come up an explicit sequence of compact sets that do the trick, especially in infinite dimensions. We note that the condition (QR1) is equivalent to the tightness of a certain capacity defined on subsets of E cf. [Lyons and Röckner 1991], [Röckner and Schmuland 1992], [Schmuland 1992]. The following result is based on Proposition 3.1 of [Röckner and Schmuland 1992], and it shows when a Dirichlet form of gradient type defined on a Banach space is quasi-regular (see also [Ma and Röckner 1992, Chapter IV, Section 4b]).

EXAMPLE 17. Gradient-type forms on an infinite dimensional Banach space.

Let E be a (real) separable Banach space, and m a finite measure on $\mathcal{B}(E)$ which charges every weakly open set. Define a linear space of functions on E by

(9) $\qquad \mathcal{F}C_b^\infty = \{f(l_1, \ldots, l_m) : m \in \mathbb{N}, f \in C_b^\infty(\mathbb{R}^m), l_1, \ldots, l_m \in E'\}.$

Here $C_b^\infty(\mathbb{R}^m)$ denotes the space of all infinitely differentiable functions on $\mathbb{R}^m$ with all partial derivatives bounded. By the Hahn-Banach theorem, $\mathcal{F}C_b^\infty$ separates the points of E. The support condition on m means that we can regard $\mathcal{F}C_b^\infty$ as a subspace of $L^2(E;m)$, and by a monotone class argument you can show that it is dense in $L^2(E;m)$. Define for $u \in \mathcal{F}C_b^\infty$ and $k \in E$,

(10) $\qquad \dfrac{\partial u}{\partial k}(z) := \dfrac{d}{ds} u(z + sk)|_{s=0}, \; z \in E.$

Observe that if $u = f(l_1, \ldots, l_m)$, then

(11) $\qquad \dfrac{\partial u}{\partial k} = \sum_{i=1}^{m} \dfrac{\partial f}{\partial x_i}(l_1, \ldots, l_m) \, _{E'}\langle l_i, k\rangle_E,$

which shows us that $\partial u/\partial k$ is again a member of $\mathcal{F}C_b^\infty$. Also let us assume that there is a separable real Hilbert space $(H, \langle , \rangle_H)$ densely and continuously embedded into E. Identifying H with its dual H' we have that

(12) $\qquad E' \subset H \subset E \quad \text{densely and continuously,}$

and $_{E'}\langle , \rangle_E$ restricted to $E' \times H$ coincides with $\langle , \rangle_H$. Observe that by (11) and (12), for $u \in \mathcal{F}C_b^\infty$ and fixed $z \in E$, the map $k \to (\partial u/\partial k)(z)$ is a continuous linear functional on H. Define $\nabla u(z) \in H$ by

(13) $\qquad \langle \nabla u(z), k\rangle_H = \dfrac{\partial u}{\partial k}(z), \quad k \in H.$

Define a bilinear form on $L^2(E;m)$ by

(14) $\qquad \begin{aligned} \mathcal{E}(u,v) &= \int \Gamma(\nabla u(z), \nabla v(z)) \, m(dz), \\ D(\mathcal{E}) &= \mathcal{F}C_b^\infty, \end{aligned}$

where $\Gamma(\cdot, \cdot)$ is a positive semi-definite, continuous, bilinear form on H, which satisfies the sector condition. A common example is to let $\Gamma(\cdot, \cdot) = \langle \cdot, \cdot\rangle_H$, but in general Γ is not assumed to be symmetric. Then $\mathcal{E}$ is also a densely defined, positive semi-definite, bilinear form on $\mathcal{F}C_b^\infty$, satisfying the sector condition, because $\mathcal{E}$

inherits those properties from Γ. If such an $\mathcal{E}$ is closable in $L^2(E;m)$, then its closure $(\mathcal{E}, D(\mathcal{E}))$ is a Dirichlet form.

PROPOSITION 18. *The form $(\mathcal{E}, D(\mathcal{E}))$, defined as the closure of $(\mathcal{E}, \mathcal{F}C_b^\infty)$ above, is quasi-regular.*

The bulk of the proof is dedicated to showing (QR1), which requires coming up with compact sets that exhaust the space in the appropriate manner. It eventually goes back to a special choice of functions in $D(\mathcal{E})$ which are shown to converge in $\tilde{\mathcal{E}}_1^{1/2}$-norm. An application of Lemma 12 gives a nest $(F_k)_{k\in\mathbb{N}}$ on which the convergence is uniform, and topological considerations are used to conclude that the members of this nest are in fact compact. The interested reader is referred to [Röckner and Schmuland 1992] or [Ma and Röckner 1992] for details.

This proof has also been generalized [Röckner and Schmuland 1993] to "square-field operator"-type Dirichlet forms, which are similar to the one seen in equation (14). Such Dirichlet forms can be used to construct and study a variety of infinite dimensional processes; such as infinite dimensional Ornstein-Uhlenbeck processes [Schmuland 1990, 1993], diffusions on loop space [Albeverio, Léandre, and Röckner 1992], [Driver and Röckner 1992], [Getzler 1989], and certain Fleming-Viot (measure-valued) processes [Overbeck, Röckner, and Schmuland 1993], [Schmuland 1991]. Once the quasi-regularity of the form has been established and the associated process constructed, you can use the Dirichlet form to analyze the behaviour of the sample paths. Determining what parts of the state space the process visits and what parts it avoids, helps you to better understand the model. We conclude by presenting two concrete examples, where a Dirichlet form is used to study an infinite dimensional process. These two examples are extracted from [Schmuland 1993] and [Overbeck, Röckner, and Schmuland 1993] respectively, where the reader can find more details. I hope that these give a flavor of the type of results one can get using Dirichlet forms, and the kind of calculations required to get them.

EXAMPLE 19. Walsh's stochastic model of neural response.

In his 1981 paper "A Stochastic Model of Neural Response," J. Walsh [Walsh 1981] proposed a model for a nerve cylinder undergoing random stimulation along its length. The cylinder itself is modelled using the interval $[0, L]$ while $\{X(x,t,\omega) : 0 \le x \le L, \ t \ge 0\}$ denotes the nerve membrane potential at the time t at a location x along the axis. He found that this potential could be approximated by the solution $\{X_t\}$ of the equation

$$(15) \qquad dX_t = -(I - \partial^2/\partial x^2)X_t\, dt + dW_t.$$

The Laplacian $\partial^2/\partial x^2$ is given reflecting boundary conditions at the endpoints 0 and L, and W is a white noise based on the measure $\eta(x)dx\,dt$. We set

$$(16) \qquad \begin{aligned} A &= I - \partial^2/\partial x^2, \\ H &= L^2([0, L]; dx/\eta(x)). \end{aligned}$$

The symmetric case $\eta \equiv 1$ serves as a benchmark, but we will work in a more general setting and in fact only assume that η is absolutely continuous and $\eta' \in L^2([0, L]; dx)$. We also assume that η is bounded away from zero on $[0, L]$, consequently H is just $L^2([0, L]; dx)$ equipped with a different norm.

Just to keep things straight we mention that unlabelled quantities like $\langle,\rangle$ and $\|\ \|$ are the usual inner product and norm on $L^2([0, L]; dx)$, while labelled versions $\langle,\rangle_H$ and $\|\ \|_H$ are the inner product and norm in the $L^2([0, L]; dx/\eta(x))$ sense.

The eigenvectors of A are given by $e_0 \equiv L^{-1/2}$ and $e_j(x) = \sqrt{2}L^{-1/2}\cos(\pi j x L^{-1})$ for $j \geq 1$, with $Ae_j = \lambda_j e_j = (1 + \pi^2 j^2 L^{-2})e_j$. Now the solution of (15) is a Gaussian process, in fact, an Ornstein-Uhlenbeck process in infinite dimensions. The invariant measure m for this process is also Gaussian and satisfies

$$(17) \quad \int \langle h, z\rangle_H \langle k, z\rangle_H \, m(dz) = (h, k) := \int_0^\infty \langle e^{-tA^*}h, e^{-tA^*}k\rangle_H \, dt.$$

Since $e^{-tA^*}f = \eta e^{-tA}(f/\eta)$, the above equation tells us that

$$(18) \quad \int \langle e_j, z\rangle^2 \, m(dz) = \int \langle \eta e_j, z\rangle_H^2 \, m(dz)$$
$$= \int_0^\infty \int \eta(x)\bigl(e^{-tA}(e_j)(x)\bigr)^2 \, dx \, dt$$
$$= \frac{1}{2\lambda_j} \int \eta(x) e_j^2(x) \, dx$$
$$\leq \frac{c}{\lambda_j}.$$

Summing over j gives $\int \|z\|^2 \, m(dz) < \infty$, which tells us that the invariant measure m lives on H.

The conditions on η imply that the bilinear form (A^*h, k) is continuous and satisfies $(A^*h, k) + (A^*k, h) = \langle h, k\rangle_H$. (recall (17) for the definition of the round bracket form $(,)$). We now define a form on $\mathcal{F}C_b^\infty$ as in (14) by

$$(19) \quad \mathcal{E}(u, v) = \int (A^*\nabla u, \nabla v) \, dm.$$

Here the gradient is measured in the H sense, so that the linear map $\langle h, \cdot\rangle_H$ has gradient h, but the linear map $\langle h, \cdot\rangle = \langle \eta h, \cdot\rangle_H$ has gradient ηh. From Proposition 18 and Theorem 14 we can construct the associated process $(\Omega, \mathcal{F}, \{X_t\}_{t\geq 0}, (P_z)_{z\in H})$ on H, and this process solves (15), in the sense of [Albeverio and Röckner 1991]. So now we have a function-valued process $\{X_t\}_{t\geq 0}$ which lives in $L^2([0, L]; dx)$. What kind of functions are they? What would a snapshot of the process look like? Walsh had already showed that the process, in fact, takes values in $C[0, L]$ and has continuous sample paths there. We shall show that the function $x \to X(t, x)$ is, nevertheless, quite rough, in the sense of having unbounded variation.

LEMMA 20. $P_m(x \to X(x, t)$ is of unbounded variation for all $t) = 1$.

PROOF. Properties of cosine series show us that if $f \in L^2[0, L]$ has bounded variation, then the sequence of numbers $(\langle f, e_k\rangle)_{k\in\mathbb{N}}$ is $O(1/k)$. A calculation similar to the one done in (18) shows that $(\langle e_k, \cdot\rangle)_{k\in\mathbb{N}}$ on $L^2(H; m)$ are mean-zero Gaussian random variables with covariance

$$(20) \quad \int \langle e_j, z\rangle \langle e_k, z\rangle \, m(dz) = \left(\int \eta(x) e_j(x) e_k(x) \, dx\right)/(\lambda_j + \lambda_k).$$

Notice that the variance of $\langle e_j, \cdot\rangle$ is of the order j^{-2}. In addition, the conditions on η mean that the random variables $(\langle e_k, \cdot\rangle)_{k\in\mathbb{N}}$ not too strongly correlated and as a consequence you can show that for fixed N there exist $0 < \alpha < 1$ and $c > 0$ so that

$$(21) \quad m\left(\sup_{j=1}^n |j\langle e_j, z\rangle| \leq N\right) \leq c\alpha^n.$$

Letting $n \to \infty$ gives us $m(\sup_{j=1}^{\infty} |j\langle e_j, z\rangle| < \infty) = 0$, and using the observation above about cosine series we get,

(22) $\qquad m(z : x \to z(x) \text{ is of bounded variation }) = 0.$

This gives us the fixed time result that for every $t \geq 0$,

(23) $\qquad P_m(x \to X(x,t) \text{ is of bounded variation }) = 0,$

and now we would like to use the Dirichlet form $\mathcal{E}$ to show that this is true for all t, with probability one.

Set $u_n(z) = \left(\sup_{j=1}^{n} |j\langle e_j, z\rangle| \wedge N\right)$. Then $u_n(z) \in D(\mathcal{E})$ and $u_n(z) \uparrow u(z)$ pointwise everywhere on E, where the function u satisfies $m(u \equiv N) = 1$. Calculating gradients gives

(24) $\qquad \nabla u_n(z) = (\pm k)(\eta e_k)\, \chi\left(|k\langle e_k, z\rangle| = \sup_{j=1}^{n} |j\langle e_j, z\rangle| \leq N\right).$

Here χ means indicator function, so that $\nabla u_n(z)$ is zero if the supremum of $|j\langle e_j, z\rangle|$ is bigger than N, otherwise we get plus or minus $k\eta e_k$, where the index k is the same one at which the supremum occurred. The upshot is we get a nice bound for the norm of the gradient, because at each point z the gradient has at most one non-zero coordinate, so

(25) $\qquad \|\nabla u_n(z)\|_H^2 \leq c\, k^2\, \chi\left(|k\langle e_k, z\rangle| = \sup_{j=1}^{n} |j\langle e_j, z\rangle| \leq N\right).$

Therefore, from (21) and (25), we obtain

(26) $\qquad \mathcal{E}(u_n, u_n) = 1/2 \int \|\nabla u_n(z)\|_H^2\, m(dz)$
$$\leq c\, n^2 m\left(\sup_{j=1}^{n} |j\langle e_j, z\rangle| \leq N\right)$$
$$\leq c\, n^2 \alpha^n.$$

Applying Lemma 12 and Proposition 15 we find that

(27) $\qquad P_m\bigl(t \to u(X_t) \text{ is continuous }\bigr) = P_m\bigl(u(X_t) = N \text{ for all } t\bigr) = 1,$

and since this is true for all N we conclude that

(28) $P_m\bigl(\langle X_t, e_k\rangle \neq O(1/k) \text{ for all } t\bigr)$
$\qquad = P_m(x \to X(x,t) \text{ is of unbounded variation for all } t) = 1.$ $\square$

EXAMPLE 21. The Fleming-Viot process.

Let S be a locally compact separable metric space and set $E := \mathcal{P}(S) =$ all probability measures on S. Consider the space of functions

(29)
$$\mathcal{F}C_b^{\infty} := \{F((f_1, \cdot), \ldots, (f_m, \cdot)) : m \in \mathbb{N},\ F \in C_b^{\infty}(\mathbb{R}^m),\ f_1, \ldots, f_m \in C_b(S)\},$$

where the round brackets refer to integration on S; i.e., $(f, \mu) = \int_S f(x)\, \mu(dx)$. For every $x \in S$ and $u \in \mathcal{F}C_b^{\infty}$ define

(30) $\qquad \dfrac{\partial u}{\partial \epsilon_x}(\mu) = \dfrac{d}{ds} u(\mu + s\epsilon_x)|_{s=0}.$

Consider the Fleming-Viot generator

$$(31) \quad Lu(\mu) = \frac{1}{2}\int_S\int_S \mu(dx)\bigl(\epsilon_x(dy) - \mu(dy)\bigr)\frac{\partial^2 u}{\partial\epsilon_x\partial\epsilon_y}(\mu) + \int_S \mu(dx) A\left(\frac{\partial u}{\partial\epsilon_x}(\mu)\right).$$

We will only consider bounded mutation operators A of the form

$$Af(x) = (\theta/2)\int_S \bigl(f(\xi) - f(x)\bigr)\nu_0(d\xi),$$

where $\theta > 0$ and $\nu_0 \in \mathcal{P}(S)$. S.N. Ethier and T.G. Kurtz [Ethier and Kurtz 1993] have shown that there is a probability measure m on E such that

$$(32) \quad \mathcal{E}(u,v) = \int (-Lu) v\, dm \qquad u, v \in \mathcal{F}C_b^\infty,$$

defines a symmetric bilinear form. This form is closable on $L^2(E; m)$ and its closure $(\mathcal{E}, D(\mathcal{E}))$ is a symmetric Dirichlet form. This form can be re-written in a more symmetric way on $\mathcal{F}C_b^\infty$ at least as:

$$(33) \quad \mathcal{E}(u,v) = \int \langle \nabla u(\mu), \nabla v(\mu)\rangle_\mu\, m(d\mu),$$

where $\langle f, g\rangle_\mu := \int fg\, d\mu - (\int f d\mu)(\int g d\mu)$ and, for each fixed μ, as a function of $x \in S$ we let $(\nabla u(\mu))(x) = (\partial u/\partial\epsilon_x)(\mu)$. The invariant measure m has the following form:

$$(34) \quad m(\cdot) = P\left(\sum_{i=1}^\infty \rho_i \epsilon_{\xi_i} \in \cdot\right).$$

Here $(\rho_1, \rho_2, \ldots)$ has a Poisson-Dirichlet distribution with parameter θ, and $\{\xi_i\}$ are i.i.d. S-valued random variables that are independent of $(\rho_1, \rho_2, \ldots)$ and have distribution ν_0.

Consider a linear function $u \in \mathcal{F}C_b^\infty$ given by $u(\mu) := \int_S f(x)\, \mu(dx)$, for some $f \in C_b(S)$. The average value of u is

$$(35) \quad \begin{aligned}\int_E u(\mu)\, m(\mu) &= E\left(\int_S f\, d(\Sigma\rho_i \epsilon_{\xi_i})\right) \\ &= \sum E(\rho_i) E f(\xi_i) \\ &= \int_S f(x)\, \nu_0(dx).\end{aligned}$$

Taking partial derivatives we discover $(\partial u/\partial\epsilon_x)(\mu) = f(x)$ at every point $\mu \in E$, and therefore $\|\partial u/\partial\epsilon_x\|_\mu^2 = \int f^2\, d\mu - (\int f\, d\mu)^2$. Plugging this into the Dirichlet form we discover that

$$(36) \quad \mathcal{E}(u,u) = \int_E \left(\int f^2\, d\mu - (\int f\, d\mu)^2\right) m(d\mu).$$

Using (36) and the fact that $(\mathcal{E}, D(\mathcal{E}))$ is a closed form we pass may from continuous functions f to bounded, measurable f. In particular, for any fixed Borel set $F \subseteq S$, the map $\mu \to \mu(F)$ belongs to $D(\mathcal{E})$ and is $\mathcal{E}$-quasi-continuous. Thus, if $\nu_0(F) = 0$, then $\mu(F) = 0$ m-a.e. from (35), and so $\mu(F) = 0$ $\mathcal{E}$-quasi-everywhere. We conclude that if $\nu_0(F) = 0$, then $P_m(X_t(F) = 0 \text{ for all } t) = 1$. In other words, if ν_0 fails to charge the set F, then the Fleming-Viot process will never charge it either, not even at exceptional times. More calculations in a similar vein lead to the fact that the function $u(\mu) = \sum(\text{atoms of } \mu)$ belongs to $D(\mathcal{E})$ and is $\mathcal{E}$-quasi-continuous. But (34) shows that $u(\mu) = 1$ m-a.e., and so $u(\mu) = 1$ $\mathcal{E}$-quasi-everywhere, and therefore $P_m(X_t \text{ is purely atomic for all } t) = 1$. Thus we can recover, using Dirichlet forms, a result [Ethier and Kurtz 1986, Chapter 10, Theorem 4.5] proved by Ethier and Kurtz. We can also use $\mathcal{E}$ to discover other properties of this measure-valued process [Overbeck, Röckner, and Schmuland 1993].

References

S. Albeverio and R. Høegh-Krohn, *Quasi-invariant measures, symmetric diffusion processes and quantum fields*, Les Méthodes Mathématiques de la Théorie Quantique des Champs (Marseille, 23–27 juin 1975), Colloques Internationaux du C.N.R.S., vol. 248, C.R.N.S, 1976.

____, *Dirichlet forms and diffusion processes on rigged Hilbert spaces*, Z. Wahrsch. Verw. Gebiete **40** (1977), 1–57.

____, *Hunt processes and analytic potential theory on rigged Hilbert spaces*, Ann. Inst. Henri Poincaré. Probab. Statist. **13** (1977), 269–291.

S. Albeverio, R. Léandre, and M. Röckner, *Construction of a rotational invariant diffusion on the free loop space* (preprint).

S. Albeverio and Z. M. Ma, *Diffusion processes associated with singular Dirichlet forms*, Proc. Stochastic Analysis and Applications (A. B. Cruzeiro and J. C. Zambrini, eds.), Birkhäuser, New York, 1991.

____, *Necessary and sufficient conditions for the existence of m-perfect processes associated with Dirichlet forms*, Séminaire de Probabilités XXV, Lecture Notes in Math., vol. 1485, Springer, Berlin, 1991, pp. 374–406.

S. Albeverio, Z. M. Ma, and M. Röckner, *Local property of Dirichlet forms and diffusions on general state spaces*, Math. Ann. (1992) (to appear).

____, *Quasi-regular Dirichlet forms and Markov processes*, J. Funct. Anal. **111** (1993), 118–154.

S. Albeverio and M. Röckner, *Classical Dirichlet forms on topological vector spaces — the construction of the associated diffusion process*, Probab. Theor. Relat. Fields **83** (1989), 405–434.

____, *Classical Dirichlet forms on topological vector spaces — closability and a Cameron-Martin formula*, J. Funct. Anal. **88** (1990), 395–436.

____, *Stochastic differential equations in infinite dimensions: solutions via Dirichlet forms*, Probab. Theor. Relat. Fields **89** (1991), 347–386.

N. Bouleau and F. Hirsch, *Dirichlet forms and analysis on Wiener space* (1991), Walter de Gruyter, Berlin New York.

S. Carrillo Menendez, *Processus de Markov associé a une forme de Dirichlet non symétrique*, Z. Wahrsch. Verw. Gebiete **33** (1975), 139–154.

B. Driver and M. Röckner, *Construction of diffusions on path and loop spaces of compact Riemannian manifolds*, 1992 (preprint).

S. N. Ethier and T. G. Kurtz, *Markov processes — Characterization and convergence*, John Wiley & Sons, New York, 1986.

____, *Convergence to Fleming-Viot processes in the weak atomic topology*, Stochastic Processes Appl. (1993) (to appear).

D. Feyel and A. de La Pradelle, *Dualité des quasi-résolvantes de Ray*, Sem. Th. Potentiel 4 (Paris), Lecture Notes in Math., vol. 713, Springer, Berlin, 1979, pp. 67–88.

M. Fukushima, *Dirichlet forms and Markov processes*, North Holland, Amsterdam, 1980.

E. Getzler, *Dirichlet forms on loop space*, Bull. Sci. Math. (2) **113** (1989), 151–174.

S. Kusuoka, *Dirichlet forms and diffusion processes on Banach spaces*, J. Fac. Sci. Univ. Tokyo Sect. IA Math. **29** (1982), 79–95.

Y. LeJan, *Dual Markovian semigroups and processes*, Functional Analysis in Markov Processes (M. Fukushima, ed.), Lecture Notes in Math., vol. 923, Springer, Berlin, 1982, pp. 47–75.

T. Lyons and M. Röckner, *A note on tightness of capacities associated with Dirichlet forms*, Bull. London Math. Soc. **24** (1992), 181–184.

Z. M. Ma and M. Röckner, *An Introduction to the theory of (non-symmetric) Dirichlet forms*, Springer, Berlin, 1992.

Y. Oshima, *Lectures on Dirichlet forms*, Erlangen, 1988 (preprint).

L. Overbeck and M. Röckner, *Existence of Markov processes associated with semi-Dirichlet forms* (1993) (in preparation).

L. Overbeck, M. Röckner, and B. Schmuland, *Dirichlet forms and Fleming-Viot processes* (in preparation).

M. Röckner and B. Schmuland, *Tightness of general $C_{1,p}$ capacities on Banach space*, J. Funct. Anal. **108** (1992), 1–12.

____, *Quasi-regular Dirichlet forms: Examples and counterexamples* (1993) (in preparation).

M. Röckner and T. S. Zhang, *Uniqueness of generalized Schrödinger operators and applications*, J. Funct. Anal. **105** (1992), 187–231.

B. Schmuland, *Sample path properties of ℓ^p-valued Ornstein-Uhlenbeck processes*, Canad. Math. Bull. **33** (1990), 358–366.

____, *An alternative compactification for classical Dirichlet forms on topological vector spaces*, Stochastics Stochastic Rep. **33** (1990), 75–90.

____, *A result on the infinitely many neutral alleles diffusion model*, J. Appl. Probab. **28** (1991), 253–267.

____, *Tightness of Gaussian capacities on subspaces* **14** (1992), C. R. Math. Rep. Acad. Sci. Canada, 125–130.

____, *Non-symmetric Ornstein-Uhlenbeck processes in Banach space via Dirichlet forms*, Canad. J. Math. (to appear).

M. Sharpe, *General theory of Markov processes* (1988), Academic Press, New York.

J.B. Walsh, *A stochastic model of neural response*, Adv. in Appl. Probab. **14** (1981), 231–281.

DEPARTMENT OF STATISTICS AND APPLIED PROBABILITY, UNIVERSITY OF ALBERTA, EDMONTON (ALBERTA), CANADA T6G 2G1

E-mail address: schmu@hal.stat.ualberta.ca

Stationary Distribution Problem for Interacting Diffusion Systems

Tokuzo Shiga

1. Introduction

In the theory of stochastic interacting particle systems one of fundamental problem is to seek all stationary distributions. However, as far as we know, there are only a few models such as symmetric simple exclusion processes, basic contact processes and voter models for which one can completely characterize all stationary distributions, see [Liggett 1985]. A common feature of these processes is that there corresponds a tractable dual process. Accordingly if the process has no dual process it seems difficult to seek all stationary distibutions. However if one consider the processes in $\mathbb{Z}^d$-shift invariant situation, it is possible to seek all $\mathbb{Z}^d$-shift invariant stationary distributions. Concerning this we know the results for non-symmetric simple exclusion processes [Liggett 1975], zero-range processes [Andjel 1982] and branching random walks [Dawson 1977], [Deuschel 1993]. In these cases the main tool of the proof is to use a variety of coupling processes. Even in such cases it is not easy to seek all stationary distributions. It would indeed be expected that in a broad class every stationary distribution is $\mathbb{Z}^d$-shift invariant.

Very recently Bramson, Cox and Greven [Bramson 1993] succeeded in proving this for branching Brownian particles and its spatial continuum version (so called "super Brownian motion").

For the interacting diffusion systems one can obtain all $\mathbb{Z}^d$-shift invariant stationary distributions by an analogous usage of coupling processes to [Liggett and Spitzer 1981], (see [Cox and Greven 1993] and [Shiga 1992]). As the next stage we would like to attack the problem to seek all stationary distributions. However it seems hard to consider the problem in the general setting. In Section 2 we discuss the interacting diffusion systems with state space $[0,1]^S$. In particular under a restrictive situation we will seek all stationary distributions. In Section 3 we discuss an interacting Fleming-Viot model that is a measure-valued diffusion, for which Handa [Handa 1990] discussed the ergodicity in case incorporating mutation factor. For the interacting Fleming-Viot model incorporating no mutation factor it is possible to obtain complete description of all stationary distributions since it has a nice dual process similar to that of the stepping stone diffusion model in [Shiga 1980], [Shiga 1980].

1991 *Mathematics Subject Classification.* 60K35.

Key words and phrases. Interacting diffusions, stationary distributions, infinite particle systems.

This is the final form of the paper.

Let us first formulate interacting diffusion systems and survey the previous results. Let S be a countable set. Suppose that we are given an $S \times S$-matrix $A = \{A_{ij}\}$ and a function $a: R \to R$ satisfying the following condition.

CONDITION (A).

(A.1) $A = \{A_{ij}\}$ is a bounded infinitesimal matrix of a continuous time irreducible Markov chain with state space S, i.e.

$$A_{ij} \geq 0 \text{ for } i \neq j, \sum_{j \in S} A_{ij} = 0 \text{ and } \sup_{i \in S} |A_{ii}| < \infty.$$

(A.2) $a(u)$ is locally 1/2-Hölder continuous and satisfies a linear growth condition, i.e. $|a(u)| \leq \text{const.}(1 + |u|)$ for $u \in R$.

Let us consider the following stochastic differential equation (SDE):

(IDS) $$dx_i(t) = \sum_{j \in S} A_{ij} x_j(t)\, dt + a(x_i(t)) dB_i(t), \quad i \in S,$$

where $\{B_i(t)\}_{i \in S}$ is an independent system of standard Brownian motions.

To formulate solutions we first choose a positive summable sequence $\gamma = \{\gamma_i\}_{i \in S}$ such that for some $C > 0$

$$\sum_{i \in S} \gamma_i |A_{ij}| \leq C \gamma_j, \quad j \in S.$$

The condition (A.1) guarantees existence of such a sequence $\gamma = \{\gamma_i\}_{i \in S}$.

Let $\mathbb{E} = L^2(\gamma)$ be a Hilbert space equipped with the norm $\|x\|^2 = \sum_{i \in S} \gamma_i |x_i|^2$. Then it is known that for every $x(0) = x \in \mathbb{E}$ there is a pathwise unique $\mathbb{E}$-valued solution $x(t)$, which defines a diffusion process $(\Omega, \mathcal{F}, \mathcal{F}_t, P^x, x(t))$ with the state space $\mathbb{E}$. cf. [Shiga and Shimizu 1980]. We call the diffusion process an interacting diffusion system. The transition probability induces a Feller Markov semi-group T_t acting on $C_b(\mathbb{E})$ (the totality of bounded continuous functions defined on $\mathbb{E}$) such that

$$T_t f - f = \int_0^t T_s L f\, ds, \quad f \in C_0^2(\mathbb{E}),$$

where $C_0^2(\mathbb{E})$ stands for the totality of C^2-functions defined on $\mathbb{E}$ depending on finitely many coordinates with bounded derivatives and Lf being bounded,

$$L f = \frac{1}{2} \sum_{i \in S} a(x_i)^2 \frac{\partial^2 f}{\partial x_i^2} + \sum_{i \in S} \left(\sum_{j \in S} A_{ij} x_j \right) \frac{\partial f}{\partial x_i}.$$

Let $\mathcal{P} = \mathcal{P}(\mathbb{E})$ be the totality of probability measures on $\mathbb{E}$, equipped with the topology of weak convergence. T_t induces the dual semi-group T_t^* acting on $\mathcal{P}$ by

$$\langle T_t^* \mu, f \rangle = \langle \mu, T_t f \rangle, \quad f \in C_b(\mathbb{E}),$$

where $\langle \mu, f \rangle = \int_\mathbb{E} f(x) \mu(dx)$.

Let $\mathcal{S}$ be the totality of stationary distributions of the interacting diffusion process $(\Omega, \mathcal{F}, \mathcal{F}_t, P^x, x(t))$, i.e.

$$\mathcal{S} = \{\mu \in \mathcal{P} \mid T_t^* \mu = \mu,\ t > 0\}.$$

We note that $\mathcal{S}$ is convex and simplex, see [Dynkin 1978]. Accordingly our problem is reduced to characterize the totality of extremal stationary distributions $\mathcal{S}_{\text{ext}}$.

In order to discuss the stationary distribution problem in the $\mathbb{Z}^d$-shift invariant situation we consider the following condition.

CONDITION (B). Let $S = \mathbb{Z}^d$.
(B.1) A is irreducible and $\mathbb{Z}^d$-shift invariant. i.e. $A_{i,j} = A_{0,j-i}$ for $i,j \in \mathbb{Z}^d$. Furthermore the symmetrized random walks generated by the infinitesimal matrix $\widehat{A} = (A_{ij} + A_{ji})$ is transient.
(B.2) $\limsup_{|u| \to \infty} |a(u)|/|u| < \widehat{G}(0,0)^{-1/2}$, where $\widehat{G}(i,j)$ is the potential matrix of the symmetrized random walks.

Let $\mathcal{T}(\mathbb{E})$ be the totality of $\mathbb{Z}^d$-shift invariant probability measures on $\mathbb{E}$, and let $\mathcal{T}_1 = \{\mu \in \mathcal{T} \mid \langle \mu, |x_i| \rangle < \infty, i \in \mathbb{Z}^d\}$.

Then we have the following results (cf. [Cox and Greven 1993], [Shiga 1992]).

THEOREM 1.1. *Assume the conditions* (A) *and* (B).
(a) *For each* $\theta \in R$, *there exists a unique* $\nu_\theta \in (\mathcal{S} \cap \mathcal{T}_1)_{\mathrm{ext}}$ *such that* $\langle \nu, x_i \rangle = \theta$ *for* $i \in \mathbb{Z}^d$.
(b) $(\mathcal{S} \cap \mathcal{T}_1)_{\mathrm{ext}} = \{\nu_\theta \mid \theta \in R\}$. *For every* $\nu \in \mathcal{S} \cap \mathcal{T}_1$, *there exists an* $m \in \mathcal{P}(R)$ (*the totality of probability measures on* R) *such that* $\nu = \int_R \nu_\theta m(d\theta)$.
(c) *If* $\mu \in \mathcal{T}_1$ *is* $\mathbb{Z}^d$-*shift ergodic, then*
$$\lim_{t \to \infty} T_t^* \mu = \nu_\theta \text{ with } \theta = \langle \mu, x_i \rangle.$$

Moreover for every $\mu \in \mathcal{T}_1$, $\lim_{t \to \infty} T_t^* \mu$ *exists in* $\mathcal{S} \cap \mathcal{T}_1$.

REMARK 1. In addition to the condition (A) suppose that $a(\theta_0) = 0$ for some $\theta_0 \in R$. Then one can restrict the diffusion process $(\Omega, \mathcal{F}, \mathcal{F}_t, P^x, x(t))$ into the smaller state space $\mathbb{E}_+(\theta_0) = L^2(\gamma) \cap [\theta_0, \infty)^{\mathbb{Z}^d}$. Then Theorem 1.1(b) can be refined as follows:
$$(\mathcal{S} \cap \mathcal{T})_{\mathrm{ext}} = \{\nu_\theta \mid \theta \geq \theta_0\}.$$

REMARK 2. It is obvious that if $a(\theta) = 0$ for $\theta \in R$, then $\nu_\theta = \delta_{\boldsymbol{\theta}}$, where $\delta_{\boldsymbol{\theta}}$ stands for the unit mass at $\boldsymbol{\theta} = (x_i \equiv \theta) \in \mathbb{E}$. On the other hand if $a(\theta_1) = a(\theta_2) = 0$ for some $\theta_1 < \theta_2$, the diffusion process $(\Omega, \mathcal{F}, \mathcal{F}_t, P^x, x(t))$ can be restricted into $\mathbb{E}(\theta_1, \theta_2) = [\theta_1, \theta_2]^{\mathbb{Z}^d}$, and it holds that $(\mathcal{S} \cap \mathcal{T})_{\mathrm{ext}} = \{\nu_\theta \mid \theta_1 \leq \theta \leq \theta_2\}$.

We note that the condition (B) is crucial for Theorem 1.1. If $a(0) = 0$ and the linearly growing rate of $a(u)$ is large, it can be shown that there is a unique $\mathbb{Z}^d$-shift invariant stationary distribution that coincides with $\delta_{\mathbf{0}}$. Furthermore one can prove the exponential decay of sample paths, (see Theorem 1.2 in [Shiga 1992]).

Under this $\mathbb{Z}^d$-shift invariant setting we conjecture that *all stationary distribution is* $\mathbb{Z}^d$-*shift invariant*.

2. Interacting diffusion systems on $[0,1]^S$

In this section we discuss the stationary distribution problem for an interacting diffusion system with the state space $[0,1]^S$. Recall that we are assuming the following condition.

CONDITION (C). Let S be a countable set.
(C.1) $A = \{A_{ij}\}$ is an irreducible $S \times S$-matrix satisfying that
$$A_{ij} \geq 0 \text{ for } i \neq j, \quad \sum_{j \in S} A_{ij} = 0, \text{ and } \sup_{i \in S} |A_{ii}| < \infty,$$
(C.2) $a(u) \colon [0,1] \to R$ is 1/2-Hölder continuous, $a(0) = a(1) = 0$, and $a(u) \neq 0$ for $0 < u < 1$.

Then we have a diffusion process $(\Omega, \mathcal{F}, \mathcal{F}_t, P^x, x(t))$ with the state space $[0,1]^S$ governed by the equation (IDS).

In order to formulate our result we need several new notations.
- $P_t = \exp tA$: the transition matrix of a continuous time Markov chain with the state space S generated by A,
- $\mathcal{H}$: the totality of $[0,1]$-valued A-harmonic functions, i.e.

$$\mathcal{H} = \{h\colon S \to [0,1]\colon Ah = 0\},$$

- $(\boldsymbol{\xi}(t), P_{i,j}^{(2)})$: two particles motion that is a continuous time Markov chain on $S \times S$ with the transition matrix $P_t \otimes P_t$,
- $\sigma_\Delta = \inf\{t \geq 0\colon \boldsymbol{\xi}(t) \in \Delta\}$, where $\Delta = \{(i,i)\colon i \in S\} \subset S \times S$,
- $H_{ij} = P_{ij}^{(2)}(\sigma_\Delta < \infty)$,
- $F_{ij} = P_{ij}^{(2)}(\int_0^\infty I_\Delta(\boldsymbol{\xi}(t))\,dt < \infty)$, where I_Δ denotes the indicator function of Δ.
- $\mathcal{H}^*$: the totality of $h \in \mathcal{H}$ such that

(2.1) $$h(i) + h(j) - 2h(i)h(j) \leq F_{ij} \text{ for } i, j \in S.$$

The class $\mathcal{H}^*$ of A-harmonic functions appears in the voter model [Liggett t 1985] and the stepping stone diffusion model [Shiga 1980], [Shiga 1980].

Now we can state our results.

THEOREM 2.1.
(i) Let $\nu \in \mathcal{S}_{\text{ext}}$ and set $h^\nu(i) = \langle \nu, x_i \rangle$ for $i \in S$. Then $h^\nu \in \mathcal{H}^*$.
(ii) Conversely let $h \in \mathcal{H}^*$. Then there exists a $\nu \in \mathcal{S}_{\text{ext}}$ such that $\langle \nu, x_i \rangle = h(i)$ for $i \in S$.

THEOREM 2.2.

$$\mathcal{S}_{\text{ext}} = \{\delta_{\mathbf{1}}, \delta_{\mathbf{0}}\} \text{ if and only if } F_{ij} = 0 \text{ for all } i,j \in S.$$

Theorem 2.1 seems still unsatisfactory. What we really want to assert is the following uniqueness, which together with Theorem 2.1 would establishes the one-to-one correspondence between $\mathcal{S}_{\text{ext}}$ and $\mathcal{H}^*$, but it is remained as a conjecture.

CONJECTURE. Let $h \in \mathcal{H}^*$. Then there exists a unique $\nu \in \mathcal{S}_{\text{ext}}$ such that $\langle \nu, x_i \rangle = h(i)$ for $i \in S$. In other words, letting $\nu \in \mathcal{S}_{\text{ext}}$ and $\nu' \in \mathcal{S}_{\text{ext}}$, $\langle \nu, x_i \rangle = \langle \nu', x_i \rangle$ for $i \in S$ yields $\nu = \nu'$.

We next discuss the above uniqueness problem in a very restrictive situation. Now let $S = \mathbb{Z}^d$ and denote by σ_i the shift operator by $i \in \mathbb{Z}^d$, in particular

$$\sigma_k A = (A_{i+k,j+k})_{i,j \in S}.$$

We formulate the following condition:
CONDITION (D).
(i) Any limit point $\bar{A}$ of $\{\sigma_k A\}$ as $|k| \to \infty$ is a generator of a continuous time irreducible Markov chain in $\mathbb{Z}^d$.
(ii) Let $\bar{A} = \lim_{p \to \infty} \sigma_{k_p} A$ for some $\{k_p\} \subset \mathbb{Z}^d$ with $|k_p| \to \infty$. Then

(2.2) $$\limsup_{p \to \infty} F_{i+k_p, j+k_p} \leq \overline{F}_{ij},$$

(2.3) $$\limsup_{p \to \infty} H_{i+k_p, j+k_p} \leq \overline{H}_{ij},$$

where $\overline{F}_{ij}$ and $\overline{H}_{ij}$ stand for the corresponding ones to F_{ij} and H_{ij} relative to the two particles motion $(\boldsymbol{\xi}(t), \overline{P}_{ij}^{(2)})$ associated with $\overline{A}$.

THEOREM 2.3. *In addition to the condition* (C) *we assume the condition* (D). *Then for every* $h \in \mathcal{H}^*$ *there is a unique* $\nu \in \mathcal{S}_{\text{ext}}$ *such that* $\langle \nu, x_i \rangle = h(i)$ *for* $i \in S$. *Thus denoting it by* ν_h *we obtain*

$$\mathcal{S}_{\text{ext}} = \{\nu_h \colon h \in \mathcal{H}^*\}.$$

Although the condition (D) seems too artificial and hard to be checked, one finds new examples for which all stationary distributions can be sought.

COROLLARY 2.4. *In addition to the condition* (C) *one of the following three conditions is fulfilled.*
  (i) A *is* $\mathbb{Z}^d$-*shift invariant.*
  (ii) $\{\sigma_k A\}_{k \in \mathbb{Z}^d}$ *is a finite set.* (*Note that if A is $\mathbb{Z}^d$-periodic, this is true.*)
  (iii) $A = A^0 + B$, *where A^0 is a $\mathbb{Z}^d$-shift invariant generator of a continuous time irreducible random walks and for some finite subset F of $\mathbb{Z}^d$*

$$B_{ij} = 0 \text{ for } i \notin F \text{ and } j \in \mathbb{Z}^d.$$

Then the condition (D) *is fulfilled.*

We here present the proofs of these results. Main tools for the proofs of Theorem 2.1 and 2.2 are second moment calculations and Feynman-Kac formula. For the proof of Theorem 2.3 we use a coupling technique which were useful to seek all $\mathbb{Z}^d$-shift invariant stationary distributions in [Cox and Greven 1993] and [Shiga 1992].

For $\mu \in \mathcal{P}(\mathbb{E})$ let

(2.4)
$$f_\mu(i,j) = \langle \mu, x_i + x_j - 2x_i x_j \rangle,$$
$$f_\mu(t;i,j) = \langle T_t^* \mu, x_i + x_j - 2x_i x_j \rangle.$$

Let us start the proofs with the following elementary lemma.

LEMMA 2.5. *There is a non-decreasing Lipschitz continuous function* $G : [0,1] \to R$ *vanishing only at 0 such that*

$$\langle \mu, a(x_i)^2 \rangle \geq G\bigl(f_\mu(i,i)\bigr) f_\mu(i,i) \text{ for } \mu \in \mathcal{P}(\mathbb{E}) \text{ and } i \in S.$$

PROOF. First using the condition (C.2) find a Lipschitz continuous function $g \colon [0,1] \to R$ such that $g(0) = g(1) = 0$, $g(u) > 0$ for $0 < u < 1$ and $a(u)^2 \geq g(u)$ for $0 \leq u \leq 1$.

Next let $0 < \epsilon < 1/2$.

$$\langle \mu, g(x_i) \rangle \geq \min_{\epsilon \leq u \leq 1-\epsilon} g(u) \langle \mu, 2x_i(1-x_i) I(\epsilon \leq x_i \leq 1-\epsilon) \rangle$$
$$\geq \min_{\epsilon \leq u \leq 1-\epsilon} g(u) \bigl(f_\mu(i,i) - 2\epsilon\bigr).$$

Setting

$$\epsilon = \frac{f_\mu(i,i)}{4} \text{ and } G(f) = \frac{1}{2} \min_{f/4 \leq u \leq 1-f/4} g(u),$$

we get the desired inequality.

LEMMA 2.6. *Let $\mu \in \mathcal{P}(\mathbb{E})$.*

(i) $$\left| f_\mu(t;i,j) - \sum_{k \in S} \sum_{l \in S} P_t(i,k) P_t(j,l) f_\mu(k,l) \right| \leq H_{ij}, \quad i,j \in S.$$

(ii) In particular, if $\nu \in \mathcal{S}$,
$$\left| f_\nu(i,j) - \sum_{k \in S} \sum_{l \in S} P_t(i,k) P_t(j,l) f_\nu(k,l) \right| \leq H_{ij}, \quad i,j \in S.$$

(iii) $$\limsup_{t \to \infty} f_\mu(t;i,j) \leq F_{ij}, \quad i,j \in S.$$

In particular, if $\nu \in \mathcal{S}$, $f_\nu(i,j) \leq F_{ij}$, $i,j \in S$.

PROOF. By Ito's formula

(2.5) $$\frac{df_\mu(t;i,j)}{dt} = \sum_{k \in S} A_{ik} f_\mu(t;k,j) + \sum_{l \in S} A_{jl} f_\mu(t;i,l) - I_\Delta(i,j) \langle T_t^* \mu, a(x_i)^2 \rangle.$$

The condition (C.2) implies that for some $K > 0$, $|a(u)|^2 \leq 2Ku(1-u)$, so that

(2.6) $$\langle T_t^* \mu, a(x_i)^2 \rangle \leq K f(t;i,i).$$

Hence by Feyman-Kac formula

(2.7) $$f_\mu(t;i,j) = E_{ij}^{(2)} \left(f_\mu(\boldsymbol{\xi}(t)) \exp - \int_0^t I_\Delta(\boldsymbol{\xi}(s)) V_\mu(t-s, \boldsymbol{\xi}(s)) \, ds \right),$$

where $V_\mu(t;i,i) = \langle T_t^* \mu, a(x_i)^2 \rangle / f(t;i,i) \leq K$. (i) follows from this because

$$\left| f_\mu(t;i,j) - \sum_{k \in S} \sum_{l \in S} P_t(i,k) P_t(j,l) f_\mu(k,l) \right|$$
$$\leq E_{ij}^{(2)} \left(\left| 1 - \exp - \int_0^t I_\Delta(\boldsymbol{\xi}(s)) V_\mu(t-s, \boldsymbol{\xi}(s)) ds \right| \right)$$
$$\leq H_{ij}.$$

Following [Notohara and Shiga 1980] let us consider the following non-linear diffusion equation:

(2.8) $$\frac{dg(t;i,j)}{dt} = \sum_{k \in S} A_{ik} g(t;k,j) + \sum_{l \in S} A_{jl} g(t;i,l) - I_\Delta(i,j) G(g(t;i,j)) g(t;i,j)$$
$$g(0;i,j) = 1 \text{ for } i,j \in S.$$

The Lipschitz continuity of G guarantees the uniquely existence of $[0,1]$-valued solution $g(t;i,j)$. As easily seen, $g(t;i,j)$ is non-increasing in $t \geq 0$, so that the limit $\bar{g}(i,j) = \lim_{t \to \infty} g(t;i,j)$ exists and it is a stationary solution of (2.8), hence by Feynman-Kac formula

(2.9) $$\bar{g}(i,j) = E_{ij}^{(2)} \left(\bar{g}(\boldsymbol{\xi}(t)) \exp - \int_0^t I_\Delta(\boldsymbol{\xi}(s)) G(\bar{g}(\boldsymbol{\xi}(s))) \, ds \right).$$

Since $\bar{g}$ is a $P_t \otimes P_t$-excessive function, $\lim_{t\to\infty} \bar{g}(\boldsymbol{\xi}(t))$ exists $P_{ij}^{(2)}$-a.s., from (2.9) it follows

(2.10) $$\bar{g}(i,j) \leq F_{ij} \text{ for } i,j \in S.$$

Noting that the comparison of solutions holds between (2.5) and (2.8) we finally obtain
$$\limsup_{t\to\infty} f_\mu(t;i,j) \leq \bar{g}(i,j) \leq F_{ij}.$$
(See [Notohara and Shiga 1980] for the details.)

LEMMA 2.7. *Let* $h \in \mathcal{H}^*$ *and* $\lim_{n\to\infty} \frac{1}{t_n} \int_0^{t_n} T_s^* \delta_h \, ds = \nu$ *for some* $t_n \to \infty$. *Then*
$$\left|\langle \nu, x_i x_j \rangle - h(i)h(j)\right| \leq F_{ij} \wedge H_{ij} \text{ for } i,j \in S,$$
where δ_h *stands for the point mass at* $h \in \mathbb{E}$.

PROOF. Apply Lemma 2.6(i) to $\mu = \delta_h$. Since $\langle T_t^* \delta_h, x_i \rangle = h(i)$ and $\langle \delta_h, x_i x_j \rangle = h(i)h(j)$,
$$\left|\langle T_t^* \delta_h, x_i x_j \rangle - h(i)h(j)\right| \leq \frac{H_{ij}}{2},$$
hence
$$\left|\langle \nu, x_i x_j \rangle - h(i)h(j)\right| \leq \frac{H_{ij}}{2}.$$
Combining this with (2.3) and Lemma 2.6(ii) we obtain the desired inequality.

LEMMA 2.8. *Suppose that* $h \in \mathcal{H}^*$ *and* $\nu \in \mathcal{S}$ *satisfies* $\langle \nu, x_i \rangle = h(i)$ *for* $i \in S$. *Then the following three are equivalent.*

(i) $$\left|\langle \nu, x_i x_j \rangle - h(i)h(j)\right| \leq F_{ij} \wedge H_{ij}, \ i,j \in S.$$

(ii) $$\lim_{t\to\infty} \sum_{k\in S} \sum_{l\in S} P_t(i,k) P_t(j,l) \langle \nu, x_k x_l \rangle = h(i)h(j), \ i,j \in S.$$

(iii) $$\lim_{t\to\infty} \left\| \sum_{k\in S} P_t(i,k) x_k - h(i) \right\|_{L^2(\nu)}^2 = 0, \ i \in S.$$

PROOF. By Markov property of $(\boldsymbol{\xi}(t), P_{ij}^{(2)})$, it holds that for $P_{ij}^{(2)}$-a.s.
$$\lim_{t\to\infty} F_{\boldsymbol{\xi}(t)} = I\left(\int_0^\infty I_\Delta(\xi(s)) \, ds < \infty\right),$$
and
$$\lim_{t\to\infty} H_{\boldsymbol{\xi}(t)} = I\left(\sigma_\Delta(\theta_t) < \infty \text{ for every } t > 0\right).$$
Hence

(2.11) $\lim_{t\to\infty} \sum_{k\in S}\sum_{l\in S} P_t(i,k)P_t(j,l) F_{kl} \wedge H_{kl}$
$$= P_{ij}^{(2)}\left(\int_0^\infty I_\Delta(\boldsymbol{\xi}(s))\, ds < \infty \text{ and } \sigma_\Delta(\theta_t) < \infty \text{ for every } t > 0\right) = 0.$$

Accordingly (ii) follows from (i) and (2.11).

Next let assume (ii). Lemma 2.6 (i) implies
$$\left|\langle \nu, x_i x_j \rangle - \sum_{k\in S}\sum_{l\in S} P_t(i,k)P_t(j,l)\langle \nu, x_k x_l \rangle\right| \leq \frac{H_{ij}}{2}.$$

So letting $t \to \infty$, we get

$$|\langle \nu, x_i x_j \rangle - h(i)h(j)| \leq \frac{H_{ij}}{2}.$$

On the other hand by (2.3) and Lemma 2.6 (ii)

$$\left|\langle \nu, x_i x_j \rangle - h(i)h(j)\right|$$
$$= \frac{1}{2}\left|f_\nu(i,j) - \left(h(i) + h(j) - 2h(i)h(j)\right)\right|$$
$$\leq F_{ij},$$

which yields (i). It is obvious that (ii) is equivalent to (iii).

We are now in position to prove Theorem 2.1.

PROOF OF THEOREM 2.1. Let us first show (ii). Let $h \in \mathcal{H}^*$. Define a convex subset of $\mathcal{S}$ by

$$\mathcal{E}(h) = \{\nu \in \mathcal{S} : \langle \nu, x_i \rangle = h(i), |\langle \nu, x_i x_j \rangle - h(i)h(j)| \leq F_{ij} \wedge H_{ij} \text{ for } i,j \in S\}.$$

Obviously $\mathcal{E}(h)$ is compact and non-empty by Lemma 2.7. Hence $\mathcal{E}(h)_{\text{ext}}$ is also non-empty. To conclude (i) it suffices to show that $\mathcal{E}(h)_{\text{ext}} \subset \mathcal{S}_{\text{ext}}$. Suppose that a $\nu \in \mathcal{E}(h)_{\text{ext}}$ is decomposed into $\nu = (\nu' + \nu'')/2$ with $\nu', \nu'' \in \mathcal{S}$. Since ν satisfies (iii) of Lemma 2.8, so do ν' and ν'', which means $\nu' \in \mathcal{E}(h)$ and $\nu'' \in \mathcal{E}(h)$. Thus $\nu' = \nu'' = \nu$ holds and $\nu \in \mathcal{S}_{\text{ext}}$, completing the proof of part (ii).

For (i) let $\nu \in \mathcal{S}_{\text{ext}}$ and set $h(i) = \langle \nu, x_i \rangle$. By Lemma 2.6 (ii)

$$\left|h(i) + h(j) - 2\langle \nu, x_i x_j \rangle\right| \leq F_{ij}.$$

Noting that F_{ij} is $P_t \otimes P_t$-harmonic we see

(2.12) $$\left|h(i) + h(j) - 2\sum_{k \in S}\sum_{l \in S} P_t(i,k)P_t(j,l)\langle \nu, x_k x_l \rangle\right| \leq F_{ij}.$$

From this it follows

$$\left|\langle \nu, x_i x_j \rangle - \sum_{k \in S}\sum_{l \in S} P_t(i,k)P_t(j,l)\langle \nu, x_k x_l \rangle\right| \leq F_{ij}.$$

Combining this with Lemma 2.6 (i) we get

(2.13) $$\left|\langle \nu, x_i x_j \rangle - \sum_{k \in S}\sum_{l \in S} P_t(i,k)P_t(j,l)\langle \nu, x_k x_l \rangle\right| \leq F_{ij} \wedge H_{ij}$$

Note that $g(i,j) = \langle \nu, x_i x_j \rangle$ is $P_t \otimes P_t$-subharmonic, so from (2.15) it follows that

$$\lim_{t \to \infty} \sum_{k \in S}\sum_{l \in S} P_t(i,k)P_t(j,l)\langle \nu, x_k x_l \rangle \equiv \bar{h}(i,j)$$

exists and (2.13) turns to

(2.14) $$\left|\langle \nu, x_i x_j \rangle - \bar{h}(i,j)\right| \leq F_{ij} \wedge H_{ij}.$$

Using the extremality of ν one can show

(2.15) $$\bar{h}(i,j) = h(i)h(j) \text{ for } i,j \in S.$$

(The proof of this part is the same as Lemma 2.5 of [Shiga 1980], so it is omitted.) Therefore by (2.12), (2.14) and (2.15) we obtain
$$|h(i) + h(j) - 2h(i)h(j)| \leq F_{ij}$$
which concludes $h \in \mathcal{H}^*$. Thus the proof of Theorem 2.1 is complete.

PROOF OF THEOREM 2.2. Suppose that $F_{ij} = 0$ for every $i, j \in S$. Then $\mathcal{H}^* = \{0, 1\}$. Hence by Theorem 2.1 $\mathcal{S}_{\text{ext}} = \{\delta_0, \delta_1\}$.

Conversely suppose that $\mathcal{S}_{\text{ext}} = \{\delta_0, \delta_1\}$. Then it is easy to see

$$(2.16) \quad \lim_{t \to \infty} \left\langle \frac{1}{t} \int_0^t T_s^* \delta_\theta \, ds, x_i + x_j - 2x_i x_j \right\rangle = 0$$

since any limit point of $1/t \int_0^t T_s^* \delta_\theta \, ds$ as $t \to \infty$ is a stationary distribution. However by (2.7)

$$(2.17) \quad \limsup_{t \to \infty} \langle T_t^* \delta_\theta, x_i + x_j - 2x_i x_j \rangle \geq 2\theta(1-\theta) E_{ij}^{(2)} \left(\exp -K \int_0^\infty I_\Delta(\boldsymbol{\xi}(s)) \, ds \right)$$

which is positive if $F_{ij} > 0$. Therefore from (2.16) and (2.17) $\mathcal{S}_{\text{ext}} = \{\delta_0, \delta_1\}$ implies that $F_{ij} = 0$ for every $i, j \in S$, completing the proof of Theorem 2.2.

PROOF OF THEOREM 2.3. Suppose that $\nu \in \mathcal{S}_{\text{ext}}$ and $\nu' \in \mathcal{S}_{\text{ext}}$ satisfy

$$(2.18) \quad \langle \nu, x_i \rangle = \langle \nu', x_i \rangle = h(i) \text{ for } i \in \mathbb{Z}^d.$$

Notice by Theorem 2.1 that $h \in \mathcal{H}^*$.

Following [Cox and Greven 1993] and [Shiga 1992] let us consider the following SDE defining a coupling system:

$$\begin{cases} dx_i(t) = \sum_{j \in S} A_{ij} x_j(t) \, dt + a(x_i(t)) dB_i(t), \\ dy_i(t) = \sum_{j \in S} A_{ij} y_j(t) \, dt + a(y_i(t)) dB_i(t), \ i \in \mathbb{Z}^d. \end{cases}$$

Then there exists a stationary distribution λ of the coupled process $((x(t), y(t)), \bar{P}^{x,y})$ of which marginal distributions coincide with ν and ν'.

Let $\Delta_i(t) = x_i(t) - y_i(t)$. Applying Ito's formula to $f(u) = |u|$ we have

$$\overline{E}^\lambda (|\Delta_i(t)|) - \overline{E}^\lambda (|\Delta_i(0)|) = \int_0^t \sum_{j \in S} \overline{E}^\lambda \left(\operatorname{sgn} \Delta_i(s) \Delta_j(s) \right) ds,$$

where $\operatorname{sgn} u = 1$ for $u > 0$, $= 0$ for $u = 0$ and -1 for $u < 0$, so by the stationarity of λ we get

$$\sum_{j \neq i} A_{ij} \langle \lambda, (\operatorname{sgn} \Delta_i) \Delta_j - |\Delta_i| \rangle = 0.$$

Rearrange this as follows:

$$(2.19) \quad \sum_{j \neq i} A_{ij} \left(\langle \lambda, |\Delta_j| \rangle - \langle \lambda, |\Delta_i| \rangle \right)$$
$$= \sum_{j \neq i} A_{ij} \langle \lambda, (1 - \operatorname{sgn} \Delta_i \operatorname{sgn} \Delta_j) |\Delta_j| \rangle, \quad i \in \mathbb{Z}^d.$$

The argument is divided into the following two cases:

CASE 1. $\sup_{i \in \mathbb{Z}^d} \langle \lambda, |\Delta_i| \rangle = \langle \lambda, |\Delta_{i_0}| \rangle$ for some $i_0 \in \mathbb{Z}^d$.

CASE 2. $\sup_{i \in \mathbb{Z}^d} \langle \lambda, |\Delta_i| \rangle = \lim_{p \to \infty} \langle \lambda, |\Delta_{k_p}| \rangle)$ for some $k_p \in \mathbb{Z}^d$ with $|k_p| \to \infty$ as $p \to \infty$.

The proof of the Case 1 is omitted since it is easy and does not need the condition (D). Now suppose that the Case 2 holds. Taking a suitable subsequence we may assume that $\{\sigma_{k_p}\lambda\}$, $\{\sigma_{k_p}\nu\}$, $\{\sigma_{k_p}\nu'\}$, $\{\sigma_{k_p}A\}$ and $\{h_{k_p}(i) = h(i + k_p)\}$ converges to $\bar{\lambda}$, $\bar{\nu}$, $\bar{\nu}'$, $\bar{A}$ and $\bar{h}$ respectively as $p \to \infty$. Then it holds that

$$(2.20) \qquad \sup_{i \in \mathbb{Z}^d} \langle \lambda, |\Delta_i| \rangle = \langle \bar{\lambda}, |\Delta_0| \rangle = \sup_{i \in \mathbb{Z}^d} \langle \bar{\lambda}, |\Delta_i| \rangle$$

so that

$$(2.21) \quad \sum_{j \neq i} \bar{A}_{ij}(\langle \bar{\lambda}, (|\Delta_j| - \langle \bar{\lambda}, |\Delta_i| \rangle \rangle)$$
$$= \sum_{j \neq i} \bar{A}_{ij} \langle \bar{\lambda}, (1 - \operatorname{sgn}\Delta_i \operatorname{sgn}\Delta_j)|\Delta_j| \rangle, \; i \in \mathbb{Z}^d.$$

By (2.20), (2.21) and the irreducibility of $\bar{A}$ it holds that $\langle \bar{\lambda}, |\Delta_i| \rangle$ is constant in $i \in \mathbb{Z}^d$, and

$$(2.22) \qquad \bar{\lambda}(\Delta_i \geq 0 \text{ for all } \in \mathbb{Z}^d, \text{ or } \Delta_i \leq 0 \text{ for all } i \in \mathbb{Z}^d) = 1.$$

Since $\nu \in \mathcal{S}_{\text{ext}}$, by (2.14) and (2.15) in the proof of Theorem 2.1

$$\left|\langle \nu, x_i x_j \rangle - h(i)h(j)\right| \leq F_{ij} \wedge H_{ij},$$

so we see

$$\left|\langle \sigma_{k_p}\nu, x_i x_j \rangle - h_{k_p}(i)h_{k_p}(j)\right| \leq F_{i+k_p,j+k_p} \wedge H_{i+k_p,j+k_p}.$$

Using here the condition (D) we have

$$\left|\langle \bar{\nu}, x_i x_j \rangle - \bar{h}(i)\bar{h}(j)\right| \leq \bar{F}_{ij} \wedge \bar{H}_{ij},$$

and by Lemma 2.8

$$(2.23) \qquad \left\| \sum_{j \in S} \bar{P}_t(i,j) x_j - \bar{h}(i) \right\|_{L^2(\bar{\nu})} \to 0 \text{ as } t \to \infty.$$

Hence $\bar{\nu}'$ also satisfies (2.23). Noting that $\bar{\nu}$ and $\bar{\nu}'$ are marginals of $\bar{\lambda}$ we obtain

$$\langle \bar{\lambda}, |\Delta_i| \rangle = \sum_{j \in S} \bar{P}_t(i,j) \langle \bar{\lambda}, |\Delta_j| \rangle$$
$$= \left\langle \bar{\lambda}, \left| \sum_{j \in S} \bar{P}_t(i,j) \Delta_j \right| \right\rangle$$
$$\leq \left\langle \bar{\nu}, \left| \sum_{j \in S} \bar{P}_t(i,j) x_j - \bar{h}(i) \right| \right\rangle + \left\langle \bar{\nu}', \left| \sum_{j \in S} \bar{P}_t(i,j) x_j - \bar{h}(i) \right| \right\rangle$$
$$\to 0 \text{ as } t \to \infty.$$

Therefore it holds $\langle \lambda, |\Delta_i| \rangle = 0$ for all $i \in S$, which yields $\nu = \nu'$. Thus we complete the proof of Theorem 2.3.

PROOF OF COROLLARY 2.4. The condition (D) is trivially satisfied in the cases (i) and (ii), because $\sigma_{k_p}A$ is constant in p in the case (i) and it is true by taking a subsequence in the case (ii). Assume the condition (iii). If the symmetrized random walks generated by $\bar{A} + \bar{A}^* = (\bar{A}_{ij} + \bar{A}_{ji})$ is transient, then $\overline{F}_{ij} = 1$, so that it suffices to show

(2.24) $$\lim_{|k| \to \infty} H_{i+k, j+k} = \overline{H}_{ij}.$$

To see this let $\widehat{F} = (F \times \mathbb{Z}^d) \cup (\mathbb{Z}^d \times F)$ and $\widehat{F}(k) = \widehat{F} - (k,k)$. Clearly

$$P^{(2)}_{i+k, j+k}(\sigma_\Delta \leq \sigma_{\widehat{F}}) = \overline{P}^{(2)}_{i+k, j+k}(\sigma_\Delta \leq \sigma_{\widehat{F}}),$$

$$\lim_{|k| \to \infty} \overline{P}^{(2)}_{i+k, j+k}(\sigma_{\widehat{F}} < \sigma_\Delta < \infty) = \lim_{|k| \to \infty} \overline{P}^{(2)}_{i,j}(\sigma_{\widehat{F}(k)} < \sigma_\Delta < \infty) = 0,$$

where σ_A denotes the hitting time for $A \subset \mathbb{Z}^d \times \mathbb{Z}^d$. Moreover a little more consideration shows

$$\lim_{|k| \to \infty} P^{(2)}_{i+k, j+k}(\sigma_{\widehat{F}} < \sigma_\Delta < \infty) = 0.$$

Combining these relations we get (2.24).

On the other hand if the symmetrized random walks generated by $\bar{A} + \bar{A}^*$ is reccurent, it holds $\overline{F}_{ij} = 0$ for all $i, j \in \mathbb{Z}^d$. Then one can show that $F_{ij} = 0$ for all $i, j \in \mathbb{Z}^d$. Although it is not trivial but not so difficult, hence it is left as an exercise.

3. Interacting Fleming-Viot diffusions

As explained in Introduction, if the interacting diffusion system has a tractable dual process, then the stationary distribution problem is easy. In fact $a(u) = \sqrt{u(1-u)}$ is this case and characterization of all stationary distributions is obtained in [Shiga 1980a], [Shiga 1980b]. Using the same method one can characterize all stationary distributions for an interacting Fleming-Viot diffusion system.

Let K be a compact metrizable space, and let $I = \mathcal{P}(K)$ be the totality of probability measures on K, which is also compact with the weak topology. Under the condition (C.2) an interacting Fleming-Viot diffusion is well-posed, which is a diffusion process $(\Omega, \mathcal{F}, \mathcal{F}_t, P^X, X(t))$ with the state space $\mathbb{E}(I) = I^S$ governed by the following martingale problem, cf. [Handa 1990]:

(FVD) $$X_i(t) - X_i(0) = \int_0^t \sum_{j \in S} A_{ij} X_j(s) \, ds + M_i(t) \text{ for } i \in S,$$

where $\{M_i(t)\}_{i \in S}$ are $C(K)^*$-valued martingales with quadratic variation processes

$$\langle M_i, M_j \rangle(t)(dxdy) = \delta_{i,j} \int_0^t \big(X(s)(dx)\delta_x(dy) - X(s)(dx)X(s)(dy)\big) \, ds.$$

In order to characterize the extremal stationary distributions we prepare new notations.
- $\mathbb{T}_t^*$: the transition dual semi-group of the interacting Fleming-Viot diffusion system (IFVD) acting on $\mathcal{P}(\mathbb{E}(I))$,
- $\mathcal{S}(I)$: the totality of stationary distributions of the IFVD i.e.,
- $\mathcal{S}(I) = \{\nu : \mathbb{T}_t^* \nu = \nu \text{ for } t > 0\}$,
- $\mathcal{S}(I)_{\text{ext}}$: the totality of extremal elements of $\mathcal{S}(I)$,
- $\mathcal{H}(I) = \{H = \{H_i\} : S \to I : \sum_{j \in S} A_{ij} H_j = H_i, j \in S\}$,

- $\mathcal{H}(I)^* = \{H \in \mathcal{H}(I)$ such that letting $h_B(i) = H_i(B)$ for a Borel subset B of K, $h_B \in \mathcal{H}^*$ for all $B\}$.

Then we obtain

THEOREM 3.1.
(i) For every $H \in \mathcal{H}(I)^*$ there exists a unique $\nu \in \mathcal{S}(I)_{\text{ext}}$ such that $\langle \nu, X_i \rangle = H_i$ for $i \in S$, which is denoted by ν_H.
(ii) $\mathcal{S}(I)_{\text{ext}} = \{\nu_H : H \in \mathcal{H}(I)^*\}$.
(iii) Let $\mu \in \mathcal{P}(\mathbb{E}(I))$ and $H \in \mathcal{H}(I)^*$. Then $\lim_{t\to\infty} \mathbb{T}_t^* \mu = \nu_H$ if and only if

$$\lim_{t\to\infty} \left\langle \mu, \left(\sum_{j\in S} P_t(i,j) X_j(\phi) - H_i(\phi)\right)^2 \right\rangle = 0 \text{ for } \phi \in C(K), i \in S.$$

PROOF OF THEOREM 3.1. As in the stepping stone diffusion process in [Shiga 1980], [Shiga 1980] the IFVD also has a nice dual process which is defined in the state space

$$\Xi = \cup_{n=1}^{\infty} \{S \times C(K)\}^n$$

with the following infinitesimal transition law

$$\big((i_1, \phi_1), \ldots, (i_n, \phi_n)\big)$$
$$\to \big((i_1, \phi_1), \ldots, (i_p, \phi_p \phi_q), \ldots, \vee^q, \ldots, (i_n, \phi_n)\big) \text{ with rate } \frac{\delta_{i_p, i_q}}{2},$$
$$\to \big((i_1, \phi_1), \ldots, (j, \phi_p), \ldots, (i_n, \phi_n)\big) \text{ with rate } A_{i_p, j},$$
$$(1 \leq p, q \leq n, n \geq 1),$$

where $\vee^{(q)}$ means deletion of the q-th component. Let denote the associated jump Markov process by

$$\left(\boldsymbol{\xi}(t) = \big(\xi^1(t), \phi^1(t), \xi^2(t), \phi^2(t), \ldots, \xi^{N(t)}(t), \phi^{N(t)}(t)\big), \mathbb{P}_{\boldsymbol{\xi}}\right),$$

where $N(t)$ is the particle number at $t \geq 0$.

The motion of particles is interpreted as follows. Each element $\phi \in C(K)$ is the colour of the particle. By the first factor of the transition rule two particles occupying the same site may coalesce into one particle with mixed colour of the two particles, and by the second factor each particle executes the continuous time Markov chain generated by A.

Then the following duality relation holds.

LEMMA 3.2. For $X = \{X_i\} \in \mathbb{E}(I)$ and $\boldsymbol{\xi} = \big((i_1, \phi_1), \ldots, (i_n, \phi_n)\big) \in \Xi$, let $F_{\boldsymbol{\xi}}(X) = \prod_{p=1}^n X_{i_p}(\phi_p)$. Then

$$\int_{\mathbb{E}(I)} \mathbb{T}_t^* \mu(dX) F_{\boldsymbol{\xi}}(X) = \int_{\mathbb{E}(I)} \mu(dX) \mathbb{E}_{\boldsymbol{\xi}} \big(F_{\boldsymbol{\xi}(t)}(X)\big).$$

Using this duality relation together with the arguments exploited for the stepping stone diffusion processes in [Shiga 1980], [Shiga 1980], it is not difficult to reach the conclusion of Theorem 3.1, so that we omit the details.

References

[1] E.D. Andjel, *Invariant measures for the zero range process*, Ann. Probab. **10** (1982), 525-547.

[2] M. Bramson, J.T. Cox and A. Greven, *Ergodicity of critical spatial branching processes in low dimensions*, Ann. Probab. (1993) (to appear).

[3] M. Bramson, J.T. Cox and A. Greven, *In preparation* (1993).

[4] J.T. Cox and A. Greven, *Ergodic theorems for infinite systems of locally interacting diffusions*, Ann. Probab. (1993) (to appear).

[5] D.A. Dawson, *The critical measure diffusion process*, Z. Wahrsch. Verw. Gebiete **40** (1977), 125-145.

[6] J-N. Deuschel, *Algebraic L^2-decay of attractive critical process on the lattice*, Ann. Probab. (1993).

[7] E.B. Dynkin, *Sufficient statistics and extreme points*, Ann. Probab. **6** (1978), 705-730.

[8] K. Handa, *A measure-valued diffusion process describing the stepping stone model with infinitely many alleles*, Stoch. Proc. Appl. **36** (1990), 269-296.

[9] T.M. Liggett, *Ergodic theorems for the asymmetric simple exclusion process*, Trans. Amer Math. Soc. (1975), 237-261.

[10] T.M. Liggett, *Interacting Particle Systems*, Springer Verlag, 1985.

[11] T.M. Liggett and F. Spitzer, *Ergodic theorems for coupled random walks and other systems with locally interacting components*, Z. Wahrsch. Verw. Gebiete **56** (1981), 443-468.

[12] M. Notohara and T. Shiga, *Convergence to gentically uniform state in stepping stone models of population genetics*, J. Math. Bio. **10** (1980), 281-294.

[13] T. Shiga, *An interacting system in population genetics*, J. Math. Kyoto Univ. **20** (1980), 213-243.

[14] T. Shiga, *An interacting system in population genetics II*, J. Math. Kyoto Univ. **20** (1980), 723-733.

[15] T. Shiga and A. Shimizu, *Infinite-dimensional stochastic differential equations and their applications*, J. Math. Kyoto Univ. **20** (1980), 395-416.

[16] T. Shiga, *Ergodic theorems and exponential decay of sample paths for certain interacting diffusion systems*, Osaka J. Math. **29** (1992), 789-807.

TOKYO INSTITUTE OF TECHNOLOGY, DEPARTMENT OF APPLIED PHYSICS, TOKYO INSTITUTE OF TECHNOLOGY, OH-OKAYAMA, MEGURO, TOKYO 152, JAPAN

E-mail address: tshiga@cc.titech.ac.jp

On Clan-Recurrence and -Transience in Time Stationary Branching Brownian Particle Systems

Andreas Stoeckl and Anton Wakolbinger

1. Introduction and formulation of the main result

We consider a critical binary branching Brownian particle system in $\mathbb{R}^d$, in which each particle independently of all others performs Brownian motion, and after an exponentially distributed time splits into two particles with probability 1/2 and dies with probability 1/2. It is well known that for $d \geq 3$ and each "intensity parameter" $c > 0$ there exists a unique equilibrium distribution with intensity c times Lebesgue measure λ^d. By starting the process off "at early time" in this equilibrium and keeping track of the individual histories, we can lift the equilibrium particle system (ϕ_t) to a time stationary historical process $H = (H_t)_{t \in \mathbb{R}}$, which is the process of populations of ancestral paths.

We will briefly indicate a way of constructing (H_t) which follows closely [Dawson and Perkins 1991, Chapter 6]: For $s < t$, let the random population H_t^s on $C((-\infty, +\infty), \mathbb{R}^d)$ arise in the following way:

Start with a random population $\sum \delta_{Y^i}$ of constant paths $Y^i : (-\infty, s] \to \mathbb{R}^d$ such that $\sum \delta_{Y_s^i}$ is a Poisson population on $\mathbb{R}^d$ with Lebesgue intensity, let them develop over the time interval $[s, t]$ according to the branching Brownian dynamics, collect all paths surviving up to time t and freeze their positions during the time interval $[t, +\infty)$. For fixed $t \in \mathbb{R}$ and $p > \frac{d}{2}$, H_t^s converges, as $s \to -\infty$, in distribution in N_p towards an infinitely divisible distribution Q_t. (Here and in the sequel N_p denotes the set of integer valued measures in M_p, cf. [Dawson and Perkins 1991, p. 95] for this notation.)

The distributions Q_t, $t \in \mathbb{R}$, constitute an entrance law for the (historical) branching dynamics. Therefore there exists a random Markovian path $H = (H_t)_{t \in \mathbb{R}}$ which takes its values in N_p, follows the branching dynamics and is time stationary in the sense that $\Theta_s H_t$ equals H_{t+s} in distribution (where Θ_s acts by shifting a path (X_r) into (X_{r+s})).

Infinite divisibility of the distributions Q_t is inherited by the law of H, which we denote by Q. The Poisson cluster decomposition of H is given by $H = \sum_{j \in J} T^j$, $T^j = (T_t^j)_{t \in \mathbb{R}}$, where each T_t^j is an integer valued *clan measure* in the sense of [Dawson and Perkins 1991, p. 106], i.e. consists of eventually backwards coalescing paths. We will address $\sum_{j \in J} \delta_{T^j}$ as the *population of clans* in the time stationary branching particle system H. Note that $\sum_{j \in J} \delta_{T^j}$ is a Poisson system with intensity measure R, where R is the canonical measure of Q.

1991 *Mathematics Subject Classification.* Primary: 60J80, 60K35; Secondary: 60J65, 60G57.
This is the final form of the paper.

Clearly, the clan decomposition of the historical system H renders a clan decomposition of the particle system at each time t: If $T_t^j = \sum_{i \in I_t^j} \delta_{X^i}$, then we write $\phi_t^{T^j} := \sum_{i \in I_t^j} \delta_{X_t^i}$. Let B be some ball in $\mathbb{R}^d$. We say that the clan T^j *populates* B *at time t* if $\phi_t^{T^j}(B) > 0$.

THEOREM 1. a) Let $d = 3$ or 4. Then, with probability 1, all clans in a time stationary critical binary branching Brownian particle system populate a fixed ball in $\mathbb{R}^d$ at arbitrarily late and early times.

b) Let $d \geq 5$. Then with probability 1, each clan in a time stationary critical binary branching Brownian particle system populates a fixed ball in $\mathbb{R}^d$ only over a finite time horizon.

Remark. a) In [Matthes, Kerstan, and Mecke 1978, Prop. 3.3.2], it is shown that there exists a complete separable metric on the set of locally finite, non-void populations on $\mathbb{R}^d$ which makes a subset G of populations bounded iff there exists a bounded $B \subset \mathbb{R}^d$ such that $\phi(B) > 0$ for all $\phi \in G$, i.e. all populations in G have individuals in a common bounded region. (Note, however, that this metric differs from the one referred to in [Dawson and Perkins 1991, p. 108]).

This concept of "boundednes" in the space of non-void populations justifies to subsume the statement of Theorem 1 as "clan recurrence" in dimensions $d = 3$ or 4 and "clan transience" in dimensions $d \geq 5$.

b) In [Wu 1993] it is proved by analytic methods that 2-level superbrownian motion (which is the measure-valued limit of a particle system in which not only the individual particles but also the clans perform a binary critical branching) suffers local extinction iff $d \leq 4$. This together with our Theorem 1 suggests a strong connection between local extinction and "clan recurrence" in a 2-level critical finite variance branching process.

c) Before turning to the proof of Theorem 1, we will give an outline of its idea. Consider a clan T "distributed" according to the measure R. We will derive the "almost everywhere" statement of Theorem 1 by conditioning T to populate a site $x \in \mathbb{R}^d$ at time 0. (Formally, this is achieved by disintegrating the "Campbell type" measure $R(dT)\phi_0^T(dx)$ with respect to its second, i.e., location component, which yields the Palm type distribution R_x^0, cf. (2.1) below.) The conditioned clan can be represented by a "backward tree" picture (Proposition 1). We will then prove that exactly in dimensions $d = 3$ or 4 infinitely many of the sidetrees emerging from the backbone of the backward tree ever populate a given ball B in $\mathbb{R}^d$, which allows to conclude that for $d \geq 5$, R-almost all clans populate B only over a bounded time set, whereas for $d = 3$ or 4, R-almost all clans populate B for a time set of infinite Lebesgue measure. Exploiting the time stationarity of R, we will finally conclude that for R-almost every clan the set of time points in which it populates B is unbounded from below and above.

2. Proof of Theorem 1

We define the *Palm type distributions* R_x^0, $x \in \mathbb{R}^d$, by the formula

(2.1) $$\int_F \Phi_0^T(A) R(dT) = \int_A R_x^0(F) \lambda^d(dx)$$

(where F is a measurable set of N_p-valued paths, and A is a Borel subset of $\mathbb{R}^d$). As mentioned in part b) of the remark at the end of Section 1, R_x^0 can be viewed as the distribution of a clan, given that it populates site x at time zero.

A key tool in the proof of Theorem 1 is the following probabilistic representation of the Palm type distributions R_x^0:

PROPOSITION 1. *For $x \in \mathbb{R}^d$, consider the following ingredients:*
  (i) *a Brownian path $(W_t)_{t \geq 0}$ starting in x, leading to the random "trunk" $X_{(-\infty, 0]}$, where $X_t := W_{-t}$, $t \leq 0$,*
  (ii) *a homogeneous Poisson process $t_1 \leq t_2 \leq \ldots$ on $\mathbb{R}^+$, whose intensity parameter is that of the exponentially distributed individual lifetime (yielding the birthtimes along the backward trunk),*
  (iii) *an i.i.d. sequence $\Sigma_0, \Sigma_1, \ldots$ of critical branching Brownian trees (starting from one mother individual each), and a sequence $S_0, S_1, \ldots$, where S_n is obtained by shifting Σ_n in space and time such that its root is in the space-time point $(-t_n, W_{t_n})$ (where we put $t_0 := 0$).*

Then, for λ^d-a. all x, R_x^0 arises as the distribution of the random clan $T^{x,0}$, which is defined as the superposition of the "trunk" $X_{(-\infty, 0]}$, the "offspring tree" S_0 and the "sidetrees" $S_1, S_2, \ldots$

The proof of Proposition 1, which combines the results of [Chauvin, Rouault, and Wakolbinger 1991, Theorem 2] and [Gorostiza and Wakolbinger 1991, Lemma 5.1] is omitted here.

For the rest of the section let B be a fixed ball in $\mathbb{R}^d$.

LEMMA 1. *Let $x \in \mathbb{R}^d$, $W = (W_t), t_0, t_1, \ldots, S_0, S_1, \ldots$ be as in Proposition 1. Then if $d = 3$ or 4, with probability 1 only finitely many of the S_n hit the ball B, whereas, if $d \geq 5$, infinitely many of the S_n hit the ball B.*

PROOF. Let us put, for $y \in \mathbb{R}^d$,

$p(y) := \text{Prob}[\text{a critical binary branching Brownian tree}$
$\text{starting in } y \text{ ever populates } B]$.

By the lemma of Borel-Cantelli it suffices to show:

$$\text{If } d = 3 \text{ or } 4, \text{ then } \sum_{n=0}^{\infty} p(W_{t_n}) = \infty, \text{ for a. all } W.$$

$$\text{If } d \geq 5, \text{ then } \sum_{n=0}^{\infty} p(W_{t_n}) < \infty \text{ for a. all } W.$$

Since $(t_1, t_2, \ldots)$ is a homogeneous Poisson process independent of W, this amounts to showing:

(2.2) $\quad$ If $d = 3$ or 4, then $\int_0^{\infty} p(W_t) dt = \infty$ for a. all W.

(2.3) $\quad$ If $d \geq 5$, then $\int_0^{\infty} p(W_t) dt < \infty$ for a. all W.

It is well known ([Sawyer and Fleischman 1979], cf. also [Iscoe 1988] and [Dynkin 1993]) that, as $r := \|y\| \to \infty$, $p(y) \sim 1/r^2$ if $d = 3$, $\sim 1/(r^2 \log r)$ if $d = 4$, and $\sim 1/r^{d-2}$ if $d \geq 5$.

By the law of the iterated logarithm (cf. [Aldous 1990, p. 182–183]), for suitable constants c_1, c_2 depending on W there holds for a. all W:

$$c_1 \frac{\sqrt{t}}{\log t} \leq \|W_t\| \leq c_2 \sqrt{t \log \log t} \text{ for } t \text{ large enough}.$$

Hence (2.2) and (2.3) follow from the elementary facts that for all suitably large $t_0 > 0$ there holds

$$\int_{t_0}^{\infty} \frac{1}{t(\log \log t) \log \sqrt{t \log \log t}} dt = \infty$$

and

$$\int_{t_0}^{\infty} t^{-(d-2)/2} (\log t)^{-d-2} dt < \infty.$$

PROOF OF THE SECOND PART OF THEOREM 1. Assume $d \geq 5$. Then for λ^d-almost all x and for R_x^0-almost all T there holds: T does not populate B at late and early times. (In fact, this is clear from Lemma 1, since each of the trees S_n has a finite height.) Now let F be the set of clans which populate B at arbitrarily late and early times. Putting $A := \mathbb{R}^d$ in (2.1), we obtain that $R(F) = 0$.

In order to prove also the first part of Theorem 1, we introduce the following notion:

A space-time point $(t, y) \in \mathbb{R} \times B$ is called a *B-immigrant* in the clan T, if T_t charges a path X with $X_t = y$ and $X_s \notin B$ for all $s < t$. In this case, t is called a *B-immigration time* (in T).

PROPOSITION 2. *If $d = 3$ or 4, then R-almost all T contain infinitely many B-immigrants.*

PROOF. Let F consist of all those clans which contain only finitely many B-immigrants. Then it is clear from Lemma 1 that for λ^d-almost all x there holds $R_x^0(F) = 0$. Hence, putting $A := \mathbb{R}^d$ in (2.1), we infer that $R(F) = 0$.

PROOF OF THE FIRST PART OF THEOREM 1. Assume $d = 3$ or 4.

Since for R-almost all T the set $I(T)$ of B-immigration times has no accumulation point (otherwise we would get a contradiction to the fact that for R-a. all T the population ϕ_t^T is locally finite for all t), it follows that for R-almost all t the set $I(T)$ is unbounded. It remains to show that in this case $I(T)$ is unbounded both from above and below.

To this end, let us first state that

(2.4) $$\int_{\mathbb{R}} 1_{\{\phi_t^T(B) > 0\}} \lambda(dt) = \infty \text{ for } R\text{-almost all } T.$$

(This follows from the fact that each path immigrating into B spends a.s. a positive amount of time in B immediately after the immigration time.)

We denote by L the (Markovian) measure on the population-valued paths which arises from R under the mapping $T \to (\phi_t^T)_{t \in \mathbb{R}}$.

Let G be the set of all populations on $\mathbb{R}^d$ having individuals in B, and denote by $\tau := \tau(\Phi)$ and $\eta := \eta(\Phi)$ the first entrance and last exit time of a population valued path $\Phi = (\Phi_t)_{t \in \mathbb{R}}$ into and from G, respectively. All we have to show is that $\tau(\Phi) = -\infty$ and $\eta(\Phi) = +\infty$ for L-almost all Φ.

1. Suppose that $L(\{\tau > -\infty\}) > 0$.
   It follows from (2.4) that

   (2.5) $$\int_{\tau(\Phi)}^{\infty} 1_{\{\Phi_t \in G\}} \lambda(dt) = \infty \text{ for } L\text{-almost all } \Phi.$$

   Let $\lambda \times \rho$ be the measure on $\mathbb{R} \times G$ obtained by first restricting L to $\{\tau > -\infty\}$ and then transporting it under the mapping (τ, Φ_τ) (note that the resulting measure is indeed of the stated product form due to time stationarity of L), and write E_φ for the expectation w.r.to L_φ, the latter denoting a regular disintegration of L with respect to $\Phi_0 = \varphi$.
   We then have

   $$L(\Phi_0 \in G) = L(\Phi_0 \in G, \Phi_\tau \in G, \tau(\Phi) \in (-\infty, 0])$$
   $$= \int_{\mathbb{R}^+} \lambda(dt) \int_G \rho(d\varphi) L_\varphi(\Phi_t \in G)$$
   $$= \int_G \rho(d\varphi) E_\varphi \left[\int_{\mathbb{R}^+} 1_{\{\Phi_t \in G\}} \lambda(dt) \right].$$

   Now $L(\Phi_0 \in G)$ is nothing but the parameter of the Poisson number of clans in H which populate B at time 0, and therefore $L(\Phi_0 \in G)$ is finite. Consequently, $\int_{\mathbb{R}^+} 1_{\{\Phi_t \in G\}} \lambda(dt)$ is finite for L_φ-almost all φ and ρ-almost all φ. This, however, contradicts (2.5), revealing that $R(\{\tau > -\infty\}) = 0$.

2. By using "time reversed" arguments, one concludes in the same way as in (a). that $L(\{\eta < +\infty\}) = 0$.

This proves also the first part of Theorem 1.

3. The case of "$1 + \beta$-stable" branching

If, instead of binary branching, the particles perform a critical branching with offspring generating function $f_\beta(s) = s + \frac{1}{2}(1-s)^{1+\beta}$, $\beta \in (0,1)$, then (cf. [Gorostiza and Wakolbinger 1991]) the critical dimension for the existence of a non-trivial equilibrium distribution with uniform intensity is $\frac{2}{\beta}$. (Note that in this case the offspring distribution is in the domain of normal attraction of a stable law with exponent $1 + \beta$ and has moments only of order less than $1 + \beta$, smaller β therefore leading to a bigger "clumping effect".) Now again the question arises for which dimensions $d > 2/\beta$ the clans in the time stationary branching particle system with offspring generating function f_β are recurrent in the sense of the remark at the end of Section 1. The following result gives a partial answer:

THEOREM 2. *Let $d > \frac{2}{\beta} + 2$. Then with probability 1, each clan in a time stationary critical branching Brownian particle system with offspring generating function f_β populates a fixed ball in $\mathbb{R}^d$ only over a finite time horizon.*

The *proof* of Theorem 1 is based on the adequate modifications of Proposition 1 and Lemma 1 above: instead of one sidetree, now a random number Z_n of sidetrees emerge from the space-time point $(-t_n, W_{t_n})$, $n = 1, 2, \ldots$, where the Z_n are independent copies of a random variable Z with distribution $P[Z = k] = p_{k+1}(k+1)$, and (p_k) denotes the offspring distribution.

Let B be a ball in $\mathbb{R}^d$ and put, for $y \in \mathbb{R}^d$,

$$p(y) := \text{Prob [an offspring tree starting in } y \text{ ever populates } B]$$

and

$q(y) :=$ Prob[at least one of the Z trees starting in y ever populates B].

Then there holds:

$$q(y) = 1 - \sum_{k=0}^{\infty} P[Z=k]\bigl(1-p(y)\bigr)^k$$
$$= 1 - \sum_{k=0}^{\infty} p_{k+1}(k+1)\bigl(1-p(y)\bigr)^k$$
$$= 1 - f'\bigl(1-p(y)\bigr) = \frac{1+\beta}{2} p(y)^\beta.$$

Since $p(y) \leq E[$Brownian motion starting in y ever hits $B]$ and

$$E[\text{Brownian motion starting in } y \text{ ever hits } B] \sim \text{const.} \frac{1}{\|y\|^{d-2}} \text{ as } \|y\| \to \infty$$

for $d \geq 3$ (cf. [Port and Stone 1978, p. 56]), it follows from the law of iterated logarithm in the same way as in the proof of Lemma 1 that

$$\int_{\mathbb{R}^+} q(W_t) dt < \infty \text{ for a. all } W.$$

Hence, by the lemma of Borel-Cantelli, only finitely many of the sidetrees emerging from the backbone W ever populate B. Now the same argument as in the proof of the second part of Theorem 1 completes the proof of Theorem 2.

4. An application to critical branching Brownian motion conditioned to non-extinction

Critical branching Brownian motion starting at time zero from one mother individual at site $x \in \mathbb{R}^d$ and *conditioned to non-extinction* arises as follows (cf. [Evans 1993, Overbeck 1993, Roelly-Coppoletta and Rouault 1989]): Take, as a "trunk" (or path of the "immortal particle") a Brownian path $W = (W_t)_{t \geq 0}$ starting in x, and let the "sidetrees" $S_1, S_2, \ldots$ (all of which are independent copies of the original branching process) grow out of the space-time points $(t_1, W_{t_1}), (t_2, W_{t_2})$, $\ldots$, where $t_1 \leq t_2 \leq \cdots$ are the points of a homogeneous Poisson process on $\mathbb{R}^+$.

An immediate consequence of Lemma 1 is the following

THEOREM 3. *Critical binary branching Brownian motion in* $\mathbb{R}^d$, *conditioned to non-extinction, in dimension* 3 *and* 4 *is* recurrent *in the sense that it comes back to populate a ball* $B \subseteq \mathbb{R}^d$ *at arbitrarily large times, and in dimension* ≥ 5 *is* transient *in the sense that it eventually ceases to populate* B.

Acknowledgement. We are grateful to D. Dawson for a discussion which has led to the question answered in this note, to K. Matthes for helping us with the last part in the proof of Theorem 1, and to S. Evans, whose talk at the Montreal conference on superprocesses in autumn 1992 stimulated the application given in Section 4.

NOTE ADDED IN PROOF. By using a result of [Iscoe 1988, p. 217] on the asymptotics of $p(y)$ as defined in the proof of Theorem 2, and with the arguments in our proof Theorem 1, one can in fact show the following complement of Theorem 2: For $\frac{2}{\beta} < d \leq \frac{2}{\beta} + 2$, each clan in a time stationary branching Brownian particle system with offspring generating function f_β populates a fixed ball in $\mathbb{R}^d$ at arbitrarily late and early times. □

Bibliography

D. Aldous, *Probability Approximations via the Poisson Clumping Heuristics*, Springer, 1990.

B. Chauvin, A. Rouault and A. Wakolbinger, *Growing conditioned trees*, Stochastic Process. Appl. **39** (1991), 117–130.

D. Dawson and E. Perkins, *Historical Processes*, Mem. Amer. Math. Soc., vol. 93, 1991.

E. B. Dynkin, *Superprocesses and partial differential equations*, Ann. Probab. (to appear).

S. N. Evans, *Two representations of conditioned superprocesses*, Proc. Roy. Soc. Edinburgh Sect. A (to appear).

L. Gorostiza and A. Wakolbinger, *Persistence criteria for a class of critical branching particle systems in continous time*, Ann. Probab. **19** (1991), 266–288.

I. Iscoe, *On the supports of measure-valued critical branching Brownian motion*, Ann. Prob. **16** (1988), 200–221.

S. Port and C. Stone, *Brownian Motion and Classical Potential Theory*, Academic Press, 1978.

A. Liemant, K. Matthes, and A. Wakolbinger, *Equilibrium distributions of branching processes*, Akademie Verlag, Berlin, and Kluwer Academic Publishers, Dordrecht, 1988.

K. Matthes, J. Kerstan and J. Mecke, *Infinitely Divisible Point Processes*, Wiley, 1978.

L. Overbeck, *Conditioned superbrownian motion*, Probab. Theory Relat. Fields **96** (1993), 545–570.

S. Roelly-Coppoletta and A. Rouault, *Processus de Dawson-Watanabe conditionné par le futur lointain*, C. R. Acad. Sci. Paris Sér. I Math. **309** (1989), 867–872.

S. Sawyer and J. Fleischman, *Maximum range of a mutant allele considered as a subtype of a Brownian branching random field*, Proc. Nat. Acad. Sci. U.S.A. **76** (1979), 872–875.

Y. Wu, *Asymptotic behavior of two level measure branching processes*, Ann. Probab. (to appear).

ABSTRACT. It is proved that each "clan" (of mutually related particles) in a time stationary critical binary branching Brownian particle system in $\mathbb{R}^d$ populates a ball $B \subseteq \mathbb{R}^d$ at arbitrarily late and early times if $d = 3$ or 4, whereas the first and last times of populating B are a.s. finite for each clan if $d \geq 5$. A corresponding result is obtained for processes conditioned to non-extinction, and "clan transience" for $d > 2/\beta+2$ is also proved for "$1+\beta$-stable" branching particle systems.

INSTITUT FÜR MATHEMATIK, JOHANNES KEPLER UNIVERSITÄT, A-4040 LINZ, AUSTRIA
E-mail address: k315970@alijku11.edvz.uni-linz.ac.at

FACHBEREICH MATHEMATIK, JOHANN WOLFGANG GOETHE-UNIVERSITÄT, D-60054 FRANKFURT/M., GERMANY
E-mail address: wakolbin@math.uni-frankfurt.de

Tagged Particle Problem for an Infinite Hard Core Particle System in $\mathbb{R}^d$

Hideki Tanemura

Introduction

In this paper we consider a system of infinitely many hard balls with the same diameter r (> 0) moving discontinuously in $\mathbb{R}^d$, ($d \geq 2$). The system is described as follows:
(1) There are always infinitely many hard balls with diameter r in $\mathbb{R}^d$.
(2) A ball with center x waits a random time which depends on the configuration of balls and chooses the position y according to a probability measure $p(|x-y|)\,dy$, where $p(\cdot)$ is a given nonnegative function on $[0,\infty)$ such that $\int_{\mathbb{R}^d} p(|x|)\,dx = 1$.
(3) If there are no centers of balls in the open r-neighborhood of y, the ball jumps to y ; otherwise it remains at x.

We denote the configuration space of hard balls by $\mathfrak{X}$:

$$\mathfrak{X} = \{\xi = \{x_i\} \subset \mathbb{R}^d : |x_i - x_j| \geq r, i \neq j\},$$

the position of a ball being represented by its center. The system is completely specified by the measure $c(x, dy, \xi)$ which gives the rate of the movement of the ball at the position x to the position y when the entire configuration is ξ. We shall consider the case, where $c(x, dy, \xi)$ is given by

$$c(x, dy, \xi) = \exp\left\{-\sum_{v \in \xi \setminus \{x\}} \Phi(|y-v|)\right\} p(|x-y|)\,dy,$$

where Φ is a measurable function on $[0, \infty)$ which is bounded from below and satisfies the following properties:
(Φ.1) $\Phi(\alpha) = \infty$ if and only if $\alpha \in [0, r)$,
(Φ.2) $\Phi(\alpha) = 0$ if $\alpha \in [r_0, \infty)$,
for some positive constants r and r_0 with $r \leq r_0$. Φ is regarded as a hard core pair potential which is rotation invariant, stable and of finite range. We construct the Markov process ξ_t which describes our system. This process has the Gibbs state μ associated with the potential Φ as a reversible measure.

We showed the ergodicity of the stationary process in [Tanemura 1989] and the central limit theorem for the tagged particle of the process in [Tanemura 1993] under the assumption that the density of the balls is sufficiently low. In this paper we study these problems in a general case. The behavior of the system can depend on the density of balls. Denote by $\mathcal{G}(z)$ the set of all Gibbs states with respect to

---

1991 *Mathematics Subject Classification.* 60K35.
This is the final form of the paper.

the activity $z > 0$ and the potential Φ. In the case where the activity z is large, that is, the density of balls is high, it is not known whether the set $\mathcal{G}(z)$ is a singleton or not. The set $\mathcal{G}(z)$ is convex and any element of $\mathcal{G}(z)$ is represented by the extreme points of $\mathcal{G}(z)$. In this paper we show that the stationary process is ergodic in the case where the stationary measure is an extreme point of $\mathcal{G}(z)$, $z > 0$. This improves the earlier result [Tanemura 1989].

To prove the ergodicity it is important to develop a topological argument on the configuration space. Denote by $\Lambda(K, n, \xi)$ the set of configurations of n balls in a compact set K with boundary condition $\xi \cap K^c$, and by $\mathfrak{X}(K, n, \xi)$ the set of configurations $\zeta \in \mathfrak{X}$ such that $\zeta \cap K \in \Lambda(K, n, \xi)$ and $\zeta \cap K^c = \xi \cap K^c$. Then the ergodicity can be derived from the property that for μ-almost all ξ, any configuration of $\mathfrak{X}(K, n, \xi)$ can be attained from any other configuration of $\mathfrak{X}(K, n, \xi)$ by means of a finite number of jumps of magnitude less than h, where h is a positive constant depending on $p(\cdot)$. In the case where the density of balls is sufficiently small, we proved this property and showed the ergodicity in [Tanemura 1989].

If the space $\Lambda(K, n, \xi)$ were connected, any configuration of $\mathfrak{X}(K, n, \xi)$ would be attained from any other configuration of $\mathfrak{X}(K, n, \xi)$ by moving balls in K continuously. However, $\Lambda(K, n, \xi)$ is not always connected. In this paper we prove that for μ-almost all ξ there exist a compact set $K'(\supset K)$ and a positive integer $n' (\geq n)$ such that the configuration space $\{\zeta \cap K' : \zeta \in \mathfrak{X}(K, n, \xi)\}$ is a subset of a connected component of $\Lambda(K', n', \xi)$, that is, any configuration of $\mathfrak{X}(K, n, \xi)$ can be attained from any other configuration of $\mathfrak{X}(K, n, \xi)$ by moving balls in K' continuously. From this result we show the ergodicity in the general case.

The tagged particle of the process is described as an additive functional of the environment process seen from the tagged particle. To prove the central limit theorem of the tagged particle we apply an invariant principle of additive functionals of ergodic reversible Markov processes given by [De Masi, Ferrari, Goldstein and Wick 1989]. In [Tanemura 1993] we used the condition of low density to prove the ergodicity of the environment process and the nondegeneracy of the diffusion matrix. By the same way as above we can show the ergodicity of the environment process in a general case and so we also improve the result [Tanemura 1993].

In Section 1 we state our main theorems. In Section 2 we introduce a continuum percolation model and obtain estimates of cluster size of this model. From the estimates we show a property of the configuration space which is a key part to prove the theorems. The proof of the theorems is given in Section 3.

1. Statement of results

Let $\mathfrak{M}$ be the set of all countable subsets ξ of $\mathbb{R}^d$ satisfying $N_K(\xi) < \infty$ for any compact subset K, where $N_A(\xi)$ is the number of points of ξ in $A \subset \mathbb{R}^d$ ($d \geq 2$). We regard $\xi \in \mathfrak{M}$ as a non-negative integer valued Radon measure on $\mathbb{R}^d$: $\xi(\cdot) = \sum_{x \in \xi} \delta_x(\cdot)$ and accordingly equip $\mathfrak{M}$ with the vague topology, where δ_x denotes the δ-measure at x. For any $\xi \in \mathfrak{M}$ and $A \subset \mathbb{R}^d$ we denote by ξ_A the restriction of ξ to A; ξ_A is a Radon measure on A; however, we regard it as an element of $\mathfrak{M}$ in a natural way.

The configuration space $\mathfrak{X}$ of hard balls introduced in the introduction is a compact subset of $\mathfrak{M}$ with the vague topology. We define σ-fields $\mathcal{B}(\mathfrak{X})$ and $\mathcal{B}_K(\mathfrak{X})$ by $\mathcal{B}(\mathfrak{X}) = \sigma(N_A; A \in \mathcal{B}(\mathbb{R}^d))$ and $\mathcal{B}_K(\mathcal{X}) = \sigma(N_A; A \in \mathcal{B}(\mathbb{R}^d), A \subset K)$, respectively.

The σ-field $\mathcal{B}(\mathfrak{X})$ coincides with the topological Borel field of $\mathfrak{X}$.

Let $\mathbb{C}(\mathfrak{X})$ be the space of all real valued continuous functions on $\mathfrak{X}$ and $\mathbb{C}_0(\mathfrak{X})$ be the set of functions of $\mathbb{C}(\mathfrak{X})$ each of which depends only on the configurations in some compact set K. We define an operator L on $\mathbb{C}_0(\mathfrak{X})$ by

$$Lf(\xi) = \sum_{x \in \xi} \int_{\mathbb{R}^d} \{f(\xi^{x,y}) - f(\xi)\} \chi(y \mid \xi \setminus \{x\}) p(|x-y|) \, dy, \quad f \in \mathbb{C}_0(\mathfrak{X}),$$

where

$$\xi^{x,y} = \begin{cases} (\xi \setminus \{x\}) \cup \{y\}, & \text{if } x \in \xi, \, y \notin \xi, \\ \xi, & \text{otherwise,} \end{cases}$$

$p(\cdot)$ is a non-negative Borel function on $[0, \infty)$ satisfying
 (p.1) $\int_{\mathbb{R}^d} p(|x|) \, dx = 1$,
 (p.2) $\int_{\mathbb{R}^d} |x|^2 p(|x|) \, dx < \infty$,
 (p.3) ess. $\inf_{\alpha \in [0,c)} p(\alpha) > 0$ for any $c \in (0, h)$ for some $h \in (0, \infty]$,
and χ is the function defined by

$$\chi(x_1, x_2, \ldots, x_n \mid \xi) = \exp\left\{ -\sum_{1 \leq i < j \leq n} \Phi(|x_i - x_j|) - \sum_{i=1}^{n} \sum_{y \in \xi} \Phi(|x_i - y|) \right\},$$

for $x_1, x_2, \ldots, x_n \in \mathbb{R}^d$, $n \in \mathbb{N}$, $\xi \in \mathfrak{M}$. Here Φ is a given measurable function on $[0, \infty)$ which is bounded from below and satisfies (Φ.1) and (Φ.2) in the introduction. The measure $\chi(y \mid \xi \setminus \{x\}) p(|x-y|) \, dy$ gives the rate of the movement of the ball at the position x to the position y when the entire configuration is ξ. From the property (Φ.1) the ball of the system moves by random jump under the hard core condition.

By a slight modification of a Liggett's theorem in [Liggett 1985] we see that the smallest closed extension $(\bar{L}, \mathcal{D}(\bar{L}))$ of $(L, \mathbb{C}_0(\mathfrak{X}))$ generates a unique strong continuous Markov semigroup T_t on $\mathbb{C}(\mathfrak{X})$. Since T_t is a Feller semigroup, for each initial distribution μ there exists a Markov process (ξ_t, P_μ) with semigroup T_t which is right continuous and has left limits.

For any compact subset $K \subset \mathbb{R}^d$ and $n \in \mathbb{N}$, we denote by $\mathfrak{M}(K)$ and $\mathfrak{M}(K, n)$ the set of all finite subsets of K and the set of all subsets of K having n points, respectively. Note that $\chi(\cdot \mid \xi)$ is permutation invariant and can be regarded as a function on $\mathfrak{M}(K)$. We denote by $\lambda_{K,z}$ a Poisson distribution on $\mathfrak{M}(K)$ with intensity measure $z \, dx$, that is, for disjoint sets $A_1, A_2, \ldots, A_m \in \mathcal{B}(K)$, $N_{A_1}(\xi), \ldots, N_{A_m}(\xi)$ are independent random variables on the probability space $(\mathfrak{M}(K), \mathcal{B}(\mathfrak{M}(K)), \lambda_{K,z})$ and

$$\lambda_{K,z}(N_{A_i} = n) = \frac{(z|A_i|)^n}{n!} \exp(-z|A_i|), \quad i = 1, 2, \ldots, m, n \in \mathbb{N},$$

where $|A| = \int_A dx$ for $A \in \mathcal{B}(\mathbb{R}^d)$.

Now, we are going to define a Gibbs state.

DEFINITION 1.1. A probability measure μ on $\mathfrak{X}$ is called a Gibbs state with respect to the activity $z > 0$ and the potential Φ, if μ satisfies the DLR equation: for any compact subset K of $\mathbb{R}^d$

$$\mu(\cdot | \mathcal{B}_{K^c}(\mathfrak{X}))(\xi) = \mu_{K,\xi,z}(\cdot), \quad \mu\text{-a.s. } \xi,$$

where $\mu_{K,\xi,z}$ is the probability measure on $\mathfrak{M}(K)$ defined by

$$\mu_{K,\xi,z}(d\mathbf{x}) = \frac{1}{Z_{K,\xi,z}}\chi(\mathbf{x} \mid \xi_{K^c})\lambda_{K,z}(d\mathbf{x}),$$

$$Z_{K,\xi,z} = \int_{\mathfrak{M}(K)} \chi(\mathbf{x} \mid \xi_{K^c})\lambda_{K,z}(d\mathbf{x}).$$

Denote by $\mathcal{G}(z,\Phi)$ the set of all Gibbs states with respect to the activity $z > 0$ and the potential Φ, and by $\mathcal{G}_\Theta(z,\Phi)$ the set of all elements of $\mathcal{G}(z,\Phi)$ which are translation invariant. The set $\mathcal{G}(z,\Phi)$ is convex and compact with respect to the topology of weak convergence. Any element of $\mathcal{G}(z,\Phi)$ is represented by the extreme points of $\mathcal{G}(z,\Phi)$. We denote by $\mathrm{ex}\mathcal{G}(z,\Phi)$ the set of extreme points of $\mathcal{G}(z,\Phi)$. If $\sharp\mathcal{G}(z,\Phi) = 1$ and $\mu \in \mathcal{G}(z,\Phi)$, then μ is rotation invariant, translation invariant and ergodic under translation. (See [Ruelle 1969], [Georgii 1988]).

Our first main result is the following theorem.

THEOREM 1. *If $\mu \in \mathrm{ex}\mathcal{G}(z,\Phi), z \in (0,\infty)$, then the process (P_μ, ξ_t) is an ergodic reversible Markov process.*

We also study the behavior of a tagged particle in the process. In order to follow the motion of the tagged particle it is convenient to regard the process ξ_t as a Markov process $(x(t), \eta_t)$ on the locally compact space $\mathbb{R}^d \times \mathfrak{X}_0$, where

$$\mathfrak{X}_0 = \{\eta \in \mathfrak{X} : \eta \cap U_r(0) = \varnothing\}.$$

In the above (and also in the sequel) $U_r(A)$ stands for the open r-neighborhood of $A \subset \mathbb{R}^d$ and $U_r(x)$ is the abbreviated form for $U_r(\{x\})$. The process $x(t)$ is the position of the tagged particle and η_t is the entire configuration seen from the tagged particle. We define $\mathbb{C}(\mathfrak{X}_0)$ and $\mathbb{C}_0(\mathfrak{X}_0)$ by the same way as $\mathbb{C}(\mathfrak{X})$ and $\mathbb{C}_0(\mathfrak{X})$, respectively. We can see that η_t is a Markov process whose generator $\bar{\mathcal{L}}$ is the smallest closed extension of the operator $\mathcal{L}$ on $\mathbb{C}_0(\mathfrak{X}_0)$ given by

$$\mathcal{L}f(\eta) = \sum_{x \in \eta}\int_{U_r(0)^c}\{f(\eta^{x,y}) - f(\eta)\}\chi(y \mid \eta \setminus \{x\})p(|x-y|)\,dy$$
$$+ \int_{\mathbb{R}^d}\{f(\tau_{-u}\eta) - f(\eta)\}\chi(u \mid \eta)p(|u|)\,du, \quad f \in \mathbb{C}_0(\mathfrak{X}_0),$$

where $\tau_u A = \{x + u : x \in A\}$, $A \subset \mathbb{R}^d, u \in \mathbb{R}$. We denote by $(\Omega, \mathcal{F}, P_\nu^0, \eta_t)$ the Markov process with generator $\bar{\mathcal{L}}$ and initial distribution ν. For any $\mu \in \mathcal{G}_\Theta(z,\Phi)$ we define a probability measure μ_0 on $\mathfrak{X}_0$ by

$$\mu_0(d\eta) = \frac{1}{\int_\mathfrak{M}\chi(0 \mid \eta)\mu(d\eta)}\chi(0 \mid \eta)\mu(d\eta).$$

Then μ_0 is a reversible probability measure for η_t.

The process $x(t)$ is driven by the process η_t in the following way. We define a σ-finite random measure N by

$$N\big((0,t] \times A\big) = \sum_{s \in (0,t]}\mathbb{I}_{\Xi(A)}(\eta_{s-},\eta_s), t > 0, \quad A \in \mathcal{B}(\mathbb{R}^d),$$

where $\mathbb{I}_R$ stands for the indicator function of a set R and $\Xi(A)$ is the measurable subset of $\mathfrak{X}_0 \times \mathfrak{X}_0$ defined by

$$\Xi(A) = \{(\zeta,\eta) \in (\mathfrak{X}_0 \times \mathfrak{X}_0) \setminus \Delta : \zeta = \tau_{-u}\eta \text{ for some } u \in A\}, \quad A \in \mathcal{B}(\mathbb{R}^d),$$

$$\Delta = \{(\eta,\eta) : \eta \in \mathfrak{X}_0\} \cup \left(\{\eta \in \mathfrak{X}_0 : \eta = \tau_{-u}\eta \text{ for some } u \in \mathbb{R}^d \setminus \{0\}\}^2\right).$$

Then,

$$x(t) = x(0) + \int_0^t \int_{\mathbb{R}^d} u N(ds\,du).$$

Our second main result is the following theorem.

THEOREM 2. *There exists $z_0 \in (0,\infty]$ such that if $\mu \in \text{ex}\,\mathcal{G}(z,\Phi) \cap \mathcal{G}_\Theta(z,\Phi)$, $z \in (0,z_0)$, then the process $\varepsilon x(t/\varepsilon^2)$ on $(\Omega, \mathcal{F}, P_{\mu_0}^0)$ converges to $D(\mu_0)B(t)$ as $\varepsilon \to 0$ in distribution with respect to J_1-topology on Skorohod's function space $\mathbf{D}[0,\infty)$, where $D(\mu_0)$ is a positive definite $d \times d$ matrix. In particular, $z_0 = \infty$ in case $h = \infty$.*

2. Estimates of cluster size

In this section we introduce a continuum percolation model and estimate the cluster size of the model. For $\xi \in \mathfrak{X}, r' \geq r$ and disjoint regions A_1 and A_2 in $\mathbb{R}^d$, we say that a continuous curve γ is an occupied connection of A_1 and A_2 with respect to (ξ, r') if $\gamma \cap A_1 \neq \emptyset, \gamma \cap A_2 \neq \emptyset$ and $\gamma \subset U_{r'/2}(\xi)$. Define a region in $\mathbb{R}^d$:

$$W(A,\xi,r') = \{x : \text{exists an occupied connection of } A \text{ and } \{x\}$$
$$\text{with respect to } (\xi,r')\},$$

i.e. the occupied cluster which intersects A. Given a probability measure μ on $\mathfrak{X}$, we obtain a continuum percolation model. We consider the case of $\mu \in \mathcal{G}(z), z \in (0,\infty)$. The purpose of this section is to show the following proposition.

PROPOSITION 2.1. *Let $\mu \in \mathcal{G}(z), z \in (0,\infty)$. Then, there exist positive constants ε_1, c_1 and c_1' such that*

$$\mu(\text{diam}\,W(I(x), r+\varepsilon_1) \geq \ell) \leq c_1 \exp(-c_1'\ell), \quad \ell > 0, x \in \mathbb{R}^d,$$

where $I(x)$ is the cube with center x and edge length $3r$;

$$I(x) = \prod_{i=1}^d \left[x^i - \frac{3r}{2}, x^i + \frac{3r}{2}\right), \quad x = (x^1, x^2, \ldots, x^d) \in \mathbb{R}^d,$$

and $\text{diam}\,A = \sup_{x,y \in A} |x-y|$ for a bounded subset A of $\mathbb{R}^d$.

To show this proposition we introduce a site percolation model. Put $S = 3r\mathbb{Z}^d$ and call an element $x \in S$ a site. For each $x = (x^1,\ldots,x^d), y = (y^1,\ldots,y^d) \in S$ we write $\langle x,y \rangle$ and say that x is adjacent to y if

$$\max_{1 \leq i \leq d} |x^i - y^i| = 3r.$$

Put $Y = \{0,1\}^S$ and equip Y with the σ-field generated by $\{\zeta \in Y : \zeta(x) = 1\}$, $x \in S$. For $\zeta \in Y$ a site $x \in S$ is said to be occupied (or vacant), if $\zeta(x) = 1$ (or $\zeta(x) = 0$). We write $x \leftrightarrow y$ if there is an occupied path from x to y; there is a

sequence $x_0 = x, x_1, \ldots, x_n = y$ of occupied sites such that $\langle x_{m-1}, x_m \rangle, 1 \leq m \leq n$. We define
$$C(x, \zeta) = \{y \in S : x \leftrightarrow y\} \cup \{x\}.$$

For $r' \geq r$ we introduce the measurable map $\pi_{r'}$ from $\mathfrak{X}$ to Y defined by

$$(\pi_{r'} \xi)(x) = \begin{cases} 1, & \text{if } \min_{y \in \xi_{I(x)}} \min_{v \in \xi \setminus y} |y - v| < r', \\ 0, & \text{otherwise}. \end{cases}$$

Then we see that

$$\bigcup_{y \in C(x, \pi_{r'} \xi)} I(y) \supset W(I(x), \xi, r'), \quad x \in S, \xi \in \mathfrak{X}.$$

Thus, Proposition 2.1 is derived from the following lemma.

LEMMA 2.2. *Let $\mu \in \mathcal{G}(z), z \in (0, \infty)$. Then, there exist positive constants ε_2, c_2 and c_2' such that*

$$\pi_{r+\varepsilon_2} \mu(\operatorname{diam} C(0) \geq \ell) \leq c_2 \exp(-c_2' \ell), \quad \ell > 0,$$

where $\pi_{r+\varepsilon_2} \mu$ is the image measure of μ under the map $\pi_{r+\varepsilon_2}$.

PROOF. First we show that for any $p \in (0, 1)$ there exists $\varepsilon > 0$ such that

(2.1) $\quad \mu_{I(x),\xi,z}\left(\mathbf{x} \in \mathfrak{M}(I(x)) : \pi_{r+\varepsilon}(\mathbf{x} \cup \xi_{I(x)^c})(x) = 1\right) \leq p, \quad x \in S, \xi \in \mathfrak{X}.$

From the definition we obtain

$$\mu_{I(x),\xi,z}\left(\mathbf{x} \in \mathfrak{M}(I(x)) : \pi_{r+\varepsilon}(\mathbf{x} \cup \xi_{I(x)^c})(x) = 1\right)$$
$$= \frac{1}{Z_{I(x),\xi,z}} \int_{\mathfrak{M}(I(x))} \chi(\mathbf{x} \mid \xi_{I(x)^c}) \left(\chi_r(\mathbf{x} \mid \xi_{I(x)^c}) - \chi_{r+\varepsilon}(\mathbf{x} \mid \xi_{I(x)^c})\right) \lambda_{I(x),z}(d\mathbf{x})$$
$$\leq M_0 \int_{\mathfrak{M}(I(x))} \left(\chi_r(\mathbf{x} \mid \xi_{I(x)^c}) - \chi_{r+\varepsilon}(\mathbf{x} \mid \xi_{I(x)^c})\right) \lambda_{I(x),z}(d\mathbf{x}),$$

where

$$\chi_{r'}(\{x_1, x_2, \ldots, x_n\} \mid \xi) = \chi_{r'}(x_1, x_2, \ldots, x_n \mid \xi)$$
$$= \prod_{1 \leq i < j \leq n} \mathbb{I}_{[r', \infty)}(|x_i - x_j|) \prod_{i=1}^{n} \prod_{y \in \xi} \mathbb{I}_{[r', \infty)}(|x_i - y|),$$

for $x_1, x_2, \ldots, x_n \in \mathbb{R}^d, n \in \mathbb{N}, \xi \in \mathfrak{X}$ and M_0 is a positive constant satisfying

$$\chi(x_1, x_2, \ldots, x_n \mid \xi) \leq M_0, \quad x_1, x_2, \ldots, x_n \in \mathbb{R}^d, n \in \mathbb{N}, \xi \in \mathfrak{X}.$$

Put $\{w_1, \ldots, w_m\} = \xi \cap (U_{r+\varepsilon}(I(x)) \setminus I(x))$. From the hard core condition $m \leq$

$m_0 = (\frac{3}{2}\sqrt{d})^d$. Then we have

$$\chi_r(x_1, x_2, \ldots, x_n \mid \xi_{I(x)^c}) - \chi_{r+\varepsilon}(x_1, x_2, \ldots, x_n \mid \xi_{I(x)^c})$$

$$= \prod_{1 \leq i < j \leq n} \mathbb{I}_{[r,\infty)}(|x_i - x_j|) \prod_{i=1}^{n} \prod_{k=1}^{m} \mathbb{I}_{[r,\infty)}(|x_i - w_k|)$$

$$- \prod_{1 \leq i < j \leq n} \mathbb{I}_{[r+\varepsilon,\infty)}(|x_i - x_j|) \prod_{i=1}^{n} \prod_{k=1}^{m} \mathbb{I}_{[r+\varepsilon,\infty)}(|x_i - w_k|)$$

$$\leq \sum_{1 \leq i < j \leq n} \mathbb{I}_{[r,r+\varepsilon)}(|x_i - x_j|) + \sum_{\substack{1 \leq i \leq n \\ 1 \leq k \leq m}} \mathbb{I}_{[r,r+\varepsilon)}(|x_i - w_k|),$$

and

$$\int_{I(x)^n} (\chi_r(x_1, x_2, \ldots, x_n \mid \xi_{I(x)^c}) - \chi_{r+\varepsilon}(x_1, x_2, \ldots, x_n \mid \xi_{I(x)^c})) \, dx_1 \, dx_2 \ldots dx_n$$

$$\leq n(n - 1 + m_0)|I(x)|^{n-1}|U_{r+\varepsilon}(0) \setminus U_r(0)|.$$

Hence

$$\mu_{I(x),\xi,z}\left(\mathbf{x} \in \mathfrak{M}(I(x)) : \pi_{r+\varepsilon}(\mathbf{x} \cup \xi_{I(x)^c})(x) = 1\right)$$

$$\leq M_0 |U_{r+\varepsilon}(0) \setminus U_r(0)|(|I(0)|z^2 + m_0 z) \exp(|I(0)|z),$$

which implies (2.1).

Let ν_p be the Bernoulli measure with parameter $p \in (0,1)$. By [Hammersley 1957, van den Berg and Kesten 1985] it was shown that there exists $p_c \in (0,1)$ such that for any $p < p_c$

(2.2) $\qquad \nu_p(\operatorname{diam} C(0) \geq \ell) \leq c(p) \exp(-c'(p)\ell), \quad \ell > 0,$

for some positive constants $c(p)$ and $c'(p)$. Using (2.1) and a comparison argument, we obtain Lemma 2.2 from (2.2). □

The following result is easily obtained from Lemma 2.1.

COROLLARY 2.3. *Let $\mu \in \mathcal{G}(z)$, $z \in (0,\infty)$. Then, for μ-almost all ξ there exists an increasing sequence $\{A_i\}_{i \in \mathbb{N}}$ of bounded open connected subsets of $\mathbb{R}^d$ such that $\bigcup_{i \in \mathbb{N}} A_i = \mathbb{R}^d$ and $U_{r/2}(\xi) \cap (U_{\varepsilon_1}(A_i) \setminus A_i) = \varnothing$, $i \in \mathbb{N}$, where ε_1 is the positive constant in Proposition 2.1.*

3. Proof of Theorems

The purpose of this section is to prove Theorem 1 and Theorem 2. First we show Theorem 1. Let $\widehat{T}_t$ be the strong continuous semigroup on $L^2(\mathfrak{X}, \mu)$ associated with ξ_t. To prove the ergodicity of the process (ξ_t, P_μ) it is enough to show the following condition (C.1) :

(C.1) If $f \in L^2(\mathfrak{X}, \mu)$ satisfies $\widehat{T}_t f = f$ for any $t \geq 0$, then f is constant.

We shall prove that the condition (C.1) holds for $\mu \in \operatorname{ex} \mathcal{G}(z, \Phi)$, $z \in (0, \infty)$.

Let $\mathcal{E}_\infty(\mathfrak{X})$ be the σ-field defined by

$$\mathcal{E}_\infty(\mathfrak{X}) = \mu\text{-completion of} \bigcap_{K:\text{compact}} \sigma(N_K, \mathcal{B}_{K^c}(\mathfrak{X})),$$

where $\sigma(N_K, \mathcal{B}_{K^c}(\mathfrak{X}))$ is the σ-field generated by N_K and $\mathcal{B}_{K^c}(\mathfrak{X})$. Georgii [Georgii 1979] proved that if $\mu \in \mathrm{ex}\,\mathcal{G}(z, \Phi)$, then $\mu(\Lambda) = 1$ or 0 for any $\Lambda \in \mathcal{E}_\infty(\mathfrak{X})$. The following Lemma 3.1 follows from this result immediately. For $\ell > 0$ we put

$$K_\ell = \{x \in \mathbb{R}^d : |x^i| \le \ell, i = 1, 2, \ldots, d\}.$$

For $m, n \in \mathbb{N}$ and $\xi \in \mathfrak{X}$, we denote by $\Lambda(K_m, n, \xi)$ the interior of the configuration space $\{\mathbf{x} \in \mathfrak{M}(K_m, n) : \chi(\mathbf{x} \mid \xi_{K_m^c}) \ne 0\}$.

LEMMA 3.1. *Suppose that $\mu \in \mathrm{ex}\,\mathcal{G}(z, \Phi)$, $z \in (0, \infty)$. If $f \in L^2(\mathfrak{X}, \mu)$ and if f satisfies*

$$\int_{\Lambda(K_m,n,\xi)} \int_{\Lambda(K_m,n,\xi)} |f(\mathbf{x} \cup \xi_{K_m^c}) - f(\mathbf{y} \cup \xi_{K_m^c})| \lambda_{K_m,z}(d\mathbf{x}) \lambda_{K_m,z}(d\mathbf{y}) = 0,$$

for μ-almost all ξ and all $m, n \in \mathbb{N}$, then f is constant.

On the other hand, in [Tanemura 1989] we have shown the following results.

LEMMA 3.2. *Suppose that $\mu \in \mathcal{G}(z, \Phi)$, $z \in (0, \infty)$ and Λ is a connected component of $\Lambda(K_m, n, \xi)$, for some $m, n \in \mathbb{N}$. If $f \in L^2(\mathfrak{X}, \mu)$ satisfies $\widehat{T}_t f = f$ for any $t \ge 0$ then f satisfies*

$$\int_\Lambda \int_\Lambda |f(\mathbf{x} \cup \xi_{K_m^c}) - f(\mathbf{y} \cup \xi_{K_m^c})| \lambda_{K_m,z}(d\mathbf{x}) \lambda_{K_m,z}(d\mathbf{y}) = 0,$$

for μ-almost all ξ.

Let $m, m', n' \in \mathbb{N}$ with $m < m'$ and Λ be a measurable subset of $\Lambda(K_{m'}, n', \xi)$. From the definition of Gibbs states a measurable function f satisfies

$$|f(\mathbf{x} \cup \xi_{K_m^c}) - f(\mathbf{y} \cup \xi_{K_m^c})| \mathbb{I}_\Lambda(\mathbf{x} \cup \xi_{K_{m'} \setminus K_m}) \mathbb{I}_\Lambda(\mathbf{y} \cup \xi_{K_{m'} \setminus K_m}) = 0,$$

for μ-almost all ξ and $\lambda_{K_m,z}$-almost all $\mathbf{x}, \mathbf{y}$, if and only if

$$|f(\widehat{\mathbf{x}} \cup \xi_{K_{m'}^c}) - f(\widehat{\mathbf{y}} \cup \xi_{K_{m'}^c})| \mathbb{I}_\Lambda(\widehat{\mathbf{x}}) \mathbb{I}_\Lambda(\widehat{\mathbf{y}}) = 0,$$

for μ-almost all ξ and $\lambda_{K_{m'},z}$-almost all $\widehat{\mathbf{x}}, \widehat{\mathbf{y}}$. Then, by virtue of Lemma 3.1 (C.1) is derived from Lemma 3.2 by using the following lemma, which is a key part of the proof of Theorem 1.

LEMMA 3.3. *Suppose that $\mu \in \mathcal{G}(z, \Phi)$, $z \in (0, \infty)$ and $m, n \in \mathbb{N}$. Then, for μ-almost all ξ, there exists $m' \in \mathbb{N}$ and a connected component Λ of $\Lambda(K_{m'}, n', \xi)$ such that*

$$\Lambda[K_{m'}, K_m, n, \xi] \subset \Lambda, \quad n \in \mathbb{N},$$

where $n' = n + N_{K_{m'} \setminus K_m}(\xi)$ and $\Lambda[K_{m'}, K_m, n, \xi]$ is the configuration space defined by

$$\Lambda[K_{m'}, K_m, n, \xi] = \{\mathbf{x} \cup \xi_{K_{m'} \setminus K_m} : \mathbf{x} \in \Lambda(K_m, n, \xi)\}.$$

PROOF. We introduce the subset $\widetilde{\mathfrak{X}}$ of $\mathfrak{X}$ defined by

$$\widetilde{\mathfrak{X}} = \{\xi = \{x_i\} : |x_i - x_j| > r, i \ne j\}.$$

Then $\mu(\widetilde{\mathfrak{X}}) = 1$ and $\Lambda[K_{m'}, K_m, n, \xi] \subset \Lambda(K_{m'}, n', \xi)$, for $m, n, m' \in \mathbb{N}$ with $m' > m$ and $\xi \in \widetilde{\mathfrak{X}}$, where $n' = n + N_{K_{m'} \setminus K_m}(\xi)$. Let ε_1 be the positive constant in Proposition 2.1 and k be a positive integer with $k > (6m + 4r)/\varepsilon_1$. From Corollary

2.3, for μ-almost all ξ there exists a sequence $\{A_j\}_{j=1}^{dk}$ of bounded open subsets of $\mathbb{R}^d$ which satisfies the following conditions:

(3.1) $$A_j \supset K_{4m+2r}, \quad 1 \le j \le dk,$$
(3.2) $$A_{j+1} \supset U_{\varepsilon_1}(A_j), \quad 1 \le j \le dk - 1,$$
(3.3) $$U_{r/2}(\xi) \cap (U_{\varepsilon_1}(A_j) \setminus A_j) = \varnothing, \quad 1 \le j \le dk - 1.$$

Suppose that $\xi \in \widetilde{\mathfrak{X}}$ satisfies (3.3) for some sequence $\{A_j\}_{j=1}^{dk}$ with (3.3) and (3.2). We choose $m' \in \mathbb{N}$ such that $K_{m'} \supset U_{\varepsilon_1}(A_{dk})$. Then, it is enough to show the following for the proof of Lemma 3.3:

(3.4) $$\Lambda(\mathbf{x}) = \Lambda(\mathbf{y}), \quad \mathbf{x}, \mathbf{y} \in \Lambda[K_{m'}, K_m, n, \xi],$$

where $\Lambda(\mathbf{x})$ is the connected component of $\Lambda(K_{m'}, n', \xi)$ containing $\mathbf{x}$.

Put $H_+^i = \{x \in \mathbb{R}^d : x^i \ge m\}$, $H_-^i = \{x \in \mathbb{R}^d : x^i \le -m\}$, $1 \le i \le d$ and put

$$A_j^+ = A_j \cap H_+^i, \quad A_j^- = A_j \cap H_-^i, \quad a_j = \left(0, \ldots, 0, \underset{i\text{-th}}{\frac{\varepsilon_1}{2}}, 0, \ldots, 0\right),$$

$(i-1)k < j \le ik$, $1 \le i \le d$. We put $\mathbf{x}^0 = \mathbf{x}$, $\mathbf{y}^0 = \mathbf{y}$ and define sequences $\{\mathbf{x}^j\}_{j=0}^{dk}$ and $\{\mathbf{y}^j\}_{j=0}^{dk}$ inductively by the following rule:

$$\mathbf{x}^j = (\mathbf{x}^{j-1} \cap (A_j^+ \cup A_j^-)^c) \cup \tau_{a_j}(\mathbf{x}^{j-1} \cap A_j^+) \cup \tau_{-a_j}(\mathbf{x}^{j-1} \cap A_j^-),$$
$$\mathbf{y}^j = (\mathbf{y}^{j-1} \cap (A_j^+ \cup A_j^-)^c) \cup \tau_{a_j}(\mathbf{y}^{j-1} \cap A_j^+) \cup \tau_{-a_j}(\mathbf{y}^{j-1} \cap A_j^-),$$

$1 \le j \le dk$. From (3.3) we see that $\mathbf{x}^j, \mathbf{y}^j \in \Lambda(K_{m'}, n', \xi)$ and $\Lambda(\mathbf{x}^{j-1}) = \Lambda(\mathbf{x}^j)$, $\Lambda(\mathbf{y}^{j-1}) = \Lambda(\mathbf{y}^j)$, $1 \le j \le dk$, and so

(3.5) $$\Lambda(\mathbf{x}) = \Lambda(\mathbf{x}^{dk}), \quad \Lambda(\mathbf{y}) = \Lambda(\mathbf{y}^{dk}).$$

Note that $\mathbf{x}^{dk}$ and $\mathbf{y}^{dk}$ satisfy

$$\mathbf{x}^{dk} \cap K_{4m+2r}^c = \mathbf{y}^{dk} \cap K_{4m+2r}^c \equiv \mathbf{w},$$
$$\mathbf{x}^{dk} \cap (K_{4m+2r} \setminus K_m) = \mathbf{y}^{dk} \cap (K_{4m+2r} \setminus K_m) = \varnothing.$$

Put $\widehat{\mathbf{x}} = 4(\mathbf{x}^{dk} \cap K_m) \cup \mathbf{w}$ and $\widehat{\mathbf{y}} = 4(\mathbf{y}^{dk} \cap K_m) \cup \mathbf{w}$, where $\alpha A = \{\alpha x : x \in A\}$ for $A \subset \mathbb{R}^d$, $\alpha > 0$. Then we see that $\widehat{\mathbf{x}}, \widehat{\mathbf{y}} \in \Lambda(K_{m'}, n', \xi)$ and

(3.6) $$\Lambda(\widehat{\mathbf{x}}) = \Lambda(\mathbf{x}^{dk}), \quad \Lambda(\widehat{\mathbf{y}}) = \Lambda(\mathbf{y}^{dk}).$$

We introduce a notation about configurations. Let D be a convex subset of $\mathbb{R}^d$ and $\ell \in \mathbb{N}$. A configuration $\{u_1, u_2, \ldots, u_\ell\}$ is said to be standard in D if

$$\min_{1 \le i < j \le \ell} |u_i - u_j| > 4r \quad \text{and} \quad \inf_{v \in D^c} \min_{1 \le i \le \ell} |v - u_i| > 2r.$$

It is easy to see that $\widehat{\mathbf{x}} \cap K_{4m+2r}$ and $\widehat{\mathbf{y}} \cap K_{4m+2r}$ are standard in K_{4m+2r}. Then, by Lemma 3.1 in [Tanemura 1989] we see that

(3.7) $$\Lambda(\widehat{\mathbf{x}}) = \Lambda(\widehat{\mathbf{y}}).$$

Combining (3.5),(3.6) and (3.7), we obtain (3.4). This completes the proof. □

Next we give the proof of Theorem 2. By the construction of η_t, $x(t)$ is an anti-symmetric additive functional of η_t, that is,

(3.8) $\quad x(t)(\eta_{t/2-}.) = -x(t)(\eta.), \qquad t \geq 0,$

(3.9) $\quad x(t+s)(\eta.) = x(t)(\eta.) + x(s)(\eta_{t+}.), \quad t, s \geq 0.$

[De Masi, Ferrari, Goldstein and Wick 1989] proved an invariance principle for anti-symmetric additive functionals of ergodic reversible Markov processes, which is a generalization of the theorem of Kipnis and Varadhan [Kipnis 1986]. Applying their invariance principle, we obtain that if $(\eta_t, P^0_{\mu_0})$ is an ergodic reversible Markov process, $\varepsilon x(t/\varepsilon^2)$ converges to $D(\mu_0)B(t)$ as $\varepsilon \to 0$, where $D(\mu_0)$ is a non-negative definite $d \times d$ matrix determined by

$$(D(\mu_0)^2)_{i,j} = \lim_{t \to \infty} \frac{1}{t} E^0_{\mu_0}[x_i(t)x_j(t)], \quad 1 \leq i, j \leq d.$$

Here $E^0_{\mu_0}$ denotes the expectation with respect to $P^0_{\mu_0}$.

Using the same argument as Theorem 1, we have the following lemma, which improves Lemma 4.1 in [Tanemura 1989].

LEMMA 3.4. *If $\mu \in \mathrm{ex}\,\mathcal{G}(z, \Phi) \cap \mathcal{G}_\Theta(z, \Phi)$, $z \in (0, \infty)$, then $(\eta_t, P^0_{\mu_0})$ is an ergodic reversible Markov process.*

Thus, the only assertion left to be proved is the non-degeneracy of $D(\mu_0)$. For any rotation θ on $\mathbb{R}^d$ with $\theta(0) = 0$,

$$\bigl(x(t), P^0_\eta\bigr) = \bigl(\theta(x(t)), P^0_{\theta^{-1}(\eta)}\bigr),$$

in the sense of distribution. Then, for any $\mu \in \mathcal{G}_\Theta(z, \Phi)$ there exists $\mu' \in \mathcal{G}_\Theta(z, \Phi)$ such that

$$\bigl(x(t), P^0_{\mu'}\bigr) = \bigl(\theta(x(t)), P^0_\mu\bigr),$$

in the sense of distribution. Hence, it is enough to show the following lemma for the proof of the non-degeneracy of the matrix.

LEMMA 3.5. *There exist $z_0 \in (0, \infty]$ and a positive function $c_0(z)$ on $(0, z_0)$ such that if $\mu \in \mathcal{G}_\Theta(z, \Phi)$, $z \in (0, z_0)$, then $\bigl(D(\mu_0)^2\bigr)_{11} \geq c_0(z)$. In particular, $z_0 = \infty$ in case $h = \infty$.*

The proof of this lemma is given in [Tanemura 1993].

References

Berg, J. van den, Kesten H., *Inequalities with applications to percolation and reliability*, J. Appl. Prob. **22** (1985), 556–569.

De Masi, A., Ferrari, P.A., Goldstein, S., Wick, D.W., *An invariance principle for reversible Markov processes. Applications to random motions in random environments*, J. Statist. Phys. **55** (1989), 787–855.

Georgii, H.O., *Canonical and grand canonical Gibbs states for continuum systems*, Comm. Math. Phys. **48** (1979), 31–51.

———, *Gibbs measures and phase transitions*, Walter de Gruyter, 1988.

Hammersley, J. M., *Percolation processes. Lower bounds for the critical probability*, Ann. Math. Statist. **28** (1957), 790–795.

Kipnis, C. and Varadhan, S. R. S., *Central limit theorems for additive functionals of reversible Markov processes and applications to simple exclusions*, Comm. Math. Phys. **104** (1986), 1–19.

Liggett, T. M., *Interacting Particle Systems*, Springer, 1985.

Ruelle, D., *Statistical Mechanics*, W.A. Benjamin, 1969.

Tanemura, H., *Ergodicity for an infinite particle system in $\mathbb{R}^d$ of jump type with hard core interaction*, J. Math. Soc. Japan **41** (1989), 681–697.

——, *Central limit theorem for a random walk with random obstacles in $\mathbb{R}^d$*, Ann. Probab. **21** (1993), 936–960.

DEPARTMENT OF MATHEMATICS, FACULTY OF SCIENCE, CHIBA UNIVERSITY, CHIBA, 263 JAPAN

E-mail address: tanemura@neumann.s.chiba-u.ac.jp

A Three Level Particle System and Existence of General Multilevel Measure-Valued Processes

Yadong Wu

Branching processes and birth-death processes have been studying for many years. In a wide sense, a branching process or a birth-death process can be viewed as a stochastic model for the evolution of a population. Hereinafter, we refer to its representatives as individuals or particles. Particles can independently replicate or die according to certain probability laws which govern the evolution. Note that all the particles in the classical branching processes (birth and death processes) belong to the same hierarchical level. That is the reason that we refer to this classical class of particle systems as "one level" particle models.

The more interesting case is related to the situation in which the system consists of more than one level particles instead of the only one level case. In this development, during the last few years, considerable interest has developed in dynamic multilevel particle models motivated by both biological and information systems' problems. By "multilevel" we mean that the system consists of objects of different hierarchical levels. At any given level, it is assumed that each object (particle) of that level can be deleted or copied. Collections of objects (particles) at one level comprise of objects of the next higher level. Once an object is copied, it is assumed that subsequent alterations can cause it differ from the object from which it was copied (its parent). We say that a level k particle has the size i if it consists of exactly i level $k-1$ particles.

Such multilevel measure-valued processes were introduced and studied in [Dawson, Hochberg, and Wu 1990, Dawson and Hochberg 1991, Wu 1991, Wu 1993a, Wu 1993b]. Multilevel processes arise in population biology as the natural generalization of the one level processes. They can model mitochondrial DNA, where sampling takes place at both the individual and organelle levels. They also arise in models describing the transfer of information over computer networks, where each directory (a group of files) can be deleted or transferred, and each single file can be deleted and transferred as well.

Dynamic multilevel particle or information systems and multilevel measure-valued processes have been a very challenging subject. Mathematical analysis of the properties of such multilevel processes is much more complicated than that in the one level situation, not only due to the intrinsic fact that several processes at different levels are affecting numbers and sizes of individual particles, but due to

---

1991 *Mathematics Subject Classification*. Primary: 60J80, 60J57, 60J60, 60J35; Secondary: 60E10.

Key words and phrases. Multilevel birth-death system, generator, multilevel measure-valued process, non-explosion.

This is the final form of the paper.

the fact that particles no longer exhibit independent behavior, since higher level birth-death and branching affect groups of particles simultaneously.

In this paper we study more general multilevel particle systems and multilevel measure-valued processes. In Section 2, we introduce a three level birth-death particle system which generalizes the multilevel models given in [Dawson, Hochberg, and Wu 1990] and [Wu 1993b]. In Section 3, we consider the problem of existence of the multilevel processes and obtain a non-explosion criterion for the general multilevel birth-death measure-valued process.

Let us consider a system which consists of particles of three different levels. Thus, we are dealing with populations of individuals or particles which not only undergo, individually, birth-death process, but which also undergo additional level two and level three birth-death processes which act upon level two particles (each level two particle is a collection of level one particles) and level three particles (each level three particle is a group of level two particles, i.e., a group of collections of individuals) respectively. If we think of a level 3 particle as a superparticle, then the three level birth-death particle system can be described as a random walk on $N(Z^+)$ of superparticles. Our model generalizes the two level birth-death measure-valued process studied in [Wu 1993b]. In order to describe the structure of the three level birth-death particle system, we introduce some notation. Let

- $N(Z^+)$ denote the class of integer-valued measures on Z^+;
- $N(N(Z^+))$ denote the class of integer-valued measures on $N(Z^+)$;
- $N_F(N(Z^+))$ denote the class of finite integer-valued measures on $N(Z^+)$.

The entire system can be represented as a integer-valued measure on $N(Z^+)$:

$$(0.1) \qquad \sum_{i=1}^{\infty} \sum_{k=1}^{n_i} \delta_{\sum_{r=1}^{i} \delta_{x_{i,k,r}}} + n_0 \delta_{\delta_\phi} \in N(N(Z^+)),$$

where n_i is the number of level 3 particles of size i, $x_{i,k,r}$ denotes the size of rth level 2 particle in the kth level 3 particle of size i and n_0 denotes the number of null superparticles where the term "null superparticle" means that it does not contain any level 2 particle.

We define three level birth and death rates as follows:
- $b_1(n,j)$ denotes birth rate of a level 1 particle in any level 2 particle of size j (i.e, the rate with which any level 2 particle changes the size from j to $j+1$), where n is the number of level 2 particles of size j;
- $d_1(n,j)$ denotes death rate of a level 1 particle in any level 2 particle of size j (i.e, the rate with which any level 2 particle changes the size from j to $j-1$), where n is the number of level 2 particles of size j;
- $b_2(n,j)$ denotes birth rate of a level 2 particle, where j is the size of the level 2 particle and n is the number of level 2 particles of size j;
- $d_2(n,j)$ denotes death rate of a level 2 particle, where j is the size of the level 2 particle and n is the number of level 2 particles of size j;
- $b_3(n_i,i)$ denotes birth rate of a level 3 particle, which depends on the size i of the level 3 particle and the number n_i of level 3 particles of the size i;
- $d_3(n_i,i)$ denotes death rate of a level 3 particle, which depends on the size i of the level 3 particle and the number n_i of level 3 particles of the size i.

We suppose that each level 3 particle lives the exponentially distributed time with mean $[b_3(n_i,i) + d_3(n_i,i)]^{-1}$ and then either splits into two particles or dies out with the rates $b_3(n_i,i)$ and $d_3(n_i,i)$ respectively. After branching, each newly born level 3 particle is again subject to the same rules of evolution as its predecessor. Each level 2 particle performs a level 2 birth-death process with birth and death rates $b_2(n,k)$ and $d_2(n,k)$ respectively, and during the lifetime of level 1, 2, 3 particles, each level 2 particle changes its size due to the births and deaths of its elements (level 1 particles). We also assume that each level 1 particle performs a level 1 birth-death process with birth and death rates $b_1(n,j)$ and $d_1(n,j)$ respectively. Thus, all the possible transitions of states in the process can be described as follows:

(a) Level one birth:

$$(0.2) \quad \delta_{\sum_{r=1}^{i} \delta_{x_i,k,r}} \to \delta_{\sum_{r \neq r_1} \delta_{x_i,k,r} + \delta_{j+1}} \quad \text{at rate } b_1(n,j),$$

where $x_{i,k,r_1} = j$ and n is the number of level 2 particles of size j. It is easily seen that the size of a level 2 particle δ_{i,k,r_1} changes from j to $j+1$ due to the birth of a level 1 particle in it.

(b) Level one death:

$$(0.3) \quad \delta_{\sum_{r=1}^{i} \delta_{x_i,k,r}} \to \delta_{\sum_{r \neq r_1} \delta_{x_i,k,r} + \delta_{j-1}} \quad \text{at rate } d_1(n,j),$$

where $x_{i,k,r_1} = j$ and n is the number of level 2 particles of size j. It is easily seen that the size of a level 2 particle δ_{i,k,r_1} changes from j to $j-1$ due to the death of a level 1 particle in it.

(c) Level two birth:

$$(0.4) \quad \delta_{\sum_{r=1}^{i} \delta_{x_i,k,r}} \to \delta_{\sum_{r \neq r_1} \delta_{x_i,k,r} + 2\delta_j} \quad \text{at rate } b_2(n,j),$$

where $x_{i,k,r_1} = j$. It is easily seen that the size of a level 3 particle $\delta_{\sum_{r=1}^{i} \delta_{x_i,k,r}}$ changes from i to $i+1$, the number of level 3 particles of size i changes from n_i to $n_i - 1$, the number of level 3 particles of size $i+1$ changes from n_{i+1} to $n_{i+1} + 1$ and the number of level 2 particles of size j changes from n to $n+1$ due to the birth of a level 2 particle $\delta_{x_{i,k,r_1}}$ in the level 3 particle $\delta_{\sum_{r=1}^{i} \delta_{x_i,k,r}}$. Note that the newly born level 2 particle has the same size as its parent, $\delta_{x_{i,k,r_1}}$.

(d) Level two death:

$$(0.5) \quad \delta_{\sum_{r=1}^{i} \delta_{x_i,k,r}} \to \delta_{\sum_{r \neq r_1} \delta_{x_i,k,r}} \quad \text{at rate } d_2(n,j),$$

where $x_{i,k,r_1} = j$. It is easily seen that the size of a level 3 particle $\delta_{\sum_{r=1}^{i} \delta_{x_i,k,r}}$ changes from i to $i-1$, the number of level 3 particles of size i changes from n_i to $n_i - 1$, the number of level 3 particles of size $i-1$ changes from n_{i-1} to $n_{i-1} + 1$ and the number of level 2 particles of size j changes from n to $n-1$ due to the death of a level 2 particle $\delta_{x_{i,k,r_1}}$ in the level 3 particle $\delta_{\sum_{r=1}^{i} \delta_{x_i,k,r}}$. Note that if $i = 1$, then the superparticle contains only one level 2 particle. Therefore, the superparticle $\delta_{\delta_{x_{1,k,1}}}$ becomes a null superparticle due to the death of the only level 2 particle $\delta_{x_{1,k,1}}$ in it.

(e) Level three birth:

$$(0.6) \quad \sum_{k=1}^{n_i} \delta_{\sum_{r=1}^{i} \delta_{x_{i,k,r}}} \to \sum_{k \neq k_1} \delta_{\sum_{r=1}^{i} \delta_{x_{i,k,r}}} + 2\delta_{\sum_{r=1}^{i} \delta_{x_{i,k_1,r}}} \quad \text{at rate } b_3(n_i, i),$$

where n_i denotes the number of superparticles of size i. It is easily seen that the number of superparticles of size i changes from n_i to $n_i + 1$ due to the birth of a superparticle $\delta_{\sum_{r=1}^{i} \delta_{x_{i,k_1,r}}}$ of size i. Note that the newly born superparticle contains exactly the same number of identical level 2 particles as $\delta_{\sum_{r=1}^{i} \delta_{x_{i,k_1,r}}}$.

(f) Level three death:

$$(0.7) \quad \sum_{k=1}^{n_i} \delta_{\sum_{r=1}^{i} \delta_{x_{i,k,r}}} \to \sum_{k \neq k_1} \delta_{\sum_{r=1}^{i} \delta_{x_{i,k,r}}} \quad \text{at rate } d_3(n_i, i),$$

where n_i denotes the number of superparticles of size i. It is easily seen that the number of superparticles of size i changes from n_i to $n_i - 1$ due to the death of a superparticle $\delta_{\sum_{r=1}^{i} \delta_{x_{i,k_1,r}}}$ of size i.

This three level birth-death particle system can be characterized as a $N(N(Z^+))$-valued Markov process $X(t)$:

$$X(t) = \sum_{i=1}^{\infty} \sum_{k=1}^{n_i(t)} \delta_{\sum_{r=1}^{i} \delta_{x_{i,k,r}(t)}} + n_0(t) \delta_{\delta_\phi},$$

where $n_i(t)$ denotes the number of level 3 particles of size i at time t, $x_{i,k,r}(t) \in Z^+$ is the size of rth level 2 particle in kth level 3 particle of size i at time t and $n_0(t)$ denotes the number of null superparticles at time t.

Let us consider the test function $F(\nu)$ on $N(N(Z^+))$ of the following form:

$$F(\nu) = f(\langle\langle h_1(\langle h_2, \cdot\rangle), \nu\rangle\rangle),$$

where $\nu \in N(N(Z^+))$, $h_2 \in C_c^2(Z^+)$, $f, h_1 \in C_b^2(R)$ and $\langle\langle g(\cdot), \nu\rangle\rangle \equiv \int g(\mu) \nu(d\mu)$. Let

- $\nu[2, x_{i,k,r}]$ denotes the number of the level 2 particles of size $x_{i,k,r} \in Z^+$ in ν;
- $\nu[2] = \sum_i n_i i$ denotes the total number of level 2 particles in ν;
- $\nu[3, i]$ denotes the number of level 3 particles of size i in ν;
- $\nu[3] = \sum_i n_i$ denotes the total number of level 3 particles in ν.

Remark 0.1. In order to give a better explanation of our notation, we consider a three level particle system consisting of three level 3 particles and review the notation in Figure 1:

According to the definitions we get that the entire system can be represented as

$$\nu_0 = \sum_i \sum_{k=1}^{n_i} \delta_{\sum_{r=1}^{i} \delta_{x_{i,k,r}}} = \delta_{\delta_3 + \delta_5 + \delta_6} + \delta_{\delta_3 + \delta_4 + 2\delta_5} + \delta_{\delta_2 + \delta_4 + \delta_5 + \delta_6},$$

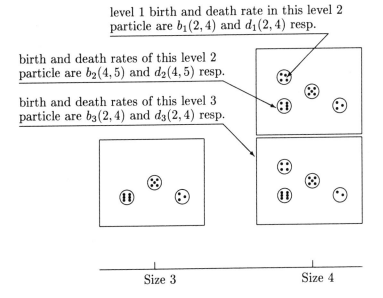

FIGURE 1. Three level particle system.

where the number n_3 of level 3 particles of size 3 is 1 and the number n_4 of level 3 particles of size 4 is 2,

$$\nu_0[2,2] = 1, \quad \nu_0[2,3] = 2, \quad \nu_0[2,4] = 2, \quad \nu_0[2,5] = 4, \quad \nu_0[2,6] = 2, \quad \nu_0[2] = 11,$$
$$\nu_0[3,3] = 1, \quad \nu_0[3,4] = 2, \quad \nu_0[3] = 3.$$

In general situation, if we let $\nu = \sum_{i=1}^{\infty} \sum_{k=1}^{n_i} \delta_{\sum_{r=1}^{i} \delta_{x_{i,k,r}}}$, then the generator of the three level birth-death process, denoted by G, is given by

$$G(\nu) = \sum_{i=1}^{\infty} \sum_{k=1}^{n_i} \sum_{r_1=1}^{i} b_1(\nu[2, x_{i,k,r_1}], x_{i,k,r_1})$$
$$\left[f\left(\left\langle\left\langle h_1(\langle h_2, \cdot \rangle), \nu - \delta_{\sum_{r=1}^{i} \delta_{x_{i,k,r}}} + \delta_{\sum_{r \neq r_1} \delta_{x_{i,k,r}} + \delta_{x_{i,k,r_1}+1}}\right\rangle\right\rangle\right) - f(\langle\langle h_1(\langle h_2, \cdot \rangle), \nu\rangle\rangle) \right]$$
$$+ \sum_{i=1}^{\infty} \sum_{k=1}^{n_i} \sum_{r_1=1}^{i} d_1(\nu[2, x_{i,k,r_1}], x_{i,k,r_1})$$
$$\left[f\left(\left\langle\left\langle h_1(\langle h_2, \cdot \rangle), \nu - \delta_{\sum_{r=1}^{i} \delta_{x_{i,k,r}}} + \delta_{\sum_{r \neq r_1} \delta_{x_{i,k,r}} + \delta_{x_{i,k,r_1}-1}}\right\rangle\right\rangle\right) - f(\langle\langle h_1(\langle h_2, \cdot \rangle), \nu\rangle\rangle) \right]$$

$$+ \sum_{i=1}^{\infty} \sum_{k=1}^{n_i} \sum_{r_1=1}^{i} b_2(\nu[2, x_{i,k,r_1}], x_{i,k,r_1})$$
$$\left[f\left(\left\langle\left\langle h_1(\langle h_2, \cdot \rangle), \nu - \delta_{\sum_{r=1}^{i} \delta_{x_{i,k,r}}} + \delta_{\sum_{r \neq r_1} \delta_{x_{i,k,r}} + 2\delta_{x_{i,k,r_1}}}\right\rangle\right\rangle\right) \right.$$
$$\left. - f(\langle\langle h_1(\langle h_2, \cdot \rangle), \nu\rangle\rangle) \right]$$

$$+ \sum_{i=1}^{\infty} \sum_{k=1}^{n_i} \sum_{r_1=1}^{i} d_2(\nu[2, x_{i,k,r_1}], x_{i,k,r_1})$$
$$\left[f\left(\left\langle\left\langle h_1(\langle h_2, \cdot \rangle), \nu - \delta_{\sum_{r=1}^{i} \delta_{x_{i,k,r}}} + \delta_{\sum_{r \neq r_1} \delta_{x_{i,k,r}}}\right\rangle\right\rangle\right) \right.$$
$$\left. - f(\langle\langle h_1(\langle h_2, \cdot \rangle), \nu\rangle\rangle) \right]$$

$$+ \sum_{i=1}^{\infty} \sum_{k=1}^{n_i} b_3(n_i, i) \left[f\left(\left\langle\left\langle h_1(\langle h_2, \cdot \rangle), \nu + \delta_{\sum_{r=1}^{i} \delta_{x_{i,k,r}}}\right\rangle\right\rangle\right) \right.$$
$$\left. - f(\langle\langle h_1(\langle h_2, \cdot \rangle), \nu\rangle\rangle) \right]$$

$$+ \sum_{i=1}^{\infty} \sum_{k=1}^{n_i} d_3(n_i, i) \left[f\left(\left\langle\left\langle h_1(\langle h_2, \cdot \rangle), \nu - \delta_{\sum_{r=1}^{i} \delta_{x_{i,k,r}}}\right\rangle\right\rangle\right) \right.$$
$$\left. - f(\langle\langle h_1(\langle h_2, \cdot \rangle), \nu\rangle\rangle) \right],$$

where the first term on the right-hand side is due to level 1 birth, the second term is due to level 1 death, the third term is due to level 2 birth, the fourth term is due to level 2 death, the fifth term is due to level 3 birth, the last term is due to level 3 death.

EXAMPLE. Three level linear growth birth-death process.

Consider a special case, namely, a three level linear growth birth and death particle system, in which a level 2 particle changes its size from j to $j+1$ due to the birth of a level 1 particle in it with level 1 birth rate $\alpha_1 j$, $\alpha_1 > 0$, a level 1 particle in a level 2 particle of size j dies out or a level 2 particle changes its size from j to $j-1$ due to the death of a level 1 particle in it with level 1 death rate $\beta_1 j$, $\beta_1 > 0$; each level 2 particle performs independently level 2 branching process with branching rate $\alpha_2 + \beta_2$, $\alpha_2, \beta_2 > 0$ and offspring distribution given by $p_0^{(2)} = \beta_2/(\alpha_2 + \beta_2)$, $p_2^{(2)} = \alpha_2/(\alpha_2 + \beta_2)$, $p_k^{(2)} = 0$ for $k \neq 0, 2$; each level 3 particle performs independently level 3 branching process with branching rate $\alpha_3 + \beta_3$, α_3, $\beta_3 > 0$ and offspring distribution given by $p_0^{(3)} = \beta_3/(\alpha_3 + \beta_3)$, $p_2^{(3)} = \alpha_3/(\alpha_3 + \beta_3)$, $p_k^{(3)} = 0$ for $k \neq 0, 2$. It is clear that in this special case we have

(0.8)
$$\begin{aligned} b_1(n, j) &= \alpha_1 j, & d_1(n, j) &= \beta_1 j; \\ b_2(n, j) &= \alpha_2, & d_1(n, j) &= \beta_2; \\ b_3(n_i, i) &= \alpha_3, & d_1(n_i, i) &= \beta_3 \end{aligned}$$

if $n, j, n_i, i > 0$.

1. Existence

The problem of existence of branching particle systems can be studied in different ways. Given a generator G, one can use the martingale problem characterization in order to show that there exists a Markov process $\{X(t)\}$ which is the unique solution of the martingale problem associated with the generator G. However, in this paper we prove the existence of the process by use of the technique of Laplace transform. The main result of this section is an existence criterion for the general multilevel birth-death measure-valued process. In particular, we prove the existence by showing that there is no explosion. We suppose that there is an M-level birth-death process $X(t)$. Each level k particle undergoes the level k birth-death process with birth and death rates $b_k(n_{k,i_k}, i_k)$ and $d_k(n_{k,i_k}, i_k)$ respectively, where i_k is the size of the level k particle and n_{k,i_k} denotes the number of level k particles of size i_k, $k = 2, 3, \ldots, M$. We also assume that each level 1 particle can be killed or copied according to the level 1 birth-death process with birth and death rates $b_1(n_{2,i_2}, i_2)$ and $d_1(n_{2,i_2}, i_2)$ respectively, where i_2 is the size of the level 2 particle which the level 1 particle belongs to and n_{2,i_2} denotes the number of level 2 particles of size i_2.

THEOREM 1.1. *Suppose that*
(i) *for $1 < k \leq M$, there exists $B_k > 0$, such that*

$$(1.1) \quad b_k(n_{k,0}, 0) \equiv 0, \quad b_k(0, i) \equiv 0 \quad \text{and} \quad b_k(n_{k,i_k}, i_k) \leq B_k,$$

for any $n_{k,0} \geq 0$, $i \geq 0$, $n_{k,i_k} > 0$ and $i_k > 0$;
(ii) $b_1(0, j) \equiv 0$, *and for each $n > 0$,*

$$(1.2) \quad \sum_j \frac{1}{b_1(n,j)} = \infty.$$

Then the M-level birth-death measure-valued process $\{X(t), t \geq 0\}$ has no explosion.

PROOF. For simplicity of notation, we only consider the worst case from the point of view of explosion, namely, the three level pure birth process. Let us define the random times between successive jumps as follows:

$$\tau_1 = \inf\{t : X(t) \neq X(0)\},$$
$$\tau_n = \inf\{t : X(t + \tau_1 + \cdots + \tau_{n-1}) \neq X(\tau_1 + \cdots + \tau_{n-1})\},$$
$$T_n = \tau_1 + \cdots + \tau_n.$$

Set

$$X(0) = \nu = \sum_{i=0}^{\infty} \sum_{k=1}^{n_i} \delta_{\sum_{r=1}^{i} \delta_{x_{i,k,r}}} \in N_F(N(Z^+)).$$

Then it is easy to see that
(a) the level 1 birth rate at time $t = 0$ is

$$\sum_{i=0}^{\infty} \sum_{k=1}^{n_i} \sum_{r=1}^{i} b_1(\nu[2, x_{i,k,r}], x_{i,k,r});$$

(b) the level 2 birth rate at time $t = 0$ is
$$\sum_{i=0}^{\infty}\sum_{k=1}^{n_i}\sum_{r=1}^{i}b_2(\nu[2,x_{i,k,r}],x_{i,k,r});$$

(c) the level 3 birth rate at time $t = 0$ is
$$\sum_{i=0}^{\infty}\sum_{k=1}^{n_i}b_3(n_i,i) = \sum_{i=0}^{\infty}n_i b_3(n_i,i).$$

Note that here we use the notation introduced in the previous section.

Since all jumps of the three level pure birth process are either level 1, level 2 or level 3 births, we can assume that the first m jumps consist of $m_1 \leq m$ jumps of the first level, $m_2 \leq m - m_1$ jumps of the second level and $m - m_1 - m_2$ jumps of the third level. Therefore, the Laplace transform of T_n is given by

$$\begin{aligned}(1.3)\quad \phi_m(\theta) &= \prod_{u=1}^{m_1}\left(1 + \frac{\theta}{\sum_{i=0}^{\infty}\sum_{k=1}^{\nu_{1,u}[3,i]}\sum_{r=1}^{i}b_1(\nu_{1,u}[2,x_{i,k,r}],x_{i,k,r})}\right)^{-1}\\ &\quad \prod_{v=1}^{m_2}\left(1 + \frac{\theta}{\sum_{i=0}^{\infty}\sum_{k=1}^{\nu_{2,v}[3,i]}\sum_{r=1}^{i}b_2(\nu_{2,v}[2,x_{i,k,r}],x_{i,k,r})}\right)^{-1}\\ &\quad \prod_{l=1}^{m-m_1-m_2}\left(1 + \frac{\theta}{\sum_{i=0}^{\infty}\sum_{k=1}^{\nu_{3,l}[3,i]}b_3(\nu_{3,l}[3,i],i)}\right)^{-1}\\ &\leq \prod_{u=1}^{m_1}\left(1 + \frac{\theta}{\sum_{i=0}^{\infty}\sum_{k=1}^{\nu_{1,u}[3,i]}\sum_{r=1}^{i}b_1(\nu_{1,u}[2,x_{i,k,r}],x_{i,k,r})}\right)^{-1}\\ &\quad \prod_{v=1}^{m_2}\left(1 + \frac{\theta}{[(\nu[2]+v)(m-m_1-m_2+1)]B_2}\right)^{-1}\\ &\quad \prod_{l=1}^{m-m_1-m_2}\left(1 + \frac{\theta}{(\nu[3]+l)B_3}\right)^{-1},\end{aligned}$$

where $\nu_{j,u}$, $j = 1, 2, 3$ denotes the state of the process $X(t)$ at the time at which the uth level j jump occurs, $\nu_{1,u}[3,i]$ is the number of level 3 particles of size i in $\nu_{1,u}$, $\nu_{2,v}[3,i]$ denotes the number of level 3 particles of size i in $\nu_{2,v}$ and $\nu_{3,l}[3,i]$ denotes the number of level 3 particles of size i in $\nu_{3,l}$.

If $m - m_1 - m_2$ goes to infinity as $m \to \infty$, then $\sum_{l=1}^{\infty}\frac{1}{(\nu[3]+l)B_3} = \infty$ would imply that
$$\prod_{l=1}^{m-m_1-m_2}\left(1 + \frac{\theta}{(\nu[3]+l)B_3}\right)^{-1} \to 0, \text{ i.e., } \phi_m(\theta) \to 0.$$

If $m - m_1 - m_2$ stays finite as $m \to \infty$ and m_2 goes to infinity, then
$$\prod_{v=1}^{m_2}\left(1 + \frac{\theta}{[(\nu[2]+v)(m-m_1-m_2+1)]B_2}\right)^{-1} \to 0$$

would imply that $\phi_m(\theta) \to 0$.

Finally, we suppose that the numbers of both level 2 and level 3 jumps are finite, i.e., there exists $N > 0$, such that $m - m_1 \leq N$ for any m. Then from (1.2) we have

$$(1.4) \quad \phi_m(\theta) \leq \max_{1 \leq n \leq (\nu[2]+N)N} \prod_{u=1}^{m_1} \left(1 + \frac{\theta}{\sum_{i=0}^{\infty} \sum_{k=1}^{\nu_{1,u}[3,i]} \sum_{r=1}^{i} b_1(n, x_{i,k,r})}\right)^{-1}$$

$$\prod_{v=1}^{m_2}\left(1 + \frac{\theta}{[(\nu[2]+v)(m-m_1-m_2+1)]B_2}\right)^{-1} \prod_{l=1}^{m-m_1-m_2}\left(1 + \frac{\theta}{(\nu[3]+l)B_3}\right)^{-1}$$
$$\to 0, \text{ as } m \to \infty.$$

Therefore, we get that $T_m \to \infty$ as $m \to \infty$, a.e. This means that the three level pure birth process has at most finite number of jumps at any finite time interval. In turn, this means that the process $\{X(t); t \geq 0\}$ has no explosion. The proof of Theorem 1.1 is complete. □

Remark 1.1. Conditions (1.1) and (1.2) are sufficient but not necessary for existence of multilevel measure-valued processes. However, it is easy to see that these conditions are satisfied in the most interesting cases. In particular, Theorem 1.1 implies the existence of the three level linear growth birth-death process.

Acknowledgement. The author would like to thank Professor D. A. Dawson for many helpful suggestions, in particular, the method of the proof of Theorem 1.1.

Bibliography

N. R. Bhattacharya and E. C. Waymire, *Stochastics processes with applications*, Wiley New York, 1990.

D. A. Dawson, *Stochastics evolution equations and related measure-valued processes*, J. Multivariate Anal. **5** (1975), 1–52.

_____, *The critical measure diffusion*, Z. Wahrsch. Verw. Gebiete **40** (1977), 125–145.

D. A. Dawson and K. J. Hochberg, *A multilevel branching model*, Adv. in Appl. Probab. **23** (1991), 701–715.

D. A. Dawson, K. J. Hochberg, and Y. Wu, *Multilevel branching systems*, White Noise Analysis: Mathematics and Applications, World Scientific Publ., Singapore, 1990, pp. 93–107.

L. G. Gorostiza and J. A. Lopez Mimbela, *The multitype measure branching process*, Adv. in Appl. Probab. **22** (1990), 49–67.

S. Watanabe, *On two dimensional Markov processes with branching property*, Trans. Amer. Math. Soc. **136** (1969) 447–466.

Y. Wu, *Dynamic particle systems and multilevel measure branching processes*, Ph.D. thesis, Carleton University, Ottawa, 1991.

_____, *Asymptotic behavior of the two level measure branching process*, Ann. Probab. **22**, No 1, (1993) forthcoming.

_____, *Multilevel birth and death particle system and its continuous diffusion*, Adv. in Appl. Probab. **25**, No. 3, (1993), 549–569.

ABSTRACT. We introduce a hierarchical (three level) dynamic stochastic model and also prove the existence of the general multilevel birth-death measure-valued process.

DEPARTMENT OF MATHEMATICS AND STATISTICS, CARLETON UNIVERSITY, OTTAWA (ONTARIO), CANADA K1S 5B6

E-mail address: ywu@ccs.carleton.ca